Fundamentals of
GRID GENERATION

Fundamentals of GRID GENERATION

Patrick Knupp
Ecodynamic Research Associates
Albuquerque, New Mexico

Stanly Steinberg
University of New Mexico
Albuquerque, New Mexico

CRC Press
Boca Raton Ann Arbor London Tokyo

Library of Congress Cataloging-in-Publication Data

Knupp, Patrick M.
The fundamentals of grid generation / Patrick M. Knupp and Stanly Steinberg.
p. cm.
Includes bibliographical references and index.
ISBN 0-8493-8987-9
1. Numerical grid generation (Numerical analysis). I. Steinberg, Stanly. II. Title.
QA377.K58 1993
519.4-dc20 93-11639
CIP

Direct all inquiries to CRC Press, Inc., 2000 Corporate Blvd., N.W., Boca Raton, Florida, 33431.

International Standard Book Number 0-8493-8987-9

Library of Congress Card Number 93-11639

Printed in the United States of America 1 2 3 4 5 6 7 8 9 0

Printed on acid-free paper

To a Sense of Space, Montana

Preface

This book has two parts, reflecting the fact that the authors have two major goals. The first part of this book, Chapters 1 through 5, is intended to be used as a textbook for a course covering the mathematical fundamentals, the basic algorithms, and some applications of structured grid generation. The second part, Chapters 6 through 11, presents results from the authors' original research in grid generation. The two parts of the book are written in the same style and the transition from textbook to monograph is gradual.

Chapters 1 through 5 provide a uniform mathematical treatment of fundamental grid generation algorithms. The mathematical orientation of the book is a result of our desire to provide the first systematic treatment of this material in textbook form. Although mathematically oriented, the book proves few theorems so as to keep the book accessible to a large audience. Chapters 1 through 5 should benefit engineers and other users of grid generation software by providing a simple explanation of current algorithms. It should also be of interest to mathematicians, who will be quick to note gaps in the underlying mathematical theory.

The second part of this book (Chapters 6 through 11) turns to more theoretical aspects of grid generation, primarily of interest to those who do research on the algorithms of grid generation. These later chapters summarize the authors' original research results. This part is not considered definitive since active research into both the practical and theoretical aspects of grid generation is currently underway. Certain approaches and algorithms are emphasized over others, reflecting the authors' orientation. Since grid generation is an open research topic, it is understandable that others having different objectives may disagree with the emphases and interpretations given. For example, generation of grids as solutions to partial differential equations is emphasized more than algebraic methods even though the latter certainly constitutes an effective approach to grid generation. Similarly, variational grid generation is emphasized over more classical partial differential equation techniques. It is hoped that in the long run, the mathematical approach to the subject will help identify truly robust methods which will serve as the basis of a final solution to the problem of grid generation on complicated objects.

The book does not introduce readers to the practical details of generating grids on complicated objects requiring patching of multiple blocks, unstructured grids, interactive grid generation, etc. Adequate discussion of these important topics would double the size of this book and carry us far afield from the basic ideas we wish to present; a separate book on these practical matters would be a useful companion to the present volume.

In recent times, there has been considerable interest in unstructured grids (see Carey [24]). For very complex geometries, satisfactory unstructured grids are easier to generate than structured grids. On the other hand, the unstructured approach has limitations. For example, there is a nontrivial computational cost in using unstructured grids. The Efficiency Note on page 44 (in Section 5.1.3) of the Thinking Machines FORTRAN manual (Thinking [207]) makes clear that, for the CM5, programs using structured grids are more efficient than ones using unstructured grids. For problems involving relatively simple geometries, structured grid algorithms may be preferable. There has also been skepticism concerning the applicability of unstructured meshes to problems of viscous flow (Mavriplis [137]). To be consistent with the goal of being an introduction to grid generation, this book is restricted to regular quadrilateral meshes and simply connected domains. Most of the key ideas of structured grid generation are met within this restriction; generalizations to composite meshes and unstructured grids require some additional ideas.

The background needed to understand the first part of this book is matrix theory, vector calculus, the elementary theory of partial differential equations, introductory finite-difference techniques, and elementary numerical programming techniques. The only advanced mathematical tools used are the calculus of variations and differential geometry. Sufficient background material for the advanced tools is provided in this book. However, reading additional material from an applied text will help the reader. A symbol manipulator such as Derive, MACSYMA, Maple, Mathematica, or Reduce will be useful for doing some of the more complex algebraic manipulations (see Steinberg [195]).

This book is not intended as a text on finite difference methods, so the reader not familiar with this topic will need to do some background reading. For example, the reader should be able to ascertain if certain approximations have second-order truncation error. Any of the textbooks, Birkhoff and Lynch [19], Celia and Gray [35], Fletcher [73], Forsythe and Wasow [76], Golub and Ortega [85], Hall and Porsching [89], Peyret and Taylor [151], Sod [180], or Strikwerda [202] provide an introduction to finite difference methods.

Exercises form an integral part of the text and provide important extensions of the material in the narrative. After studying the first part of this book, the reader will be able to generate grids on a wide range of geometric regions and solve differential equations defined on those regions using the generated grids. To increase the utility of the book, computer codes implementing the basic algorithms are provided on the floppy disk included with this book.

CHAPTER SUMMARIES

We believe a highly satisfactory solution to the problem of organizing the many interrelated ideas of grid generation has been found for this book, but it is recognized that other presentations of the material may seem more logical to some readers.

Grids are viewed in this book as discretized transformations and therefore Chapter 1, *Preliminaries*, introduces the concept of transformation and discusses the invertibility of transformations, a property critical to their use in generating grids. Examples of transformations that can be used to generate grids are given; these include transformations such as polar and spherical coordinates. The most useful and fastest general grid generator, transfinite interpolation, is introduced.

Chapter 2, *Application to Hosted Equations*, shows how to use boundary-fitted grids to solve steady-state boundary value problems for differential equations in one and two dimensions. To use a grid to solve a differential equation, the differential equation is transformed to general coordinates and then discretized. The differential equations that describe physical problems are called hosted equations. The *type* of a partial differential equation determines the kind of physics that it describes. It is important that the transformation being used to generate a grid have a positive Jacobian; in particular, the type of a differential equation is preserved by transformations with positive Jacobian. This chapter ends with a project to solve a nontrivial boundary-value problem in two dimensions using grid-generation techniques. The solution of this problem using other techniques such as those described in Birkhoff and Lynch [19] is nontrivial and more difficult than using grid-generation techniques.

Chapter 3, *Grid Generation on the Line*, discusses one-dimensional grid generation. This rather simple problem provides a context for introducing many of the ideas that are critical to the rest of the book. A fundamental problem is to generate a grid that has specified lengths of segments between nodes. This problem is completely solved for the one-dimensional case; the problem is considerably more difficult in higher dimensions. The most important classical generators use Laplace or Poisson partial differential equations to generate the grid, so the one-dimensional analogs of these generators are described. The next goal is to introduce variational ideas to generate grids with specified segment lengths. In one dimension, all of the described grid generators are intimately related, so these relationships are described. Numerical algorithms for the one dimensional generators are given in detail, providing a model for solving higher dimensional problems. The chapter ends with a discussion of how to use the weighted grid generator to create solution-adapted grids.

Before proceeding to higher dimensions, a short detour away from grid generation is taken in Chapter 4, *Vector Calculus and Differential Geometry*, to present basic geometric ideas. Tangent vectors to coordinate curves and normal vectors to coordinate surfaces are introduced. Coordinate relationships between maps and their inverse are presented. Concepts from elementary differential geometry such as the metric tensor are defined.

In Chapter 5, *Classical Planar Grid Generation*, important nonvariational planar grid generators are described. These generators form the core of most structured grid-generation software currently in use. The material is divided into two parts: nonelliptic and elliptic generators. The nonelliptic generators include algebraic techniques, conformal and quasiconformal mappings, and orthogonal, hyperbolic, parabolic, and biharmonic partial differential equation methods. Two standard elliptic generators, introduced by Amsden and Hirt [3] and Winslow [235] are described. Inhomogeneous grid generation with elliptic grid generators to provide interior control and a more refined technique for controlling the grid near the boundary are discussed. Planar solution-adaptive grid generation is outlined, and finally, a novel nonclassical approach to adaptive grid generation is described.

In Chapter 6, *Variational Planar Grid Generation,* the Steinberg-Roache variational theory of grid generation is described. This is preceded by an introduction to methods from the calculus of variations. Length, Area, Orthogonality, and combinations of these functionals are given. The idea of the reference grid is presented for use in inhomogeneous grid generation. A numerical algorithm for finding solutions to the Euler-Lagrange equations is given. The chapter closes with a brief presentation of the Direct Optimization Method.

In Chapter 7, *Tensor Analysis and Transformation Relationships,* the reader is introduced to concepts from continuum mechanics, such as the divergence of a tensor, which prove useful in unifying and extending variational grid-generation algorithms. In addition, these concepts are directly applicable to the problem of transforming hosted equations to general coordinate frames. General relationships for the gradient, divergence, curl, and Laplacian are given in conservative and nonconservative forms. Applications to grid generation, such as a derivation of the inverted Winslow equations, is given. General expressions for time derivatives are also noted.

In Chapter 8, *Advanced Planar Variational Grid Generation,* the planar variational principle is assumed to be a function of the elements of the metric tensor and the determinant of the Jacobian. This assumption permits a general derivation of the Euler-Lagrange equations in conservative form for this class of functionals; the equations are related to a variational principle for the model hosted equation. A numerical algorithm for the general variational principle is given. An entirely new form of the grid-generation equations, the covariant projection, is given and shown to be related to the Euler-Lagrange equations resulting from the Brackbill-Saltzman approach to variational grid generation. Connections between the Steinberg-Roache and Brackbill-Saltzman theories are outlined. A nonconservative form of the general Euler-Lagrange equations is derived.

In Chapter 9, *Grid Generation in Three Dimensions,* ideas from Chapters 5 and 8 are extended to three dimensions.

In Chapter 10, *Variational Grid Generation on Curves and Surfaces,* a general variational theory of curve and surface grid generation is presented. Two approaches to the problem of constraining grid points to the manifold are described, the parametric approach and the Lagrange Multiplier approach. Classical results from differential geometry and concepts from tensor analysis are used to derive the Euler-Lagrange equations for the curve and surface grid generators. The bifurcation problem is described and two approaches that avoid this problem (both essentially involving the covariant projection) are given. The concept of Off-Object truncation error is introduced in a discussion of numerical approaches to solving the Lagrange Multiplier form of the equations.

In Chapter 11, *Contravariant Functionals*, we discuss contravariant functionals based on an elliptic norm; the functionals are designed to perform either grid *alignment* with a given vector field or tensor *diagonalization* by appropriate choice of coordinate system. The latter permits an unambiguous interpretation of the weight functions in terms of stated local conditions on the grid tangents. Two special cases are of interest: the Winslow variable diffusion generator and the orthogonal/cell-aspect ratio grid generator. From a mathematical point of view, the Winslow variable diffusion generator is the proper weighted generalization of the homogeneous Thompson-Thames-Mastin generator in that it gives a clear interpretation of the weights and falls within the theory of harmonic mappings.

In Appendix A, *Tensor Coefficients*, detailed formulas for the matrix coefficients of the nonconservative form of the Euler-Lagrange equations are given.

Appendix B, *Fortran Code Directory*, is a guide to the software on the floppy disk.

Appendix C, *A Rogue's Gallery of Grids*, gives a large sample of grids that illustrate the performance of the important grid generators on a selection of basic planar regions.

ACKNOWLEDGMENTS

Thanks to the first class to use this text and who corrected many errors and typographical errors: A. Farhat, L. James, S. Weber. Thanks to Frank Hatfield for his help with the figures, T. Larkin for his comments and corrections, and finally Patrick J. Roache for his support and encouragement.

Contents

Notation

Symbol	*Explanation*
E^n	Euclidean space of dimension n
x, y, z	Physical space coordinates
ξ, η, ζ	Logical space coordinates
$U_k, \partial U_k$	Logical space of dimension k and its boundary
$\Omega_k^n, \partial\Omega_k^n$	k-Dimensional object in n-dimensional space and its boundary
$X_k^n\ \partial X_k^n$	Transformation from logical space to physical space, or their boundaries
$\mathcal{J}, \mathcal{J}^T$	The Jacobian matrix and its transpose
$J, \sqrt{g}$	Determinant of the Jacobian matrix
C	Auxiliary Jacobian matrix
$\mathbf{x}_\xi, \mathbf{x}_\eta, \mathbf{x}_\zeta$	Coordinate line tangents
C^j	Space of functions that have j continuous derivatives
$\nabla_\mathbf{x}$	Gradient operator with respect to physical variables
∇_ξ	Gradient operator with respect to logical variables
det	Determinant
tr	Trace operator
$\mathcal{I}$	Identity matrix
$\mathcal{G}$	Metric tensor
$g_{i,j}$	Elements of the metric tensor
$g^{i,j}$	Elements of the inverse metric tensor
$\mathbf{x}^\perp$	Vector perpendicular to the vector $\mathbf{x}$
$\Delta_\mathcal{B}$	Beltrami operator
Γ_{ij}^k	Surface or Space Christoffel Symbols
$[ij, k]$	Space Christoffel Symbol of the first kind
$\mathbf{P}_\xi, \mathbf{P}_\eta$	Algebraic projection operators
$\oplus$	Direct sum operator
$\nabla_\xi^2, \nabla_\mathbf{x}^2$	Laplace operators with respect to logical and physical variables
$\mathcal{Q}_H\mathbf{x}$	Grid generation operator
$I[\mathbf{x}]$	Variational principle
$D_cI[\mathbf{x}]$	First variation of variational principle
$D_c^2I[\mathbf{x}]$	Second variation of variational principle
$\mathcal{T}_{ij}$	Matrix coefficients in Euler-Lagrange equation
$\mathrm{div}_\xi, \mathrm{div}_\mathbf{x}$	Divergence operators with respect to logical and physical variables
$[\nabla S]\mathcal{T}$	Contraction of the tensors S and $\mathcal{T}$
$\mathrm{curl}_\xi, \mathrm{curl}_\mathbf{x}$	Curl operators with respect to logical and physical variables
$\otimes$	Tensor product operation
$\cdot$	Inner product operation
$\times$	Vector cross product operation
Df/Dt	Material or substantial derivative of f with respect to t
ϕ	Logical space weight function
w	Physical space weight function
H	Integrand of variational principle
κ, τ	Curvature and torsion
$\hat{\mathbf{n}}, \hat{\mathbf{t}}, \hat{\mathbf{b}}$	Unit vectors for curve (moving-trihedron)
$\mathcal{W}$	Curvature tensor

Chapter 1

Preliminaries

1.1 THE GOALS OF GRID GENERATION

Numerical *grid generation* arose from the need to compute solutions to the partial differential equations of *fluid dynamics* on *physical regions* with *complex geometry* (see the picture on the cover for an example). By transforming a physical region to a simpler region, one removes the complication of the shape of the physical region from the problem. Such *transformations* can be viewed as a general *curvilinear coordinate* system for the physical region. The classical techniques of transforming problems to polar, cylindrical, or spherical coordinates are special cases. A cost of using such coordinate systems is an increase in the complexity of the transformed *hosted equations*, i.e., those equations modeling the physical problem to be solved. An advantage of this technique is that the boundary conditions become easier to approximate accurately.

In many applications, it is possible to transform the physical region to a square in two dimensions or a cube in three dimensions in such a manner that the boundary of the square or cube corresponds to the boundary of the physical region (see Figure 1.1). The square or cube is called the *logical region* and the transformation gives rise to a *boundary conforming* coordinate system. The coordinate lines in this coordinate system are given by the images of uniform coordinate lines in the logical region. In such coordinate systems, it is relatively easy to make accurate implementations of *numerical boundary conditions*.

The terminology in this subject is flexible. For example the *physical region* is also called the *physical space*, the *physical domain*, or the *physical object*, with similar terminology for the *logical region*, *space*, or *domain* (object is not used here). *Transformations* are also called *maps*.

In this development, the Jacobian of the transformation is required to be nonzero, and consequently the transformation has an *inverse*. Because grids are first chosen in logical space and then mapped to physical space, it is natural to view the *transformation* as a *mapping* from logical to physical space (see Figure 1.1). In addition, when algorithms are implemented in computer codes, the computations are done in logical space and consequently it is helpful to view the

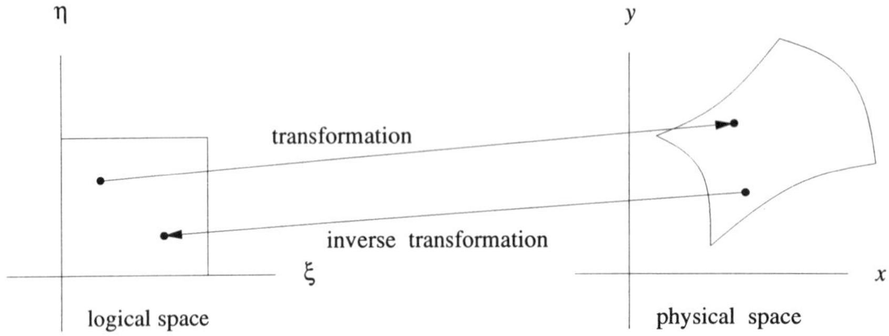

Figure 1.1 *Transformation, map, or coordinate system*

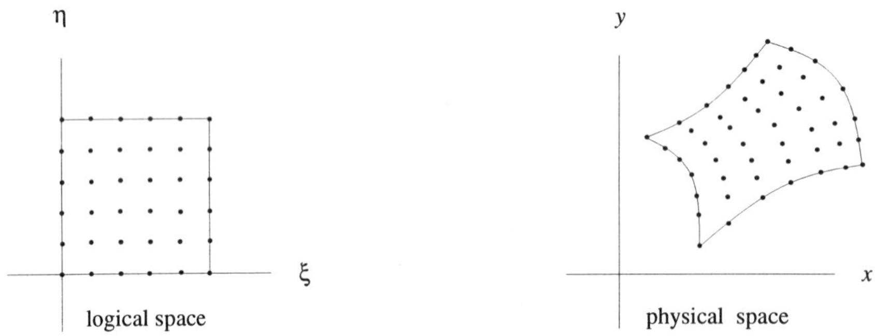

Figure 1.2 *A grid*

transformation as being from logical to physical space. The reader familiar with finite-element theory should realize that the *master element* notion from that theory is a simple variant of the notion of logical space. If a set of points is chosen in logical space, then the inverse transformation carries these points to points in physical space, where they form a *grid* (see Figure 1.2). If the points in logical space are created by dividing the square into identical rectangles or the cube into identical rectangular boxes, then the grid is *logically rectangular*; the discussion in this book is confined to logically rectangular grids.

Several attributes of a grid determine its utility in computations. If the *Jacobian* of the transformation is ever zero, the transformation fails to preserve the essential physical and mathematical properties of the hosted equations (Epstein [70]). Transformations containing a point with zero Jacobian are called *folded*; avoiding a folded transformation is a major objective of grid-generation algorithms. Also, it is well known that the *error of approximations* of the hosted equations depends not only on the derivatives of the solution of the hosted equations and the grid spacing, but also on the rate-of-change of grid spacing and the departure of the grid from orthogonality (Mastin [130] or Thompson and Mastin [213]). For a given grid spacing, smooth, *orthogonal grids* usually result in the smallest error in simple problems. Thus another major goal of grid-generation algorithms is to produce smooth grids, that is, grids where the spacing varies smoothly and the angles between grid lines do not become too small. It turns out that it is not possible to generate orthogonal grids for a wide range of problems.

If the solution of the hosted equations varies rapidly in some part of the physical region, then it is reasonable to choose a finer grid in that part of the region to reduce the error in the numerical solution. Such a grid is called *solution adapted*; it is important to be able to generate solution-adapted grids.

The main advantage of modern grid-generation algorithms is that these algorithms can be used to efficiently generate *large grids*, containing tens of thousands to a few million points, and the resulting grids allow the reduction of error and the simplified treatment of *boundary conditions*. Grid-generation algorithms have been applied to many problems in computational fluid dynamics, including *aerodynamics*, *tidal* and *estuary flow*, *plasma physics*, *electromagnetics*, and *structures*. Grid generation has been the subject of several conferences (Smith [174], Thompson [210], Ghia and Ghia [79], Haüser and Taylor [89], Sengupta et al. [167], Castillo [32], and Arcilla et al. [9]); the proceedings of these conferences are an excellent source of information about both algorithms and applications. The textbook by Thompson et al. [215] describes a wide range of algorithms. The following textbooks cover some aspects of the material in this text: George [78] (mostly unstructured grids), Celia and Gray [35], Section 2.7.2; Fletcher [72], Volume II, Chapters 12 and 13; and Peyret and Taylor [148], Section 11.4. The new text by Carey [24] provides an excellent survey of related material, with emphasis on applications to finite elements.

1.2 THE TOOLS OF GRID GENERATION

It is important to understand some background information before starting to generate complicated grids. In this text, practical algorithms are emphasized; many of these algorithms can be understood and implemented without a complete understanding of the underlying mathematical results. However, the design of the algorithms and the prediction of the shape of the grids that the algorithms generate critically rely on the underlying mathematics. The most frequently used results are reviewed, but it is assumed that the reader is familiar with elementary matrix theory and vector calculus.

One of the best ways to assess the quality of a grid-generation algorithm is to test it on a number of difficult problems. This has been done for many of the algorithms in this text; the grids generated by the algorithms are compared in Appendix C, *The Rogue's Gallery of Grids*.

In the next section, a standard notation is set up to describe transformations and their Jacobians. It is critical, as discussed later in this chapter, that the transformation be invertible. The connection between the invertibility of the transformation and the Jacobian is discussed in detail. Next, some classical trans-

n	Logical Space	Boundary ∂U_k^n
1	$U_1 = \{\xi \in E^1\,;\, 0 \le \xi \le 1\}$	2 points
2	$U_2 = \{(\xi, \eta) \in E^2\,;\, 0 \le \xi, \eta \le 1\}$	4 segments 4 points
3	$U_3 = \{(\xi, \eta, \zeta) \in E^3\,;\, 0 \le \xi, \eta, \zeta \le 1\}$	6 faces 12 segments 8 points

Table 1.1 *Logical space*

formations like polar coordinates are introduced and used to generate grids. At this stage the reader needs to know a scientific programming language such as FORTRAN, Pascal, basic, or C to complete the exercises. It is also important to have computer graphics software and hardware appropriate for plotting grids. Next a fast method of generating grids on simple regions, known as transfinite interpolation, is introduced. An exercise requires the reader to build a useful grid-generation code based on transfinite interpolation.

1.3 MAPPINGS AND INVERTIBILITY

This book is primarily concerned with mappings from one geometric object to another. The domain of the mapping is referred to as the *logical* space, while the range of the mapping is referred to as the *physical* space. In keeping with the practical origins of this subject, these spaces are limited to subsets of the Euclidean Spaces E^n with $n = 1$, 2, or 3. The variables x, y, and z are used as *coordinates* in physical space, while the variables ξ, η, and ζ are used as coordinates in logical space. For the description of some algorithms, it is helpful to have coordinates with indices, that is, using standard vector notation. Thus, in logical space $\boldsymbol{\xi} = (\xi_1, \xi_2, \ldots, \xi_k)$ is used, while in physical space $\mathbf{x} = (x_1, x_2, \ldots, x_n)$ is used, with the convention that $\xi_1 = \xi$, $\xi_2 = \eta$, $\xi_3 = \zeta$, and $x_1 = x$, $x_2 = y$, $x_3 = z$.

Logical space is chosen as follows: as the *unit interval* U_1 in E^1, as the *unit square* U_2 in E^2, as the *unit cube* U_3 in E^3 (see Table 1.1). The boundary of both the physical and logical regions play an important role in numerical computations. In one dimension the boundary of logical space consists of two points; in two dimensions it consists of four open segments and four corner points; while in three dimensions it consists of six open faces, twelve open segments, and eight corner points (see Table 1.1). The situation in physical space is a bit more complicated: in three dimensions it is necessary to place grids on *regions*, *surfaces*, and *curves*. The dimensions of these objects are, respectively, three, two, and one. In two dimensions it is necessary to place grids on regions and curves, and the dimensions of these objects are, respectively, two and one. In one dimension, only intervals need to have grids placed on them. The dimension of the object is an important

Map	Coordinates	From	To
X_1^1	$x = x(\xi)$	interval	interval
X_1^2	$x = x(\xi),\ y = y(\xi)$	interval	curve
X_1^3	$x = x(\xi),\ y = y(\xi),\ z = z(\xi)$	interval	curve
X_2^2	$x = x(\xi,\eta),\ y = y(\xi,\eta)$	square	region
X_2^3	$x = x(\xi,\eta),\ y = y(\xi,\eta),\ z = z(\xi,\eta)$	square	surface
X_3^3	$x = x(\xi,\eta,\zeta),\ y = y(\xi,\eta,\zeta),\ z = z(\xi,\eta,\zeta)$	cube	volume

Table 1.2 *Coordinate maps*

parameter: intervals and curves are one-dimensional objects; planar regions and surfaces are two-dimensional objects; while the only three-dimensional objects of interest are regions in three dimensions. Objects in physical space have two important parameters associated with them; the dimension of the object k; and the dimension of physical space n. A k-dimensional object in n-dimensional physical space is labeled Ω_k^n and is assumed to be bounded (i.e., it does not extend to infinity).

Maps or transformations from logical to physical objects give a system of general coordinates on the physical object. Such maps have two important parameters: the dimension of the object k and the dimensions of physical space n. Thus they are labeled X_k^n:

$$X_k^n : U_k \to \Omega_k^n . \tag{1.1}$$

It is assumed that $0 < n \leq 3$ and $0 < k \leq n$. In general, such maps can be written

$$\mathbf{x} = \mathbf{x}(\boldsymbol{\xi}) ; \tag{1.2}$$

the special notation given in Table 1.2 is used when the dimension is specified. The coordinate lines in the general coordinate system in physical space are the images of the coordinate lines in logical space and are thus given by the curves where one of the components of $\boldsymbol{\xi}$ varies and all other components are held constant. In addition, it is assumed that the boundary of the physical region is the image of the boundary of the logical region under the map and, consequently, the boundary of physical space has the same structure as the boundary of logical space (see Table 1.1). These coordinates are called *boundary conforming.* Such maps can be used to generate grids by choosing a uniform grid in logical space and then transforming the grid to physical space (examples are given in the Section 1.4).

To summarize, a map or transformation from logical space to physical space produces a natural grid in physical space. However, this grid depends on the parameterization of the map: different parameterizations produce different grids. Thus grid generation can be viewed as finding useful parameterizations of maps.

The next few chapters are devoted to studying maps and the grids they generate. In Chapter 3, the simplest possible grids are studied, namely grids on intervals given by maps X_1^1 from the unit interval to an interval. The next most complicated grids are those on planar regions given by maps X_2^2 from the unit square to a planar domain; these are studied in Chapters 5 and 6. Maps X_3^3 from the unit cube to a volume in three-dimensional space are studied in Chapter 9. Grids on curves are given by maps X_1^2 from the unit interval to the plane, or by maps X_1^3 from the unit interval to three-dimensional space, while grids on surfaces are given by maps X_2^3 from the unit square to the three-dimensional space. Such grids are studied in Chapter 10. Within the context of each of these chapters, it is convenient to drop the indices, referring to the map $X : U \to \Omega$.

Before a grid can be generated, a physical object must be specified mathematically; this can be done by specifying its *boundary*. The boundary of an object can be given in three fundamentally different ways: *parametrically*, *implicitly*, or *numerically*. For example, the boundary of the *unit disk* in the plane can be given parametrically using polar coordinates,

$$x = \cos(\theta), \quad y = \sin(\theta), \quad 0 \le \theta \le 2\pi, \tag{1.3}$$

or implicitly by the equation

$$x^2 + y^2 = 1. \tag{1.4}$$

If a physical object is given parametrically, then a discrete set of points on the object can be chosen on the object by discretizing the parameter:

$$x_i = \cos(\theta_i), \quad y_i = \sin(\theta_i), \quad \theta_i = \frac{2\pi i}{M}, \quad 0 \le i \le M. \tag{1.5}$$

Numerically specified boundaries can be converted to parametrically specified boundaries by using *interpolation* (see Lancaster [119]). Interpolation of discrete grids on realistic objects can pose serious difficulties.

Exercise 1.3.1 The surface of the *unit sphere* is given implicitly by $x^2 + y^2 + z^2 = 1$. Use spherical coordinates to obtain a parametric description of this surface. Change this parameterization of the sphere to one where the unit square is mapped to the sphere. What change is needed to obtain a hemisphere? §

It is required that the maps used for generating grids carry the boundary ∂U_k of the logical region U_k to the boundary $\partial\Omega_k^n$ of the physical region Ω_k^n. This can be done by first giving the boundary parametrically using a map ∂X_k^n,

$$\partial X_k^n : \partial U_k \to \partial\Omega_k^n, \tag{1.6}$$

and then extending this map to the interior of the region U_k. Thus, if an object is given implicitly, a parametric description of the boundary must be found (most calculus books have a discussion of this point in sections on the parameterization of curves and surfaces). For many simple objects, it is easy to find a parameterization of the boundary. However, the parameterization may not use values of the parameters restricted to logical space. Typically, a proper map of the logical boundary to physical boundary can be found by rescaling the parameters. In some cases, it is important to *re-parameterize* the boundary before starting on the grid-generation problem (this is also discussed in most calculus books). Many examples of parameterizations and re-parameterization are given in this book.

It is assumed that the physical region is *connected*, that is, made up of a single piece. Technically, the physical region is assumed to be a *domain*, that is, the physical region is the closure of an open connected set. In one dimension, this means the region is an interval. In two dimensions, connected regions can be more complicated; for example, an annulus is connected (an annulus is the region between two concentric circles). In this book, it is also assumed that all regions are *simply connected*. Intuitively, a region is not simply connected if it has a hole in it. The definition of "simply connected" is given in most texts on topology (see Armstrong [13]). It is easy to see that the logical region is simply connected. In one dimension, only intervals are simply connected. In the plane, a connected region is simply connected if its complement is connected. Thus, the annulus is not simply connected. The situation in three dimensions is even more complicated. For example the region between two concentric spheres or that within a torus (doughnut) are not simply connected.

It is possible to extend the techniques presented in the book to non-simply connected regions. The number and type of holes in a region gives the topological structure of the region. It is important to realize that logical space is simply connected and that the topological structure of logical space is preserved by the maps, because it is assumed that the maps are smooth and nonsingular on all of logical space, including the boundaries. Moreover, such maps are orientation preserving and consequently the parts of the boundary of physical space must bear the same relationship to each other and to the interior of the region as do the parts of the boundary of logical space. Figure 1.3 gives an example of both permissible and nonpermissible arrangements in the plane.

The basic problem of grid generation is: if the boundary of the physical object is given by a nonsingular parametric map

$$\partial X_k^n : \partial U_k \to \partial \Omega_k^n , \tag{1.7}$$

then extend this map to a map

$$X_k^n : U_k \to \Omega_k^n \tag{1.8}$$

from the interior of logical space to the interior of the physical object. Such a mapping is termed a boundary-conforming map and the map generates a boundary-conforming continuum grid. A continuum grid can be used to generate a discrete grid by choosing a uniform discrete grid in logical space and then evaluating the continuum map at the points in the logical space grid to give the grid points in physical space.

Two undesirable situations can arise: (i) a point in logical space is mapped to a point outside of the physical object (this is referred to as *spillover* or *folding*) or (ii) two or more points in logical space are mapped to the same point on the physical object. It is of paramount importance to perform the extension of the boundary map in such a way that each point in logical space is mapped to a unique point in physical space (the map is *one-to-one*) and such that each point in physical space is the image of a point in logical space (the map is *onto*). That is, to every point $\boldsymbol{\xi} \in U_k$, there corresponds a unique point $\mathbf{x} \in \Omega_k^n$ and to every point $\mathbf{x} \in \Omega_k^n$ there corresponds a unique point $\boldsymbol{\xi} \in U_k$.

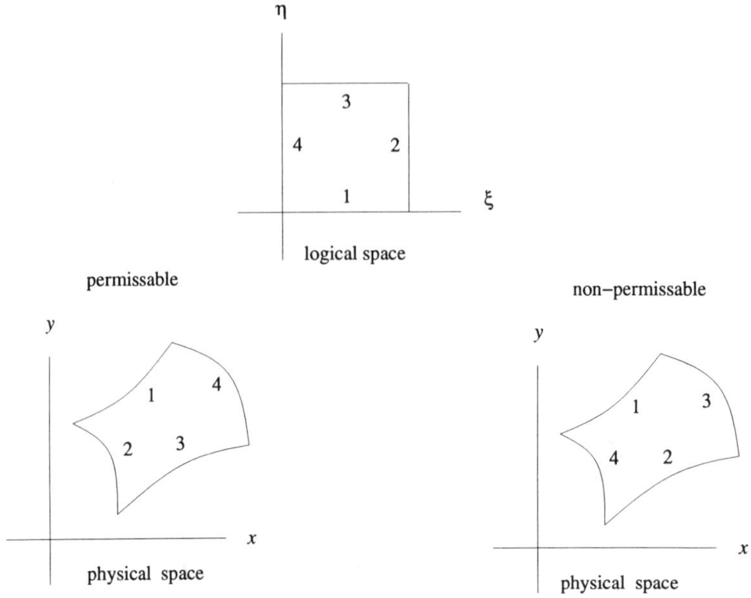

Figure 1.3 *Boundary topology*

The map X_k^n is generally required to be *smooth*, that is, each of the coordinate functions $x_i(\xi)$ must be continuous and have continuous derivatives as functions of each ξ_i on U_k. Note that U_k is defined to be a closed object, that is, it includes its boundary, which means that the map must be smooth both in the interior of U_k and on the boundary ∂U_k of U_k. In particular, this implies that the corners of logical space must map to corners of the physical object. Mathematically, the space of functions that have j continuous derivatives is written $\mathbf{C}^j$; the notation $\mathbf{C}^j$ will also be used for maps whose coordinates are in $\mathbf{C}^j$. Thus, in general, it will be assumed that $X_k^n \in \mathbf{C}^1$. If $X_k^n : U_k \to \Omega_k^n$ is one-to-one and onto, and $X_k^n \in \mathbf{C}^1$, then the map is termed a *diffeomorphism*. If the map X_k^n is restricted to the boundary ∂U_k of U_k then it is labeled ∂X_k^n. The boundary mapping ∂X_k^n is assumed to be a diffeomorphism. In some circumstances this assumption can be relaxed, particularly at a finite number of points. In other circumstances this assumption needs to be strengthened. The basic problem of grid generation can be rephrased: extend a given boundary diffeomorphism ∂X_k^n to a diffeomorphism X_k^n of U_k to Ω_k^n.

The most useful object for studying mappings is the *Jacobian matrix* $\mathcal{J}$. Since it has been assumed that $X_k^n \in \mathbf{C}^1$, the partial derivatives $\partial x_i / \partial \xi_j$ are defined and consequently the elements of $\mathcal{J}$

$$\mathcal{J}_{ij} = \frac{\partial x_i}{\partial \xi_j}, \qquad i = 1, \ldots, n, \qquad j = 1, \ldots, k \tag{1.9}$$

are defined. Note that, in general, $\mathcal{J}$ is not a square matrix. For example, a surface has a Jacobian matrix of the form

$$\mathcal{J} = \begin{pmatrix} \frac{\partial x}{\partial \xi} & \frac{\partial x}{\partial \eta} \\ \frac{\partial y}{\partial \xi} & \frac{\partial y}{\partial \eta} \\ \frac{\partial z}{\partial \xi} & \frac{\partial z}{\partial \eta} \end{pmatrix} = \begin{pmatrix} x_\xi & x_\eta \\ y_\xi & y_\eta \\ z_\xi & z_\eta \end{pmatrix} . \tag{1.10}$$

In the case $n = k$, the matrix is square and then the determinant of $\mathcal{J}$ is defined and referred to as the Jacobian J of the map X_k^k:

$$J = \det(\mathcal{J}) . \tag{1.11}$$

Exercise 1.3.2 For the parameterization of the sphere found in Exercise 1.3.1, show that rank of the Jacobian matrix is not maximal at the poles of the sphere. §

Exercise 1.3.3 Write out the Jacobian matrix for the maps $X_1^1, X_1^2, X_2^2, X_1^3, X_2^3$, and X_3^3. §

The *Inverse Mapping Theorem* is an important result relating the notion of one-to-one to the *rank* of the Jacobian matrix. The theorem is stated using the concept of *local*, meaning for all points near a given point. The proof is beyond the scope of this text.

THEOREM 1.1 (Inverse Mapping Theorem) Assume $X_k^n \in \mathbf{C}^1$. Then X_k^n is locally one-to-one at $\boldsymbol{\xi}$ in the interior of U_k, if and only if the rank of $\mathcal{J}$ is maximal (equals k) at $\boldsymbol{\xi}$. §

A mapping whose Jacobian matrix has maximal rank at $\boldsymbol{\xi}$ is referred to as a *nonsingular* map at $\boldsymbol{\xi}$.

Exercise 1.3.4 Show that the map

$$x = \cos(4\,\pi\,\xi)\,, \quad y = \sin(4\,\pi\,\xi)\,, \quad 0 \le \xi \le 1\,, \tag{1.12}$$

is locally one-to-one but not globally one-to-one. §

If the dimension of the physical object equals the dimension of physical space (the object is not a curve or surface) the Jacobian matrix is square. Square matrices have maximal rank if and only if their determinants are nonzero.

COROLLARY 1.2 Assume $X_n^n \in \mathbf{C}^1$. Then X_n^n is locally one-to-one near $\boldsymbol{\xi}$ if and only if $J(\boldsymbol{\xi}) \neq 0$.

For a proof of this corollary see Corwin [41].

There are always infinitely many diffeomorphisms from the logical domain to the physical domain, so additional criteria may be applied to select superior grids. One such criterion is to require that the map have positive Jacobian, $J > 0$ on U_k; another is to require that the map be as smooth as possible, e.g., belong to $\mathbf{C}^\infty$; another is to require orthogonality; and yet another is to require uniformity of the areas of the grid cells.

It is relatively easy to generate grids in *convex regions.* Recall that a region is convex if, when it contains two points, it contains the line segment joining the two points. Convexity can be defined more formally: $\Omega \in E^n$ is convex if and only if for every pair of points $\mathbf{x}_1$ and $\mathbf{x}_2$ in Ω, the point $\mathbf{x} = (1-\xi)\,\mathbf{x}_1 + \xi\,\mathbf{x}_2$ belongs to Ω provided $0 \le \xi \le 1$. Note that the logical domains U_k are convex. The reader should note that most of the physical regions used in this text are not convex (see Appendix C).

1.4 SPECIAL COORDINATE SYSTEMS

The simplest maps of substantial interest are the *bilinear maps* commonly used in *finite-element* methods where they are referred to as *isoparametric maps* on a quadrilateral. If physical space is one-dimensional, that is, an interval $[x_0, x_1]$, then the bilinear (in this case, actually linear) map is given by

$$x = (1-\xi)\,x_0 + \xi\,x_1\,, \quad 0 \le \xi \le 1\,. \tag{1.13}$$

For a bilinear map in two dimensions, the physical region must be a quadrilateral defined by four points $\mathbf{x}_{0,0}$, $\mathbf{x}_{1,0}$, $\mathbf{x}_{0,1}$, $\mathbf{x}_{1,1}$, where the vector notation means $\mathbf{x}_{i,j} = (x_{i,j}, y_{i,j})$. Then the map from U_2 to the quadrilateral in E^2 is given in vector notation by

$$\mathbf{x}(\xi,\eta) = (1-\xi)\,(1-\eta)\,\mathbf{x}_{0,0} + (1-\xi)\,\eta\,\mathbf{x}_{0,1} + \xi\,(1-\eta)\,\mathbf{x}_{1,0} + \xi\,\eta\,\mathbf{x}_{1,1} \tag{1.14}$$

or in coordinate notation by

$$\begin{aligned} x(\xi,\eta) &= (1-\xi)\,(1-\eta)\,x_{0,0} + (1-\xi)\,\eta\,x_{0,1} \\ &\quad + \xi\,(1-\eta)\,x_{1,0} + \xi\,\eta\,x_{1,1}\,, \\ y(\xi,\eta) &= (1-\xi)\,(1-\eta)\,y_{0,0} + (1-\xi)\,\eta\,y_{0,1} \\ &\quad + \xi\,(1-\eta)\,y_{1,0} + \xi\,\eta\,y_{1,1}\,. \end{aligned} \tag{1.15}$$

Recall that if η is fixed and ξ varies, then a coordinate line in logical space is generated. The image of this coordinate line is $\mathbf{x}(\xi)$, which is a curve in physical space. The same holds if ξ is fixed and η varies. Thus coordinate curves for the bilinear map are straight lines. In particular, the boundary of the physical region must be given by straight lines that are linearly parameterized. An example of a planar bilinear map for the corner points $(1,1)$, $(2,2)$, $(1,5)$, and $(-1,3)$ is given in Figure 1.4.

Exercise 1.4.1 Write out the general trilinear map of the cube U_3 in E^3 to a region defined by eight points in physical space. §

Computer codes for plotting one-dimensional grids are based on discretizing logical space and then mapping the logical grid to physical space. Thus if $x = x(\xi)$

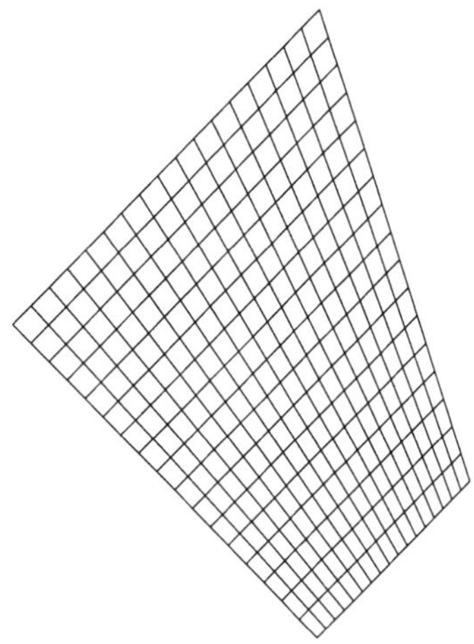

Figure 1.4 *A bilinear map*

is an analytic transformation from $U_1 = [0,1]$ to physical space and if M is a positive integer then set

$$\xi_i = \frac{i}{M}, \quad x_i = x(\xi_i), \quad 0 \le i \le M \tag{1.16}$$

to produce the grid x_i.

Exercise 1.4.2 Write a computer code for plotting linear grids in one dimension and use that code to display the points for the several linear maps. There is a sample code listed in Appendix B.1.1. §

Exercise 1.4.3 Compute the Jacobian $J = \det(\mathcal{J})$ of the bilinear map (1.14). Let J_{ij} be the Jacobian at each of the four corners. Show that $J_{11}+J_{00}-J_{10}-J_{01} = 0$. Show that the physical domain is convex if and only if J is positive at the four corners. Finally, show that J is positive everywhere in logical space if and only if J is positive at the four corners (the *corner test*). §

The answers to the previous exercise can be found in Knupp [109]. Moreover, this paper considers *trilinear isoparametric maps* from U_3 to a hexahedron in E^3 similar in form to (1.14) (see Exercise 1.4.1). Unlike the bilinear map, the corner Jacobian test for a positive Jacobian fails for the trilinear map.

Exercise 1.4.4 Show that for the two-dimensional bilinear map (1.14)

$$\mathbf{x}_{\xi\eta} = \frac{\partial^2 \mathbf{x}}{\partial \xi \, \partial \eta} = \mathbf{0} \tag{1.17}$$

if and only if the quadrilateral is a parallelogram. As usual, the derivatives of a vector are given by $\mathbf{x}_{\xi\eta} = (x_{\xi\eta}, y_{\xi\eta})$. §

There are many other coordinate systems that are used to assist in solving partial differential equations. One-dimensional transformations that exponentially stretch a grid near a point are used in the study of boundary layers. A grid is said to be *exponentially stretched* if the ratio

$$\frac{x_{i+1} - x_i}{x_i - x_{i-1}} \tag{1.18}$$

is a constant not equal to one or zero. For example, for $\lambda \neq 0$, the transformation

$$x(\xi) = \frac{e^{\lambda\xi} - 1}{e^{\lambda} - 1} \tag{1.19}$$

generates an exponentially stretched grid.

Exercise 1.4.5 Compute the ratio (1.18) for the transformation (1.19) and show that it is a constant. Consider the cases $\lambda > 0$ and $\lambda < 0$. Is this transformation singular? §

Note that the function $x = 1 - \xi$ maps the unit interval to the unit interval and has a negative Jacobian. This means that the map reverses the order of the points. Now if in (1.19) ξ is replaced by $1 - \xi$, the same grid results, but it is now in reverse order. This is just a *re-parameterization* of the interval. However, if in (1.19) x is replaced by $1 - x$, then a grid that is stretched at the opposite end of the unit interval is obtained.

Exercise 1.4.6 The transformation

$$x = 2^{1-1/\xi} \tag{1.20}$$

maps the unit interval in logical space to the unit interval in physical space. However, the transformation fails to satisfy one of the hypotheses necessary for the generation of a proper grid. Show that the Jacobian of the transformation given in (1.20) is positive for $\xi > 0$ and that the transformation is singular for $\xi = 0$. §

Another useful stretching transformation is

$$x = \tan\left(\frac{\pi}{2}\xi\right). \tag{1.21}$$

This transformation maps logical space onto half of the real line E^1, sending one of the boundary points in logical space to infinity in physical space. Note that this violates the assumption that the physical object is bounded, while the transformation given in Equation (1.20) violates the condition that the Jacobian must be bounded. Difficulties may arise at points where a transformation violates one of the basic assumptions. The transformations may still be useful, but special care must be taken when using them. Other stretching transformations are discussed in Vinokur [218].

Exercise 1.4.7 Plot a few sample grids for the transformations given in (1.19), (1.20), and (1.21) (see Appendix B.1.2). The plots should give an interval with the grid points located on it. It is also illuminating to plot the function $x = x(\xi)$ as a curve in the x-ξ plane. §

In two dimensions, there are many transformations that are useful for generating boundary-conforming coordinate systems; their use predates grid generation. For example, polar coordinates are useful in problems with circular symmetry. Similarly, in three dimensions, spherical coordinates are useful in problems with spherical symmetry. These transformations usually entail analytic formulas and are restricted to fixed physical domains with fixed boundary parameterizations. The most elementary of such coordinate systems in the plane is, of course, the Cartesian system,

$$x(\xi,\eta) = \xi\,, \quad y(\xi,\eta) = \eta\,, \tag{1.22}$$

which uniformly parameterizes the unit square. Most planar grid generators should generate this coordinate system when the identity *boundary parameterization* is specified on a physical domain, that is, the unit square. This provides a good elementary test of the algorithm.

The algorithm for generating grids in two dimensions assumes that two functions $x = x(\xi,\eta)$ and $y = y(\xi,\eta)$ defined for $0 \le \xi,\eta \le 1$ are given and that M and N are program parameters. The points in the grid are given by $(x_{i,j}, y_{i,j})$ where

$$\begin{aligned} x_{i,j} &= x\left(\frac{i}{M},\frac{j}{N}\right), \quad 0 \le i \le M\,, \\ y_{i,j} &= y\left(\frac{i}{M},\frac{j}{N}\right), \quad 0 \le j \le N\,. \end{aligned} \tag{1.23}$$

It can be very illuminating to plot the coordinate curves. They are given by drawing line segments joining certain points: the ξ curves are given by the segments

$$[(x_{i,j}, y_{i,j})\,,\ (x_{i+1,j}, y_{i,j})]\,, \quad 0 \le i \le M-1\,, \quad 0 \le j \le N\,, \tag{1.24}$$

while the η curves are given by the segments

$$[(x_{i,j}, y_{i,j})\,,\ (x_{i,j}, y_{i,j+1})]\,, \quad 0 \le i \le M\,, \quad 0 \le j \le N-1\,. \tag{1.25}$$

Exercise 1.4.8 Plot some examples of bilinear maps given by Equation (1.14) (see Appendix B.1.3). §

Four types of analytic coordinate systems for the plane are now described.

Polar Coordinates: The most famous special coordinate system is *polar coordinates*, which can be used to parameterize a circle or, more generally, a sector of an annulus. Given two radii $0 \le r_0 < r_1$ and two angles $0 \le \theta_0 < \theta_1 \le 2\pi$, the sector of the annulus between the two radii and between the two angles is given by

$$x(\xi,\eta) = r\cos(\theta)\,, \quad y(\xi,\eta) = r\sin(\theta)\,, \tag{1.26}$$

where

$$\theta = \theta_1 + (\theta_0 - \theta_1)\,\xi\,, \quad r = r_0 + (r_1 - r_0)\,\eta\,. \tag{1.27}$$

If either the domain or the boundary parameterization is changed, even slightly, this polar coordinate system is rendered useless, showing the limitations of analytic transformations. One of the main goals of grid generation is to escape this limitation. Of course, when these special transformations apply, they

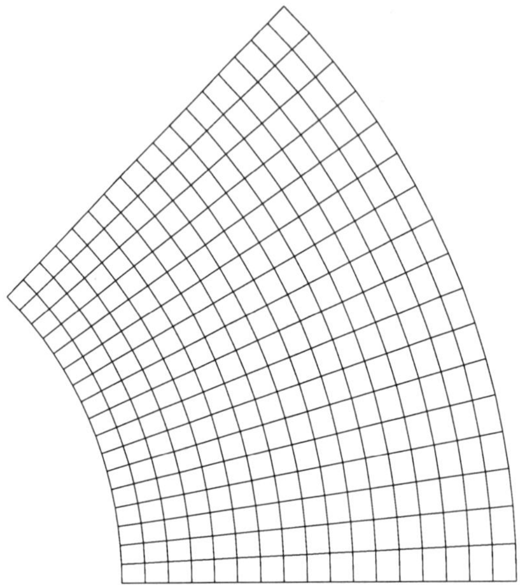

Figure 1.5 *Polar coordinates for a sector*

generally should be used since they involve analytic expressions for orthogonal coordinate systems. Grid points and grid lines for an annular sector are given in Figure 1.5.

Exercise 1.4.9 Show that for the polar transformation given in (1.26), the Jacobian is $J = r\,(\theta_1 - \theta_0)\,(r_1 - r_0)$. Note that $J > 0$ if $r > 0$. §

As with polar coordinates, the classical coordinate systems do not make use of the idea of a logical region. To make use of this idea, it is usually necessary to rescale the variables in the classical transformation, as was done in polar coordinates by going from θ and r to ξ and η variables.

Parabolic Cylinder Coordinates: Parabolic cylinder coordinates on the region in Figure 1.6 are given by the mapping

$$x(\xi,\eta) = \frac{1}{2}\,(r^2 - s^2)\,, \quad y(\xi,\eta) = r\,s\,, \tag{1.28}$$

where

$$r = 1 + \xi\,, \quad s = 1 + \eta\,. \tag{1.29}$$

Of course, altering the definitions of r and s will vary the physical region.

Elliptic Cylinder Coordinates: Elliptic cylinder coordinates on the region shown in Figure 1.7 are given by the mapping

$$x(\xi,\eta) = a\cosh(r)\cos(s)\,, \quad y(\xi,\eta) = a\sinh(r)\sin(s)\,, \tag{1.30}$$

where

$$r = 1 + \xi\,, \quad s = \pi\,\eta\,. \tag{1.31}$$

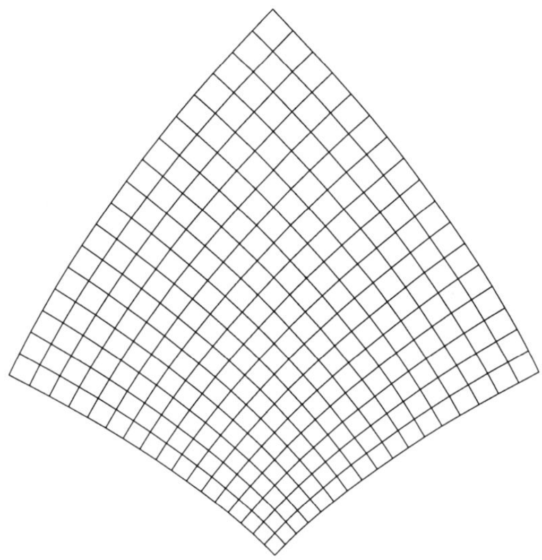

Figure 1.6 *Parabolic cylinder coordinates*

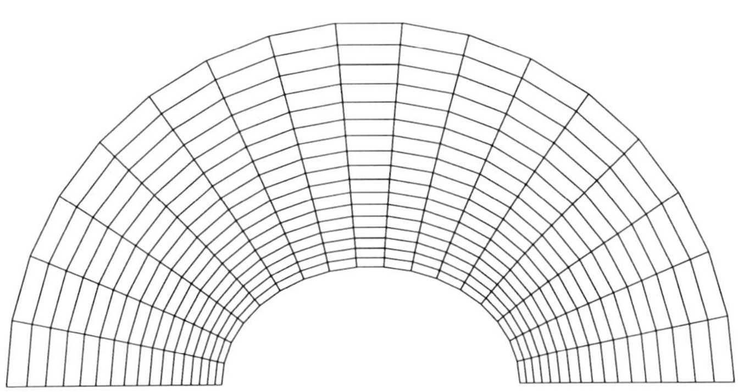

Figure 1.7 *Elliptic cylinder coordinates*

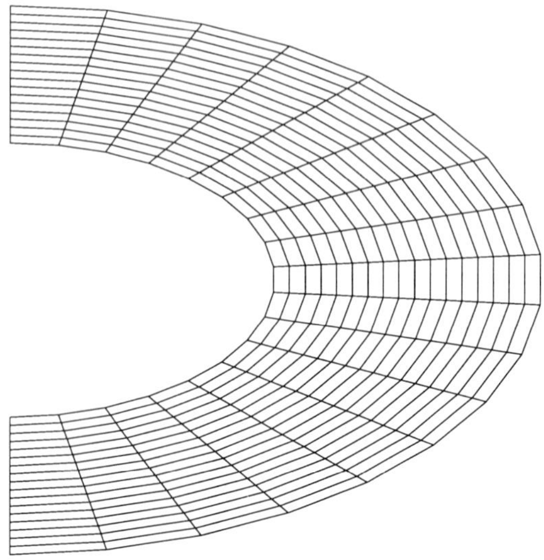

Figure 1.8 *Horseshoe*

This results in the physical domain consisting of concentric ellipses of radius $a > 0$.

Horseshoe Domains: Horseshoe-shaped domains often provide difficult test problems for grid generators. Several such domains are used in the literature. Three such domains, including the elliptic cylinder region above, are presented here. Another horseshoe-shaped region is shown in Figure 1.8 and is given by

$$x(\xi,\eta) = \rho\, r \cos(\theta)\,, \quad y(\xi,\eta) = r \sin(\theta)\,, \tag{1.32}$$

where $\rho > 0$ is the aspect ratio of the major to minor axes of the ellipses that generate the horseshoe, $0 < b_0 < b_1$, and

$$r = b_0 + (b_1 - b_0)\,\eta\,, \qquad \theta = \frac{\pi}{2}\,(1 - 2\,\xi)\,. \tag{1.33}$$

The next example, called the *Modified Horseshoe*, is shown in Figure 1.9 and is given by

$$x(\xi,\eta) = -(1+\eta)\cos(\pi\,\xi)\,, \quad y(\xi,\eta) = [1 + (2R-1)\,\eta]\sin(\pi\,\xi)\,, \tag{1.34}$$

where $R \geq 1$ is the aspect ratio.

Bipolar Coordinates: Bipolar coordinates for the physical domain shown in Figure 1.10 are given by the mapping

$$x(\xi,\eta) = \frac{a \sinh(r)}{\cosh(r) + \cos(s)}\,, \qquad y(\xi,\eta) = \frac{a \sin(s)}{\cosh(r) + \cos(s)}\,, \tag{1.35}$$

where

$$r = \xi\,, \quad s = \pi\left(\eta - \frac{1}{2}\right)\,, \tag{1.36}$$

The quantity $a > 0$ is a parameter that determines the height of the domain.

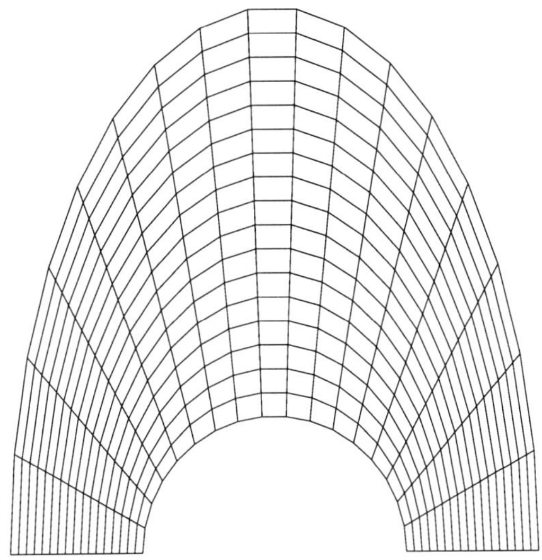

Figure 1.9 *Modified Horseshoe*

Exercise 1.4.10 Make some plots of the physical regions, grids, and grid lines generated by the previously described coordinate systems (see Appendix B.1). Vary the parameters in the map to give an idea of the range of physical regions that can be obtained with these maps. §

In three dimensions, there are two well-known coordinate systems, *cylindrical* and *spherical* coordinates.

Cylindrical Coordinates: Cylindrical coordinates are essentially polar coordinates and can be used to parameterize cylindrical shells. Given two radii $0 \le r_0 < r_1$, two angles $0 \le \theta_0 < \theta_1 \le 2\pi$, and the limits on the axial variable, $z_0 < z_1$, a cylindrical body is given by

$$x(\xi,\eta,\zeta) = r\cos(\theta)\,, \quad y(\xi,\eta,\zeta) = r\sin(\theta)\,, \quad z(\xi,\eta,\zeta) = w\,, \tag{1.37}$$

where

$$\theta = \theta_1 + (\theta_0 - \theta_1)\,\xi\,, \quad r = r_0 + (r_1 - r_0)\,\eta\,, \quad w = z_0 + (z_1 - z_0)\,\zeta\,. \tag{1.38}$$

Exercise 1.4.11 Show that for the cylindrical coordinates given in (1.37) the Jacobian is

$$J = r\,(\theta_1 - \theta_0)\,(r_1 - r_0)\,(z_1 - z_0)\,. \quad \S \tag{1.39}$$

Exercise 1.4.12 Write a computer code that can plot coordinate surfaces; that is, fix one of the variables ξ, η, or ζ at some value between 0 and 1 and vary the other two variables, then plot the resulting surface. Plot the full surface and then use a hidden surface algorithm. Can you plot the full grid? §

Spherical Coordinates: Spherical coordinates are used to describe parts of spheres. Given two radii $0 \le r_0 < r_1$, two equatorial angles $0 \le \theta_0 < \theta_1 \le 2\pi$,

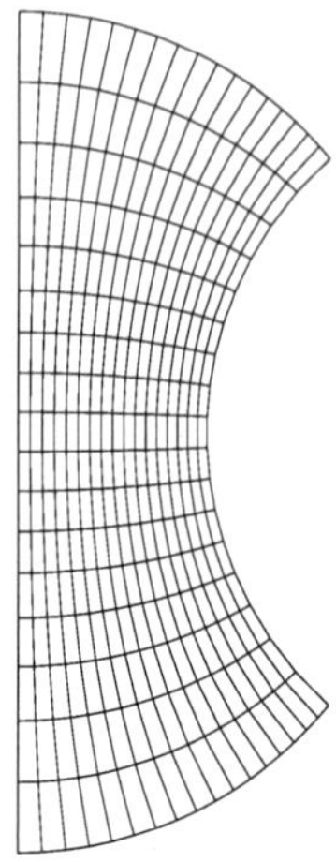

Figure 1.10 *Bipolar coordinates*

and two polar angles $0 \leq \phi_0 < \phi_1 \leq \pi$, a spherical wedge is given by

$$\begin{aligned} x(\xi,\eta,\zeta) &= r\ \cos(\theta)\ \sin(\phi)\,, \\ y(\xi,\eta,\zeta) &= r\ \sin(\theta)\ \sin(\phi)\,, \\ z(\xi,\eta,\zeta) &= r\ \cos(\phi)\,, \end{aligned} \tag{1.40}$$

where

$$\theta = \theta_1 + (\theta_0 - \theta_1)\,\xi\,, \quad r = r_0 + (r_1 - r_0)\,\eta\,, \quad \phi = \phi_1 + (\phi_0 - \phi_1)\,\zeta\,. \tag{1.41}$$

Figure 1.11 shows the coordinate variables for the spherical coordinate system.

Exercise 1.4.13 Look up some conformal mappings of the unit disk to some other region in a book on complex function theory and plot some grids generated by the mappings. §

1.5 TRANSFINITE INTERPOLATION

Grid generation methods based on *interpolation* have been extensively developed to take advantage of their two main strengths: rapid computation of the grids compared to the partial differential equation methods that will be studied later in this text, and direct control over grid point locations. These advantages are somewhat offset by the fact that interpolation methods may not generate smooth grids (see the TFI figures in Appendix C); in particular, boundary-slope discontinuities propagate into the interior. Interpolation methods are frequently called *algebraic methods.* The standard method of algebraic grid generation is known as *transfinite interpolation* (TFI) (Gordon and Hall [85]). In the one-dimensional

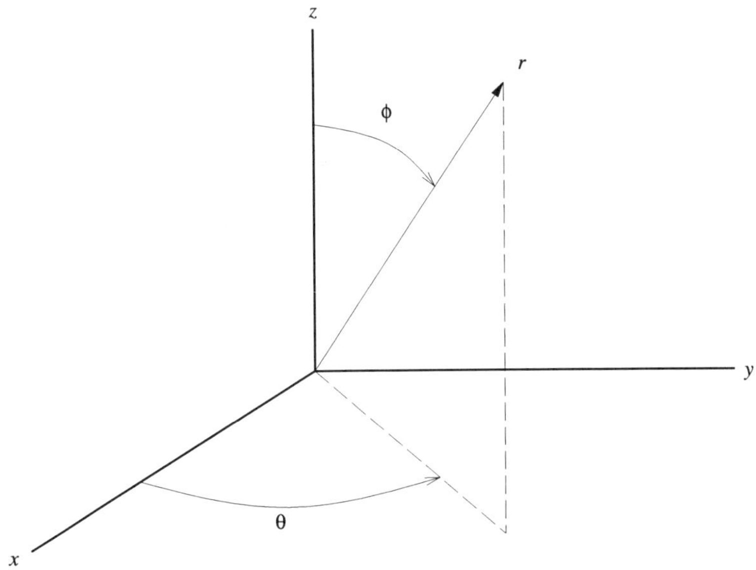

Figure 1.11 *Spherical coordinates*

case, transfinite interpolation is the same as linear interpolation, so the interesting cases begin with planar regions.

Any planar grid-generation problem begins with a description of the boundary of the region, that is, four parametric equations

$$\mathbf{x}_b(\xi), \quad \mathbf{x}_t(\xi), \; 0 \le \xi \le 1,$$
$$\mathbf{x}_l(\eta), \quad \mathbf{x}_r(\eta), \; 0 \le \eta \le 1, \tag{1.42}$$

are needed to describe each part of the boundary (see Figure 1.12). The subscripts on $\mathbf{x}$ stand for *bottom, top, left,* and *right* boundaries of the logical domain.

The simplest example of such a parameterization is the identity map, that

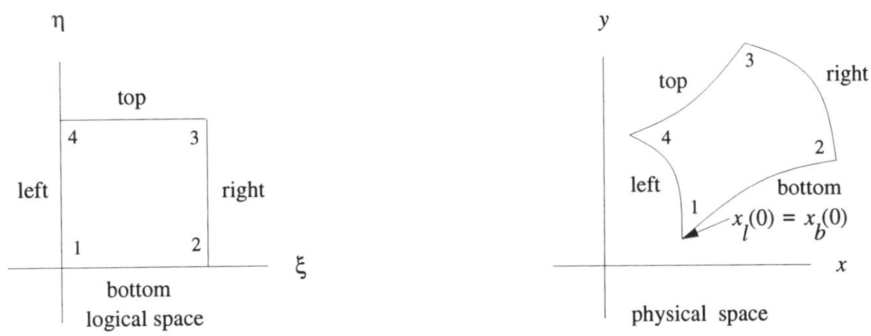

Figure 1.12 *Boundaries of planar regions*

(i)	$\mathbf{x}_b(0) = \mathbf{x}_l(0)$,
(ii)	$\mathbf{x}_b(1) = \mathbf{x}_r(0)$,
(iii)	$\mathbf{x}_r(1) = \mathbf{x}_t(1)$,
(iv)	$\mathbf{x}_l(1) = \mathbf{x}_t(0)$.

Table 1.3 *Consistency conditions*

is, where the physical region is the unit square just like the logical region. This parameterization is given by

$$\begin{aligned} \mathbf{x}_b(\xi) &= (\xi, 0)\,,\ 0 \le \xi \le 1\,, \\ \mathbf{x}_t(\xi) &= (\xi, 1)\,,\ 0 \le \xi \le 1\,, \\ \mathbf{x}_l(\eta) &= (0, \eta)\,,\ 0 \le \eta \le 1\,, \\ \mathbf{x}_r(\eta) &= (1, \eta)\,,\ 0 \le \eta \le 1\,. \end{aligned} \tag{1.43}$$

This vector notation needs to be converted to components for use in computer programs, for example,

$$\begin{aligned} x_b(\xi) &= \xi\,,\ 0 \le \xi \le 1\,, \\ y_b(\xi) &= 0\,,\ 0 \le \xi \le 1\,, \\ x_l(\eta) &= 0\,,\ 0 \le \eta \le 1\,, \\ y_l(\eta) &= \eta\,,\ 0 \le \eta \le 1\,. \end{aligned} \tag{1.44}$$

There are four important consistency checks for the boundary formulas, namely, that the four corners of the region are consistently described (see Table 1.3 and Figure 1.12). Any computer code for planar grid generation should check these conditions and give an informative error message if they are not satisfied. In a computer code, there are actually eight things to check because each of the four corner points has two components.

Exercise 1.5.1 Check the consistency conditions for the identity boundary parameterization (1.43). §

The first degree *Lagrange polynomials* $1-\xi$, ξ, $1-\eta$, and η are used as *blending* functions in the basic transfinite interpolation formula. The TFI formula is

$$\begin{aligned} \mathbf{x}(\xi,\eta) = {} & (1-\eta)\,\mathbf{x}_b(\xi) + \eta\,\mathbf{x}_t(\xi) + (1-\xi)\,\mathbf{x}_l(\eta) + \xi\,\mathbf{x}_r(\eta) \\ & - \ \{\xi\,\eta\,\mathbf{x}_t(1) + \xi\,(1-\eta)\,\mathbf{x}_b(1) \\ & + \ \eta\,(1-\xi)\,\mathbf{x}_t(0) + (1-\xi)\,(1-\eta)\,\mathbf{x}_b(0)\}\,. \end{aligned} \tag{1.45}$$

Exercise 1.5.2 Verify that $\mathbf{x}(\xi,\eta)$ in (1.45) matches the given boundary functions, that is, $\mathbf{x}(\xi,0) = \mathbf{x}_b(\xi)$ and so forth. §

The following three examples of TFI show both the power of the method and its limitations. The three examples are the Modified Horseshoe, Swan, and

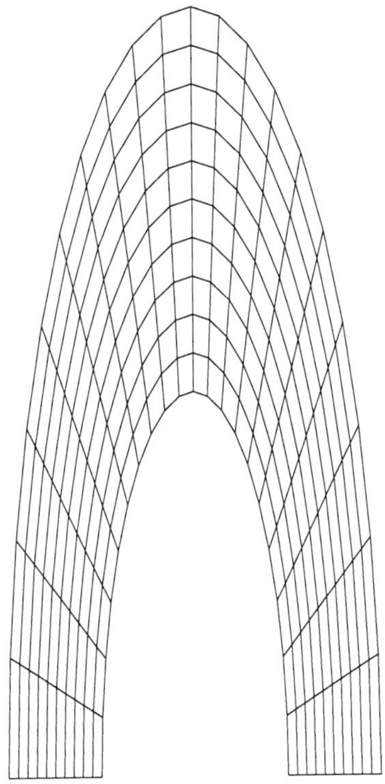

Figure 1.13 *TFI grid on the horseshoe*

Chevron. All are nonconvex regions and consequently difficult to grid nicely. The two distinct parameters ξ and η used in (1.42) are not necessary, so they are replaced by a single parameter s satisfying $0 \le s \le 1$. Additional examples of TFI grids are given in Appendix C.

A TFI grid for the Horseshoe is shown in Figure 1.13; the TFI grid is smooth and unfolded. The parameter $\rho \ge 1$ determines the eccentricity of the outer ellipse. The boundary parameterizations are obtained from (1.32):

Bottom boundary:

$$x_b(s) = \rho b_0 \cos\left\{\frac{\pi}{2}(1-2s)\right\}, \quad y_b(s) = b_0 \sin\left\{\frac{\pi}{2}(1-2s)\right\} \tag{1.46}$$

Top boundary:

$$x_t(s) = \rho b_1 \cos\left\{\frac{\pi}{2}(1-2s)\right\}, \quad y_t(s) = b_1 \sin\left\{\frac{\pi}{2}(1-2s)\right\} \tag{1.47}$$

Left boundary:

$$x_l(s) = 0, \quad y_l(s) = b_0 + (b_1 - b_0)s \tag{1.48}$$

Right boundary:

$$x_r(s) = 0, \quad y_r(s) = -y_l(s) \tag{1.49}$$

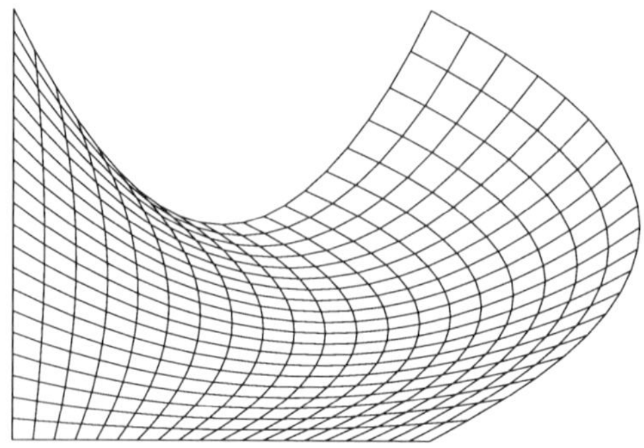

Figure 1.14 *TFI grid on the Swan*

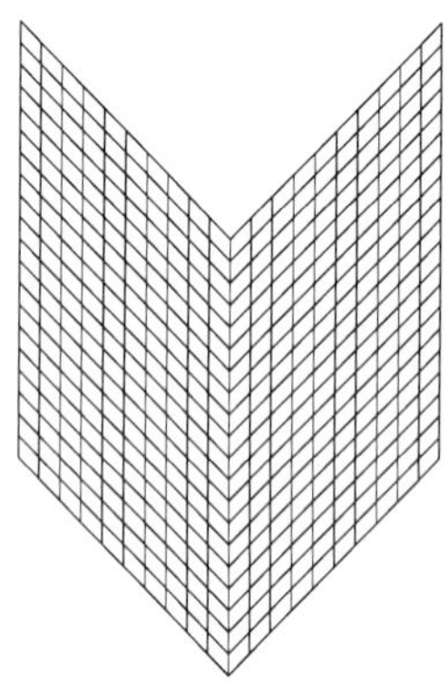

Figure 1.15 *TFI grid on the Chevron*

The TFI grid for the Swan, shown in Figure (1.14), is folded. The boundary parameterizations are as follows.

Bottom boundary:

$$x_b(s) = s\,, \quad y_b(s) = 0 \tag{1.50}$$

Top boundary:

$$x_t(s) = s\,, \quad y_t(s) = 1 - 3s + 3s^2 \tag{1.51}$$

Left boundary:

$$x_l(s) = 0\,, \quad y_l(s) = s \tag{1.52}$$

Right boundary:

$$x_r(s) = 1 + 2s - 2s^2\,, \quad y_r(s) = s \tag{1.53}$$

The TFI grid for the Chevron is shown in Figure 1.15; the slope discontinuity on the boundary of the Chevron is propagated into the interior of the physical region by TFI. The boundary parameterizations are as follows.

Bottom boundary:

$$x_b(s) = s\,, \quad y_b(s) = \begin{cases} -s, & s \le \frac{1}{2} \\ s-1, & s > \frac{1}{2} \end{cases} \tag{1.54}$$

Top boundary:

$$x_t(s) = s\,, \quad y_t(s) = \begin{cases} 1-s, & s \le \frac{1}{2} \\ s, & s > \frac{1}{2} \end{cases} \tag{1.55}$$

Left boundary:

$$x_l(s) = 0\,, \quad y_l(s) = s \tag{1.56}$$

Right boundary:

$$x_r(s) = 1\,, \quad y_r(s) = s \tag{1.57}$$

Exercise 1.5.3 Write a computer code (see Appendix B.2) to use transfinite interpolation to generate grids on a variety of planar regions. Be sure to include a computer graphics display. §

Transfinite interpolation produces excellent grids on many regions, including the Square, Trapezoid, Annulus, Modified Horseshoe, Dome, and Valley, but is inadequate on the Swan, Airfoil, Chevron (lacks smoothness), Backstep, Plow, and C. The main disadvantages of TFI are lack of smoothness and a tendency to fold grids on complex domains. Two other planar methods of algebraic grid generation are Gilding [82] and Intrinsic (Knupp [110]).

TFI can be extended in several ways. The easiest extension is to break up the region in several parts and then interpolate each part. However, there will typically be slope discontinuities at the interior boundaries of the parts. Also, TFI can be extended to use higher order polynomials as blending functions. For example, the cubic Hermite polynomials can be used to match slopes and consequently remove the slope discontinuities caused by breaking the region into parts. The full potential of algebraic grid generation is explored in Gordon [85, 86], Eiseman [58, 59, 60, 61, 68], Shih [169], and Smith [175, 176]. Algebraic grid generation is discussed further in Section 5.3.1 and in Chapter 9 of this book.

Chapter 2

Application to Hosted Equations

This chapter is intended to motivate study of the grid-generation algorithms presented in this book by illustrating the use of such grids in solving partial differential equations on irregularly shaped regions. Suppose a partial differential equation representing some physical phenomenon such as fluid flow, electromagnetic potential, or heat conduction is to be solved. Such an equation is termed the *hosted* equation; if the domain on which the hosted equation is to be solved is irregularly shaped, then a grid must be generated and the boundary conditions handled accordingly. Thus, grid-generation equations may be required in addition to the hosted equation.

There are several basic approaches to solving partial differential equations on irregular regions. For example, rectangular grids can be used to cover the region (see Birkhoff and Lynch [19]). This approach avoids the need to transform the hosted equation, but it requires major revisions to the computer program if the region is changed. It is difficult to implement the boundary conditions with second-order accuracy because the nodes of the grid do not coincide with the boundary. Other approaches include unstructured meshes as in finite elements (see George [78]), finite differences in physical space (see Heinrich [94], Samarskii et al. [163, 164]), or finite differences in logical space (see Steinberg and Roache [194]).

The latter approach is used in this book because it is simple and effective for a wide range of problems. It requires the partial differential equation to be transformed to logical coordinates. An important fact about such transformations is that the *type* of the hosted equation is invariant under a general nonsingular coordinate transformation (i.e., the physics is preserved). Second-order partial differential equations in two variables are classified as one of three types: elliptic, parabolic, or hyperbolic (see Epstein [70] for more information). Elliptic equations describe the steady states of diffusion processes and parabolic equations describe transient diffusion processes, while hyperbolic equations describe wave motion. The physical behavior modeled by a differential equation is not preserved using general transformations if the type of the transformed equation is not the same as the original equation. Only transformations that preserve type will be considered in this book. The type of a linear partial differential equation is determined by putting the differential equation in a canonical form and then computing the *discriminant* D. The type is determined by the sign of the discriminant: (1) if $D > 0$ then the equation is elliptic; (2) if $D = 0$ then the equation is parabolic; while (3) if $D < 0$ then the equation is hyperbolic. The type invariance of the model-hosted equation is returned to later in Section 7.3.6.

For illustration, the logical-space approach is applied to steady-state (elliptic) boundary value problems in one and two spatial dimensions. In Section 2.1, a one-dimensional hosted equation is transformed to general coordinates. In Section 2.2, it is shown how to discretize this equation to obtain symmetric stencils and second-order accuracy (an explanation of the reasons for the choice of differencing is beyond the scope of this book). The resulting system of discrete algebraic equations is linearized and solved using a standard linear equation solver. The two-dimensional hosted equation is transformed to general coordinates in Section 2.3 and then discretized in Section 2.4. Expressions for the gradient, divergence, curl, and Laplacian operators in general coordinates are studied in Section 7.3; other extensions and applications can be found in Steinberg and Roache [194].

By the end of this chapter the reader will be able to solve boundary value problems on nontrivial regions. The method described in this text can be used to write a single program for solving problems on a wide range of regions and the boundary conditions are, essentially automatically, second-order accurate. A wide range of assumptions can be made; the particular method used in this chapter was chosen because it is applicable to discretizing variational grid-generation equations.

2.1 TRANSFORMATION OF THE ONE-DIMENSIONAL HOSTED EQUATION

Let a, A, b, and B be given real numbers and $g = g(x)$ be a given function. Then solving the one-dimensional boundary value problem

$$f_{xx} = g\,, \quad f(a) = A\,, \quad f(b) = B\,, \quad a < x < b\,, \tag{2.1}$$

($f_{xx} = d^2f/dx^2$) for $f = f(x)$ provides an elementary example illustrating the use of grids for solving boundary value problems.

Assume that a one-to-one transformation $x = x(\xi)$ from the unit interval U_1 to the interval $[a, b]$ is given. The Jacobian matrix of this transformation is the trivial one-by-one matrix,

$$\mathcal{J} = (x_\xi) = \left(\frac{dx}{d\xi}\right), \tag{2.2}$$

and the Jacobian of this transformation is the determinant of this matrix, which is simply the entry in the matrix:

$$J = J(\xi) = x_\xi = \frac{dx}{d\xi}\,. \tag{2.3}$$

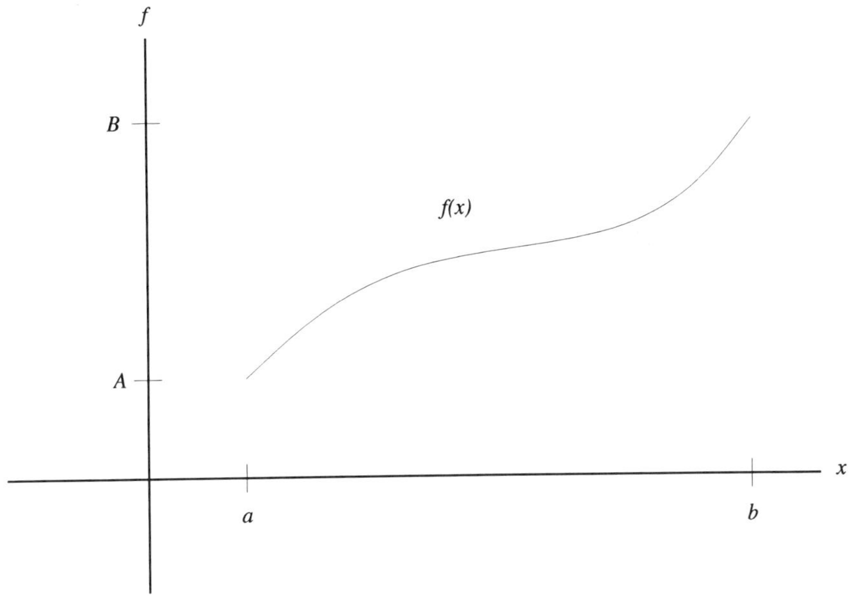

Figure 2.1 *Solution of the BVP*

The assumption that the transformation is one-to-one implies that $J \neq 0$. If the Jacobian is negative, then the transformation can be replaced by the new transformation $x = x(1 - \xi)$, whose Jacobian is $-x_\xi$ and thus is positive. From now on, the Jacobian will be assumed positive, $J > 0$.

The functions f and g in the differential equation (2.1) are transformed to functions $\tilde{f}$ and $\tilde{g}$ defined on logical space using

$$\tilde{f}(\xi) = f(x(\xi)), \quad \tilde{g}(\xi) = g(x(\xi)). \tag{2.4}$$

The *derivatives* of f can be transformed to logical space using the chain rule:

$$\tilde{f}_\xi = f_x \, x_\xi . \tag{2.5}$$

Although the Jacobian is trivial in this case, it is retained in the one-dimensional formulas, so that the latter can be easily compared to the formulas in the higher dimensional cases. Writing $J = x_\xi$ and then dividing gives

$$f_x = \frac{\tilde{f}_\xi}{J} . \tag{2.6}$$

Next, Equation (2.6) applied twice gives

$$f_{xx} = (f_x)_x = \frac{1}{J}\left(\frac{\tilde{f}_\xi}{J}\right)_\xi = \frac{1}{J^2}\tilde{f}_{\xi\xi} - \frac{x_{\xi\xi}}{J^3}\tilde{f}_\xi . \tag{2.7}$$

The last expression is the nonsymmetric form of the second derivative. Observe that the transformation produces a lower order term in f, a phenomena that occurs in higher dimensions as well.

The first part of (2.7) gives

$$J\, f_{xx} = \left(\frac{\tilde{f}_\xi}{J}\right)_\xi . \tag{2.8}$$

This is the *symmetric* form of the second derivative. To summarize, if

$$\hat{f}(\xi) = \tilde{f}(\xi), \quad \hat{g}(\xi) = J(\xi)\,\tilde{g}(\xi), \quad \hat{\alpha}(\xi) = \frac{1}{J(\xi)}, \tag{2.9}$$

then the transformed boundary value problem in symmetric form is

$$(\hat{\alpha}\,\hat{f}_\xi)_\xi = \hat{g}, \quad \hat{f}(0) = A, \quad \hat{f}(1) = B, \quad 0 < \xi < 1 . \tag{2.10}$$

Comparing the hosted equation (2.1) to the transformed equation shows that the latter has an additional coefficient $\hat{\alpha}$, a right-hand side that includes J, and holds over the logical rather than the physical domain. If at some point $J = 0$ or $J = \infty$, then the transformed differential equation is singular. This is one of the reasons that the Jacobian is assumed nonzero; the smoothness assumptions imply that the Jacobian is not infinite.

Exercise 2.1.1 In addition to the assumptions given at the beginning of this section, assume that $\alpha = \alpha(x)$ is a smooth function and $\tilde{\alpha}(\xi) = \alpha(x(\xi))$. Show that the boundary value problem

$$(\alpha\, f_x)_x = g, \quad f(a) = A, \quad f(b) = B, \quad a < x < b, \tag{2.11}$$

is transformed to the boundary value problem (2.10) where

$$\hat{\alpha}(\xi) = \frac{\tilde{\alpha}(\xi)}{J(\xi)} . \tag{2.12}$$

The differential equation in (2.11) appears in applied problems with variable material properties. The boundary value problem (2.11) is said to be invariant because its form does not change under a general change of independent variables. §

Exercise 2.1.2 Show that the solution of the boundary value problem (2.1) is given by

$$f(x) = \int_a^x (x - y)\, g(y)\, dy + K\,(x - a) + A, \tag{2.13}$$

where

$$K = \frac{B - A}{b - a} - \frac{1}{b - a}\int_a^b (b - y)\, g(y)\, dy . \tag{2.14}$$

What is the solution to the boundary value problem (2.11)? §

2.2 DISCRETIZATION OF THE ONE-DIMENSIONAL TRANSFORMED EQUATION

The goal of this section is to discretize the transformed equation (2.10) to compute numerical solutions. Two additional objectives are desired. First, the discretization must be second-order accurate (including the boundary conditions). Second, the stencils should be symmetric.

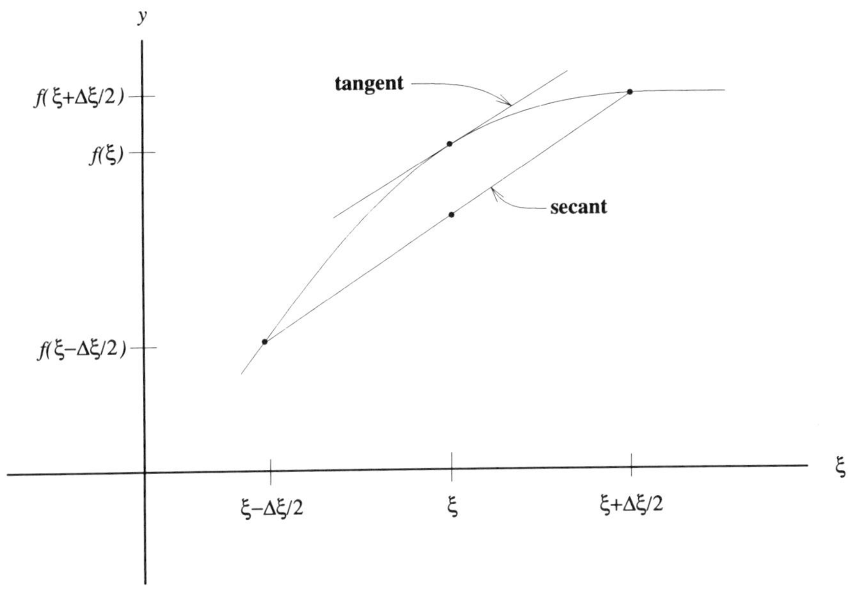

Figure 2.2 *Central differences and averages*

If $\tilde{f}(\xi)$ is a smooth function, then its derivative is approximated to second order at the center of an small interval of length $\Delta\xi$ by the *central difference*

$$\tilde{f}_\xi(\xi) \approx \frac{\tilde{f}(\xi + \frac{\Delta\xi}{2}) - \tilde{f}(\xi - \frac{\Delta\xi}{2})}{\Delta\xi}, \tag{2.15}$$

while the value of $\tilde{f}$ at the center of the interval can be approximated to second order by the *central average*

$$\tilde{f}(\xi) \approx \frac{\tilde{f}(\xi + \frac{\Delta\xi}{2}) + \tilde{f}(\xi - \frac{\Delta\xi}{2})}{2}. \tag{2.16}$$

Figure 2.2 illustrates the central-difference formula. The slope of the tangent is given by the left-hand side of (2.15), while slope of the secant line is given by the right-hand side of (2.15). The left-hand side of (2.16) gives the value of the function at the midpoint of the interval, while the right-hand side gives the height of the midpoint of the secant.

Assume that a *grid in physical space* has been created (methods for doing so are the subject of the next chapter) and has been constructed so that the *uniform grid* with $M + 1$, $M > 0$ points is set up in logical space:

$$\Delta\xi = \frac{1}{M}, \quad \xi_i = i\,\Delta\xi, \quad 0 \le i \le M. \tag{2.17}$$

The midpoints of the intervals

$$\xi_{i+\frac{1}{2}} = \left(i + \frac{1}{2}\right)\Delta\xi, \quad 0 \le i \le M - 1 \tag{2.18}$$

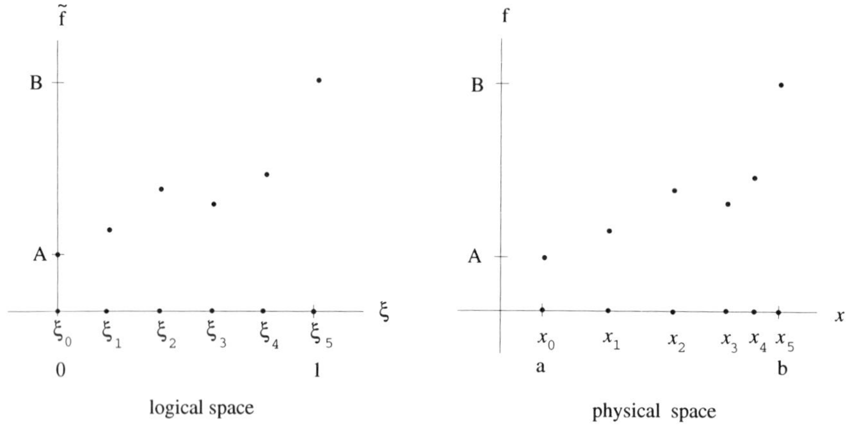

Figure 2.3 *One-dimensional discretization*

are also needed; the half indices are $1/2, 3/2, \ldots M-1/2$. If a transformation $x(\xi)$ is given, then the logical-space grid induces a grid in physical space given by

$$x_i = x(\xi_i), \quad 0 \le i \le M. \tag{2.19}$$

The discretization also induces *discrete values* for all of the transformed variables $\tilde{f}$, $\tilde{g}$, and $\tilde{\alpha}$; for example,

$$\tilde{f}_i = \tilde{f}(\xi_i) = f(x(\xi_i)) = f(x_i) = f_i, \qquad 0 \le i \le M, \tag{2.20}$$

with similar formulas for $\tilde{g}_i = g_i$ and $\tilde{\alpha}_i = \alpha_i$ (see Figure 2.3). These formulas in turn provide discrete values for the variables in the differential equation (2.10). The transformed variables are found by applying (2.9) and using formula (2.12) for the coefficient:

$$\begin{aligned} \hat{f}_i &= f_i, \quad 0 \le i \le M, \\ \hat{g}_i &= J_i g_i, \quad 1 \le i \le M-1, \\ \hat{\alpha}_i &= \frac{\alpha_i}{J_i}, \quad 0 \le i \le M, \end{aligned} \tag{2.21}$$

where

$$J_i = (x_\xi)_i. \tag{2.22}$$

The points at which these values are needed are determined by the differencing, as is seen below.

The differential equation (2.10) is differenced using the central differences given in (2.15) twice:

$$\left((\hat{\alpha}\hat{f}_\xi)_\xi\right)_i \approx \frac{(\hat{\alpha}\hat{f}_\xi)_{i+\frac{1}{2}} - (\hat{\alpha}\hat{f}_\xi)_{i-\frac{1}{2}}}{\Delta\xi}, \tag{2.23}$$

$$(\hat{\alpha}\hat{f}_\xi)_{i+\frac{1}{2}} \approx \hat{\alpha}_{i+\frac{1}{2}} \frac{\hat{f}_{i+1} - \hat{f}_i}{\Delta\xi}. \tag{2.24}$$

Substitute these approximations into the differential equation (2.10) to obtain a difference equation of the form

$$L_i \hat{f}_{i-1} + C_i \hat{f}_i + R_i \hat{f}_{i+1} = \hat{g}_i , \quad 1 \le i \le M-1 , \tag{2.25}$$

where

$$\hat{f}_0 = A , \quad \hat{f}_M = B , \tag{2.26}$$

and

$$L_i = \frac{\hat{\alpha}_{i-\frac{1}{2}}}{\Delta\xi^2} , \tag{2.27}$$

$$C_i = -(L_i + R_i) , \tag{2.28}$$

$$R_i = \frac{\hat{\alpha}_{i+\frac{1}{2}}}{\Delta\xi^2} . \tag{2.29}$$

In the one-dimensional setting, the Jacobian is the same as the derivative:

$$J_i = (x_\xi)_i , \quad J_{i+\frac{1}{2}} = (x_\xi)_{i+\frac{1}{2}} . \tag{2.30}$$

Now

$$\hat{g}_i = J_i g_i , \quad 1 \le i \le M-1 , \tag{2.31}$$

and

$$\hat{\alpha}_{i+\frac{1}{2}} = \frac{\alpha_{i+\frac{1}{2}}}{J_{i+\frac{1}{2}}} , \quad 0 \le i \le M-1 . \tag{2.32}$$

The values of $\hat{f}_i$ are to be computed for integer i, so the values of $\hat{g}_i$ are needed for integer i, while the values of $\hat{\alpha}$ are needed for half-integer i.

Since values of $\hat{g}$ are needed at integer points, derivatives of the transformation (i.e., J) are computed at integer points using the central-difference formula (2.15) with $\Delta\xi$ replaced by $2\,\Delta\xi$:

$$(x_\xi)_i = \frac{x_{i+1} - x_{i-1}}{2\,\Delta\xi} , \qquad 1 \le i \le M-1 \tag{2.33}$$

(an interval of width $2\,\Delta\xi$ is needed because it is assumed that the grid is not given at the half-integer points). Since $\hat{\alpha}$ is required at the half-integer points, derivatives of the transformation are also needed at the half-integer points; use the central-difference formula (2.15) with i replaced by $i + 1/2$:

$$(x_\xi)_{i+\frac{1}{2}} \approx \frac{x_{i+1} - x_i}{\Delta\xi} , \qquad 0 \le i \le M-1 . \tag{2.34}$$

It is assumed that the coefficient, α, may be directly computed using an analytic expression or is explicitly given at the half-integer points, i.e., there is no need for interpolation. If this is the case, then the discretization given here is second-order accurate. Additional theory for the case in which the coefficient must be interpolated is given in Steinberg and Roache [194].

There are $M-1$ unknowns in this problem, $f_i = \hat{f}_i$, $1 \le i \le M-1$; formula (2.25) provides $M-1$ equations for determining these unknowns. The L_i, C_i, and

R_i are the coefficients of the *stencil* of the finite difference equation. This algorithm has been implemented in the code described in Appendix B, Section B.3.1.

Exercise 2.2.1 Verify the formulas for L_i, C_i, and R_i. Verify that the difference equations for $\hat{f}_i$ can be written in matrix form

$$\begin{pmatrix} C_1 & R_1 & & & & \\ L_2 & C_2 & R_2 & & & \\ & L_3 & C_3 & R_3 & & \\ & & \ddots & \ddots & \ddots & \\ & & & L_{M-2} & C_{M-2} & R_{M-2} \\ & & & & L_{M-1} & C_{M-1} \end{pmatrix} \begin{pmatrix} \hat{f}_1 \\ \hat{f}_2 \\ \hat{f}_3 \\ \vdots \\ \hat{f}_{M-2} \\ \hat{f}_{M-1} \end{pmatrix} = \begin{pmatrix} G_1 \\ \hat{g}_2 \\ \hat{g}_3 \\ \vdots \\ \hat{g}_{M-2} \\ G_{M-1} \end{pmatrix} \tag{2.35}$$

where

$$G_1 = \hat{g}_1 - L_1 \hat{f}_0 , \tag{2.36}$$

$$G_{M-1} = \hat{g}_{M-1} - R_{M-1} \hat{f}_M , \tag{2.37}$$

and then verify that the coefficient matrix is $M - 1$ by $M - 1$, symmetric, and diagonally dominant but not *strictly* diagonally dominant. §

Project 2.2.2 Write a computer code (see Appendix B) for numerically solving boundary value problems of the form (2.11). A good test problem has

$$f(x) = \frac{e^{\lambda x} - 1}{e^{\lambda} - 1}, \quad \lambda > 0, \tag{2.38}$$

as a solution. To create such a problem, choose

$$\alpha(x) = 1 + x^2 \tag{2.39}$$

and then solve (2.11) for $g(x)$. Use the interval $[0, 1]$ for physical space and note that the solution f satisfies the boundary conditions $f(0) = 0$ and $f(1) = 1$.

Use the uniform grid $x(\xi) = \xi$ and $\lambda = 2$. Next, use the *stretched* grid

$$x(\xi) = \frac{1}{\nu} \ln[1 + (e^{\nu} - 1)\, \xi], \tag{2.40}$$

with $\lambda = \nu = 2$. Check that the stretched grid is given by a nonsingular transformation (including the end points).

To check your code, perform a convergence-rate test for both the uniform and stretched grids. To do this, first define the error in the approximate solution $\hat{f}$ by

$$E_M = \max_{0 \le i \le M} |\hat{f}_i - f(x_i)| . \tag{2.41}$$

Second-order difference approximations require that

$$C = \lim_{M \to \infty} M^2 E_M \tag{2.42}$$

be a finite nonzero number. For both the uniform grid and the stretched grid with $\nu = 2$, use your code to compute a table of the values of $C_M = M^2 E_M$ where $M = 2^k$, $1 \le k \le 10$, and then observe that the values in the table become constant as M increases. Hint: use double precision in your code. §

Project 2.2.3 This is a continuation of Project 2.2.2. Note that when $\lambda = \nu$, the solution f and the stretched transformation are inverses of each other and consequently $\tilde{f}(\xi) = \xi$. It seems reasonable, but is not true, that choosing $\nu = \lambda$ in the stretched grid would give minimal error. Choose $\lambda = 2$ and then find a value of ν that is "best" in the sense that the value of C in (2.42) is minimal. Explain why $\nu = \lambda$ is not the optimal choice. §

2.3 TRANSFORMATION OF THE TWO-DIMENSIONAL HOSTED EQUATION

A two-dimensional, second-order, steady-state hosted equation is transformed into general coordinates. Let Ω be a region in the physical plane and $\partial\Omega$ be the boundary of Ω. Also, let $\alpha = \alpha(x,y)$, $\beta = \beta(x,y)$, $\gamma = \gamma(x,y)$, and $g = g(x,y)$ be functions defined in Ω. Then a steady state $f = f(x,y)$ of a process occurring in an inhomogeneous media satisfies a boundary value problem of the following form:

$$(\alpha f_x)_x + (\beta f_x)_y + (\beta f_y)_x + (\gamma f_y)_y = g\,, \quad f\,|_{\partial\Omega} = 0. \tag{2.43}$$

The condition $f|_{\partial\Omega} = 0$ means that the function f is zero on the boundary $\partial\Omega$. This is a *Dirichlet* boundary condition. More general boundary conditions such as inhomogeneous Dirichlet, Neumann, or mixed boundary conditions can be treated as well (see Steinberg and Roache [194]). The differential equation was chosen to have the form given in equation (2.43) because this is the form that frequently occurs in physical problems with variable material properties such as heat conduction, electric potential, and fluid flow in porous media.

Exercise 2.3.1 Let $\mathcal{T}$ be the symmetric matrix

$$\mathcal{T} = \begin{pmatrix} \alpha & \beta \\ \beta & \gamma \end{pmatrix}, \tag{2.44}$$

and show that the differential equation (2.43) is given by

$$\nabla \cdot (\mathcal{T}\, \nabla f) = g\,, \tag{2.45}$$

that is, the equation is given by the divergence of the flux ($\mathcal{T}$ times the gradient of f) equals g. For the reader familiar with vector calculus, this gives a direct physical interpretation of the partial differential equation. §

Assume that a transformation

$$x = x(\xi,\eta)\,, \quad y = y(\xi,\eta) \tag{2.46}$$

with Jacobian

$$J = x_\xi y_\eta - x_\eta y_\xi \tag{2.47}$$

is given. To transform the partial differential equation to general coordinates, let

$$\begin{aligned}
\tilde{f}(\xi,\eta) &= f(x(\xi,\eta),\, y(\xi,\eta))\,,\\
\tilde{g}(\xi,\eta) &= g(x(\xi,\eta),\, y(\xi,\eta))\,,\\
\tilde{\alpha}(\xi,\eta) &= \alpha(x(\xi,\eta),\, y(\xi,\eta))\,,\\
\tilde{\beta}(\xi,\eta) &= \beta(x(\xi,\eta),\, y(\xi,\eta))\,,\\
\tilde{\gamma}(\xi,\eta) &= \gamma(x(\xi,\eta),\, y(\xi,\eta))\,.
\end{aligned} \tag{2.48}$$

Next, apply the chain rule to the partial derivatives in (2.43) to obtain

$$\tilde{f}_\xi = f_x\, x_\xi + f_y\, y_\xi\,, \quad \tilde{f}_\eta = f_x\, x_\eta + f_y\, y_\eta\,. \tag{2.49}$$

Because the Jacobian J of the map is assumed to be nonzero, the transformation is nonsingular and consequently (2.49) can be inverted:

$$f_x = \frac{1}{J}(\tilde{f}_\xi\, y_\eta - \tilde{f}_\eta\, y_\xi)\,, \quad f_y = \frac{1}{J}(\tilde{f}_\eta\, x_\xi - \tilde{f}_\xi\, x_\eta)\,. \tag{2.50}$$

It is an important observation that, using the product rule for derivatives and the identity $x_{\xi\eta} = x_{\eta\xi}$, equations (2.50) can be put in the form

$$f_x = \frac{1}{J}\{(\tilde{f}\, y_\eta)_\xi - (\tilde{f}\, y_\xi)_\eta\}\,, \quad f_y = \frac{1}{J}\{(\tilde{f}\, x_\xi)_\eta - (\tilde{f}\, x_\eta)_\xi\}\,. \tag{2.51}$$

This form is called the *conservative* or *symmetric* form of transformed derivatives.

The relations (2.50) can be used to obtain second derivative expressions by substituting

$$\alpha\, f_x = \frac{\tilde{\alpha}}{J}\,(\tilde{f}_\xi\, y_\eta - \tilde{f}_\eta\, y_\xi) \tag{2.52}$$

for f in the Formula (2.51) for f_x, and so forth:

$$J\,(\alpha\, f_x)_x = +\left(\tilde{\alpha}\,\frac{+\tilde{f}_\xi\, y_\eta - \tilde{f}_\eta\, y_\xi}{J}\, y_\eta\right)_\xi - \left(\tilde{\alpha}\,\frac{+\tilde{f}_\xi\, y_\eta - \tilde{f}_\eta\, y_\xi}{J}\, y_\xi\right)_\eta\,, \tag{2.53}$$

$$J\,(\beta\, f_x)_y = -\left(\tilde{\beta}\,\frac{+\tilde{f}_\xi\, y_\eta - \tilde{f}_\eta\, y_\xi}{J}\, x_\eta\right)_\xi + \left(\tilde{\beta}\,\frac{+\tilde{f}_\xi\, y_\eta - \tilde{f}_\eta\, y_\xi}{J}\, x_\xi\right)_\eta\,, \tag{2.54}$$

$$J\,(\beta\, f_y)_x = +\left(\tilde{\beta}\,\frac{-\tilde{f}_\xi\, x_\eta + \tilde{f}_\eta\, x_\xi}{J}\, y_\eta\right)_\xi - \left(\tilde{\beta}\,\frac{-\tilde{f}_\xi\, x_\eta + \tilde{f}_\eta\, x_\xi}{J}\, y_\xi\right)_\eta\,, \tag{2.55}$$

$$J\,(\gamma\, f_y)_y = -\left(\tilde{\gamma}\,\frac{-\tilde{f}_\xi\, x_\eta + \tilde{f}_\eta\, x_\xi}{J}\, x_\eta\right)_\xi + \left(\tilde{\gamma}\,\frac{-\tilde{f}_\xi\, x_\eta + \tilde{f}_\eta\, x_\xi}{J}\, x_\xi\right)_\eta\,. \tag{2.56}$$

These are called the *conservative* or *symmetric* forms for the second derivatives; there are several other forms for these derivatives that are not used here.

If the partial differential equation in (2.43) is multiplied by the Jacobian and then transformed to general coordinates using the formulas in (2.53)-(2.56), then the resulting transformed boundary value problem is

$$(\hat{\alpha}\,\hat{f}_\xi)_\xi + (\hat{\beta}\,\hat{f}_\xi)_\eta + (\hat{\beta}\,\hat{f}_\eta)_\xi + (\hat{\gamma}\,\hat{f}_\eta)_\eta = \hat{g}\,, \tag{2.57}$$

with the boundary conditions

$$\hat{f}(\xi,0) = 0\,, \quad \hat{f}(\xi,1) = 0\,, \quad \hat{f}(0,\eta) = 0\,, \quad \hat{f}(1,\eta) = 0\,, \tag{2.58}$$

where

$$\begin{aligned} \hat{f}(\xi,\eta) &= \tilde{f}(\xi,\eta)\,, \\ \hat{g}(\xi,\eta) &= J(\xi,\eta)\,\tilde{g}(\xi,\eta)\,, \end{aligned} \tag{2.59}$$

and

$$\hat{\alpha} = +\frac{1}{J}\left(+\tilde{\alpha}\,y_\eta^2 - 2\,\tilde{\beta}\,x_\eta\,y_\eta + \tilde{\gamma}\,x_\eta^2\right)\,, \tag{2.60}$$

$$\hat{\beta} = -\frac{1}{J}\left(+\tilde{\alpha}\,y_\xi\,y_\eta - \tilde{\beta}\left(x_\xi\,y_\eta + x_\eta\,y_\xi\right) + \tilde{\gamma}\,x_\xi\,x_\eta\right)\,, \tag{2.61}$$

$$\hat{\gamma} = +\frac{1}{J}\left(+\tilde{\alpha}\,y_\xi^2 - 2\,\tilde{\beta}\,x_\xi\,y_\xi + \tilde{\gamma}\,x_\xi^2\right)\,. \tag{2.62}$$

Note the simplicity of the boundary conditions (2.58) compared to statement (2.43). This is one of the advantages of boundary-conforming coordinates.

Exercise 2.3.2 Extend the boundary conditions (2.58) to inhomogeneous Dirichlet conditions. §

Exercise 2.3.3 Check one of the formulas (2.60)-(2.62). §

2.4 *DISCRETIZATION OF THE TWO-DIMENSIONAL TRANSFORMED EQUATION*

The discretization of the two-dimensional hosted equation is performed to attain second-order accuracy and symmetric stencils. The first step in approximating the transformed partial differential equation (2.57) by finite differences is to make a rectangular grid in the logical region U_2. Let M and N be positive integers and then define the grid points (ξ_i, η_j) by

$$\Delta\xi = \frac{1}{M}\,, \quad \xi_i = i\,\Delta\xi\,, \quad 0 \le i \le M\,, \tag{2.63}$$

$$\Delta\eta = \frac{1}{N}\,, \quad \eta_j = j\,\Delta\eta\,, \quad 0 \le j \le N\,. \tag{2.64}$$

As in the one-dimensional case, grid points with either whole integer i or j or

	Points	Max i	Max j
cell corners	$(i\,\Delta\xi,\ j\,\Delta\eta)$	M	N
horizontal edge centers	$\left((i+\frac{1}{2})\,\Delta\xi,\ j\,\Delta\eta\right)$	$M-1$	N
vertical edge centers	$\left(i\,\Delta\xi,\ (j+\frac{1}{2})\,\Delta\eta\right)$	M	$N-1$
cell centers	$\left((i+\frac{1}{2})\,\Delta\xi,\ (j+\frac{1}{2})\,\Delta\eta\right)$	$M-1$	$N-1$

Table 2.1 *Grid points* $(i,\ j \geq 0)$

both half-integers are used. The geometrical meaning of these points is given in Table 2.1 and Figure 2.4. The grid with integer-indexed points contains $(M+1)\times(N+1)$ points. The transformation $x = x(\xi,\eta)$, $y = y(\xi,\eta)$ carries the logical-space grid to a physical-space grid $\mathbf{x}_{i,j} = (x_{i,j},\, y_{i,j})$ where

$$x_{i,j} = x(\xi_i,\, \eta_j)\,, \quad y_{i,j} = y(\xi_i,\, \eta_j)\,, \quad 0 \leq i \leq M\,, \quad 0 \leq j \leq N\,. \tag{2.65}$$

The discretization also induces discrete values for all of the transformed variables in the boundary value problem. For example,

$$\tilde{f}_{i,j} = \tilde{f}(\xi_i,\eta_j) = f(x(\xi_i,\eta_j), y(\xi_i,\eta_j)) = f(x_{i,j},\, y_{i,j}) = f_{i,j}\,. \tag{2.66}$$

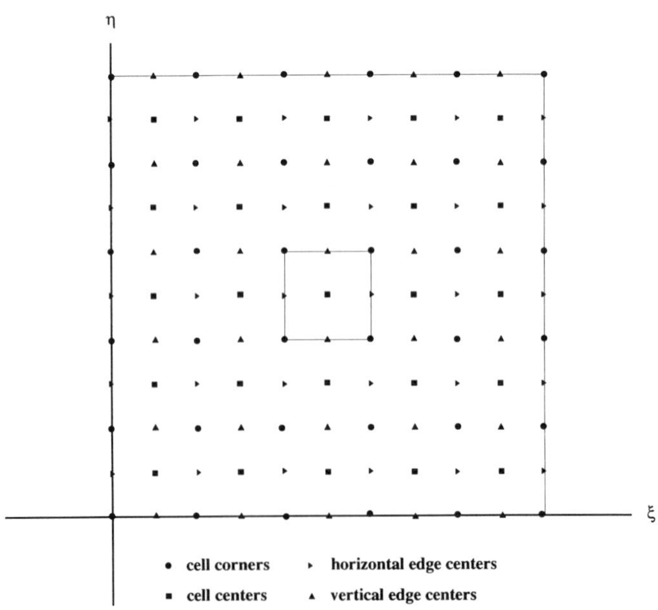

Figure 2.4 *Grid points*

Similar formulas hold for $\tilde{g}_{i,j} = g_{i,j}$, $\tilde{\alpha}_{i,j} = \alpha_{i,j}$, $\tilde{\beta}_{i,j} = \beta_{i,j}$, and $\tilde{\gamma}_{i,j} = \gamma_{i,j}$; consequently, there is no need to retain the *tilde* notation. The values of $f_{i,j}$ are needed for all points $0 \le i \le M$, $0 \le j \le N$. They are specified on the boundary by the boundary conditions:

$$\begin{aligned} \hat{f}_{0,j} &= 0\,, \quad \hat{f}_{M,j} = 0\,, \quad 0 \le j \le N\,, \\ \hat{f}_{i,0} &= 0\,, \quad \hat{f}_{i,N} = 0\,, \quad 0 \le i \le M\,. \end{aligned} \tag{2.67}$$

The $(M-1)(N-1)$ interior values $f_{i,j}$, $1 \le i \le M-1$, $1 \le j \le N-1$, are to be computed. It is assumed that the values of $g_{i,j}$ are given for the interior points, $1 \le i \le M-1$, $1 \le j \le N-1$. The coefficients $\alpha_{i,j}$, $\beta_{i,j}$, and $\gamma_{i,j}$ are needed at locations determined by the discretization of the derivatives (the formulas are derived below).

The variables that appear in the transformed differential equation are defined in terms of the tilde variables, but now the tilde has been eliminated, so

$$\begin{aligned} \hat{f}_{i,j} &= f_{i,j}\,, \quad 0 \le i \le M\,, \quad 0 \le j \le N\,, \\ \hat{g}_{i,j} &= J_{i,j}\, g_{i,j}\,, \quad 1 \le i \le M-1\,, \quad 1 \le j \le N-1\,. \end{aligned} \tag{2.68}$$

The coefficients $\hat{\alpha}_{i,j}$, $\hat{\beta}_{i,j}$, and $\hat{\gamma}_{i,j}$ are defined below using formulas (2.60) through (2.62).

Exercise 2.4.1 Generalize the above discussion to inhomogeneous Dirichlet boundary conditions $f\,|_{\partial\Omega} = p$. Assume that $\partial\Omega$ is made up of four pieces and that p is made up of functions p_k, $1 \le k \le 4$, that are defined on each piece of the boundary. Algorithms for general mixed boundary conditions can be found in Steinberg and Roache [194]. §

As before, the differential equation is differenced using second-order central differences. The approximation of the second derivative will involve nine neighboring points, as shown in Figure 2.5. The first letter in the words North, East, South, West, and Central are used to label the coefficients of the *stencil.* The diagonal terms $(\hat{\alpha}\,\hat{f}_\xi)_\xi$ and $(\hat{\gamma}\,\hat{f}_\eta)_\eta$ are differenced in a manner analogous to formula (2.23), with the stencil coefficients relabeled as in Figure 2.5. Thus

$$\left((\hat{\alpha}\hat{f}_\xi)_\xi\right)_{i,j} \approx W_{i,j}\,\hat{f}_{i-1,j} + C_{i,j}\,\hat{f}_{i,j} + E_{i,j}\,\hat{f}_{i+1,j}\,, \tag{2.69}$$

where

$$W_{i,j} = \frac{\hat{\alpha}_{i-\frac{1}{2},j}}{\Delta\xi^2}\,, \tag{2.70}$$

$$C_{i,j} = -(E_{i,j} + W_{i,j})\,, \tag{2.71}$$

$$E_{i,j} = \frac{\hat{\alpha}_{i+\frac{1}{2},j}}{\Delta\xi^2}\,. \tag{2.72}$$

Also,

$$\left((\hat{\gamma}\hat{f}_\eta)_\eta\right)_{i,j} \approx S_{i,j}\,\hat{f}_{i,j-1} + C_{i,j}\,\hat{f}_{i,j} + N_{i,j}\,\hat{f}_{i,j+1}\,, \tag{2.73}$$

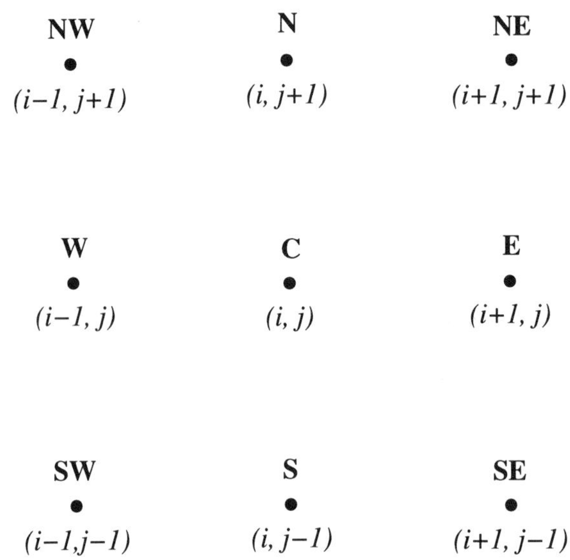

Figure 2.5 *Two-dimensional stencil*

where

$$S_{i,j} = \frac{\hat{y}_{i,j-\frac{1}{2}}}{\Delta\eta^2}\,, \tag{2.74}$$

$$C_{i,j} = -(N_{i,j} + S_{i,j})\,, \tag{2.75}$$

$$N_{i,j} = \frac{\hat{y}_{i,j+\frac{1}{2}}}{\Delta\eta^2}\,. \tag{2.76}$$

The mixed partial derivatives are approximated at cell corners by averaging (as in 2.16), using values at the four center points:

$$\left(f_\xi\right)_{i,j} \approx \frac{f_{i+\frac{1}{2},j+\frac{1}{2}} - f_{i-\frac{1}{2},j+\frac{1}{2}} + f_{i+\frac{1}{2},j-\frac{1}{2}} - f_{i-\frac{1}{2},j-\frac{1}{2}}}{2\,\Delta\xi}\,,$$

$$\left(f_\eta\right)_{i,j} \approx \frac{f_{i+\frac{1}{2},j+\frac{1}{2}} - f_{i+\frac{1}{2},j-\frac{1}{2}} + f_{i-\frac{1}{2},j+\frac{1}{2}} - f_{i-\frac{1}{2},j-\frac{1}{2}}}{2\,\Delta\eta}\,. \tag{2.77}$$

Replacing i by $i + 1/2$ and j by $j + 1/2$ gives a formula for computing derivatives at cell centers in terms of values at cell corners, for example,

$$\left(\hat{\beta}\, f_\xi\right)_{i+\frac{1}{2},j+\frac{1}{2}} \approx \hat{\beta}_{i+\frac{1}{2},j+\frac{1}{2}} \frac{f_{i+1,j+1} - f_{i,j+1} + f_{i+1,j} - f_{i,j}}{2\,\Delta\xi}\,. \tag{2.78}$$

Combining (2.78) with (2.77) gives the following stencils for the mixed partials:

$$
\begin{aligned}
\left((\hat{\beta}\hat{f}_\xi)_\eta\right)_{i,j} + \left((\hat{\beta}\hat{f}_\eta)_\xi\right)_{i,j} \approx\ & \mathrm{NE}_{i,j}\,\hat{f}_{i+1,j+1} \\
& + \mathrm{NW}_{i,j}\,\hat{f}_{i-1,j+1} \\
& + \mathrm{SE}_{i,j}\,\hat{f}_{i+1,j-1} \\
& + \mathrm{SW}_{i,j}\,\hat{f}_{i-1,j-1} \\
& + C_{i,j}\hat{f}_{i,j}\,,
\end{aligned}
\tag{2.79}
$$

where

$$\mathrm{NE}_{i,j} = +\frac{\hat{\beta}_{i+\frac{1}{2},j+\frac{1}{2}}}{4\,\Delta\xi\,\Delta\eta}\,, \tag{2.80}$$

$$\mathrm{NW}_{i,j} = -\frac{\hat{\beta}_{i-\frac{1}{2},j+\frac{1}{2}}}{4\,\Delta\xi\,\Delta\eta}\,, \tag{2.81}$$

$$\mathrm{SE}_{i,j} = -\frac{\hat{\beta}_{i+\frac{1}{2},j-\frac{1}{2}}}{4\,\Delta\xi\,\Delta\eta}\,, \tag{2.82}$$

$$\mathrm{SW}_{i,j} = +\frac{\hat{\beta}_{i-\frac{1}{2},j-\frac{1}{2}}}{4\,\Delta\xi\,\Delta\eta}\,, \tag{2.83}$$

$$C_{i,j} = -\left(\mathrm{NE}_{i,j} + \mathrm{NW}_{i,j} + \mathrm{SE}_{i,j} + \mathrm{SW}_{i,j}\right)\,. \tag{2.84}$$

If the previous formulas (2.69), (2.73), and (2.79) are combined, then an approximation for the full differential equation (2.57) is obtained:

$$
\begin{aligned}
W_{i,j}\,\hat{f}_{i-1,j} &+ \\
E_{i,j}\,\hat{f}_{i+1,j} &+ \\
S_{i,j}\,\hat{f}_{i,j-1} &+ \\
N_{i,j}\,\hat{f}_{i,j+1} &+ \\
\mathrm{NE}_{i,j}\,\hat{f}_{i+1,j+1} &+ \\
\mathrm{NW}_{i,j}\,\hat{f}_{i-1,j+1} &+ \\
\mathrm{SE}_{i,j}\,\hat{f}_{i+1,j-1} &+ \\
\mathrm{SW}_{i,j}\,\hat{f}_{i-1,j-1} &+ \\
C_{i,j}\,\hat{f}_{i,j} &= \hat{g}_{i,j}\,,
\end{aligned}
\tag{2.85}
$$

where the formulas for W and E in equations (2.70–2.72), N and S in equations

(2.74–2.76), and NE, NW, SE, and SW in equations (2.80)–(2.83) are correct, while the central stencil must be redefined as

$$C_{i,j} = -\left(N_{i,j} + W_{i,j} + S_{i,j} + E_{i,j} + \mathrm{NE}_{i,j} + \mathrm{NW}_{i,j} + \mathrm{SE}_{i,j} + \mathrm{SW}_{i,j}\right) \tag{2.86}$$

to account for all the simple stencils being added together.

To evaluate formulas (2.60)–(2.62) for the stencils, the following are used:

$$\begin{aligned}\hat{\alpha}_{i-\frac{1}{2},j} = +\frac{1}{J_{i-\frac{1}{2},j}}\Big(&+ \alpha_{i-\frac{1}{2},j}\,(y_\eta)^2_{i-\frac{1}{2},j} \\ &- 2\,\beta_{i-\frac{1}{2},j}\,(x_\eta)_{i-\frac{1}{2},j}\,(y_\eta)_{i-\frac{1}{2},j} \\ &+ \gamma_{i-\frac{1}{2},j}\,(x_\eta)^2_{i-\frac{1}{2},j}\Big),\end{aligned} \tag{2.87}$$

$$\begin{aligned}\hat{\beta}_{i-\frac{1}{2},j-\frac{1}{2}} = -\frac{1}{J_{i-\frac{1}{2},j-\frac{1}{2}}}\Big(&+ \alpha_{i-\frac{1}{2},j-\frac{1}{2}}\,(y_\xi)_{i-\frac{1}{2},j-\frac{1}{2}}\,(y_\eta)_{i-\frac{1}{2},j-\frac{1}{2}} \\ &- \beta_{i-\frac{1}{2},j-\frac{1}{2}}\Big((x_\xi)_{i-\frac{1}{2},j-\frac{1}{2}}\,(y_\eta)_{i-\frac{1}{2},j-\frac{1}{2}} \\ &\qquad + (x_\eta)_{i-\frac{1}{2},j-\frac{1}{2}}\,(y_\xi)_{i-\frac{1}{2},j-\frac{1}{2}}\Big) \\ &+ \gamma_{i-\frac{1}{2},j-\frac{1}{2}}\,(x_\xi)_{i-\frac{1}{2},j-\frac{1}{2}}\,(x_\eta)_{i-\frac{1}{2},j-\frac{1}{2}}\Big),\end{aligned} \tag{2.88}$$

$$\begin{aligned}\hat{\gamma}_{i,j-\frac{1}{2}} = +\frac{1}{J_{i,j-\frac{1}{2}}}\Big(&+ \alpha_{i,j-\frac{1}{2}}\,(y_\xi)^2_{i,j-\frac{1}{2}} \\ &- 2\,\beta_{i,j-\frac{1}{2}}\,(x_\xi)_{i,j-\frac{1}{2}}\,(y_\xi)_{i,j-\frac{1}{2}} \\ &+ \gamma_{i,j-\frac{1}{2}}\,(x_\xi)^2_{i,j-\frac{1}{2}}\Big).\end{aligned} \tag{2.89}$$

The formulas needed to compute the $\hat{\alpha}$ coefficient are

$$\begin{aligned} x_{i-\frac{1}{2},j} &= \frac{x_{i,j} + x_{i-1,j}}{2}, \\ y_{i-\frac{1}{2},j} &= \frac{y_{i,j} + y_{i-1,j}}{2}, \\ (x_\xi)_{i-\frac{1}{2},j} &= \frac{x_{i,j} - x_{i-1,j}}{\Delta\xi}, \\ (x_\eta)_{i-\frac{1}{2},j} &= \frac{x_{i,j+1} - x_{i,j-1} + x_{i-1,j+1} - x_{i-1,j-1}}{4\,\Delta\eta}, \\ (y_\xi)_{i-\frac{1}{2},j} &= \frac{y_{i,j} - y_{i-1,j}}{\Delta\xi}, \\ (y_\eta)_{i-\frac{1}{2},j} &= \frac{y_{i,j+1} - y_{i,j-1} + y_{i-1,j+1} - y_{i-1,j-1}}{4\,\Delta\eta}, \\ J_{i-\frac{1}{2},j} &= (x_\xi)_{i-\frac{1}{2},j}\,(y_\eta)_{i-\frac{1}{2},j} - (y_\xi)_{i-\frac{1}{2},j}\,(x_\eta)_{i-\frac{1}{2},j}. \end{aligned} \tag{2.90}$$

The formulas needed to compute the $\hat{\gamma}$ coefficient are

$$
\begin{aligned}
x_{i,j-\frac{1}{2}} &= \frac{x_{i,j} + x_{i,j-1}}{2}, \\
y_{i,j-\frac{1}{2}} &= \frac{y_{i,j} + y_{i,j-1}}{2}, \\
(x_\xi)_{i,j-\frac{1}{2}} &= \frac{x_{i+1,j} - x_{i-1,j} + x_{i+1,j-1} - x_{i-1,j-1}}{4\,\Delta\xi}, \\
(x_\eta)_{i,j-\frac{1}{2}} &= \frac{x_{i,j} - x_{i,j-1}}{\Delta\eta}, \\
(y_\xi)_{i,j-\frac{1}{2}} &= \frac{y_{i+1,j} - y_{i-1,j} + y_{i+1,j-1} - y_{i-1,j-1}}{4\,\Delta\xi}, \\
(y_\eta)_{i,j-\frac{1}{2}} &= \frac{y_{i,j} - y_{i,j-1}}{\Delta\eta}, \\
J_{i,j-\frac{1}{2}} &= (x_\xi)_{i,j-\frac{1}{2}}\,(y_\eta)_{i,j-\frac{1}{2}} - (y_\xi)_{i,j-\frac{1}{2}}\,(x_\eta)_{i,j-\frac{1}{2}}.
\end{aligned}
\tag{2.91}
$$

The formulas needed to compute the $\hat{\beta}$ coefficient are

$$
\begin{aligned}
x_{i-\frac{1}{2},j-\frac{1}{2}} &= \frac{x_{i,j} + x_{i,j-1} + x_{i-1,j} + x_{i-1,j-1}}{4}, \\
y_{i-\frac{1}{2},j-\frac{1}{2}} &= \frac{y_{i,j} + y_{i,j-1} + y_{i-1,j} + y_{i-1,j-1}}{4}, \\
(x_\xi)_{i-\frac{1}{2},j-\frac{1}{2}} &= \frac{x_{i,j} + x_{i,j-1} - x_{i-1,j} - x_{i-1,j-1}}{2\,\Delta\xi}, \\
(x_\eta)_{i-\frac{1}{2},j-\frac{1}{2}} &= \frac{x_{i,j} - x_{i,j-1} + x_{i-1,j} - x_{i-1,j-1}}{2\,\Delta\eta}, \\
(y_\xi)_{i-\frac{1}{2},j-\frac{1}{2}} &= \frac{y_{i,j} + y_{i,j-1} - y_{i-1,j} - y_{i-1,j-1}}{2\,\Delta\xi}, \\
(y_\eta)_{i-\frac{1}{2},j-\frac{1}{2}} &= \frac{y_{i,j} - y_{i,j-1} + y_{i-1,j} - y_{i-1,j-1}}{2\,\Delta\eta}, \\
J_{i-\frac{1}{2},j-\frac{1}{2}} &= (x_\xi)_{i-\frac{1}{2},j-\frac{1}{2}}\,(y_\eta)_{i-\frac{1}{2},j-\frac{1}{2}}, \\
&\quad -(y_\xi)_{i-\frac{1}{2},j-\frac{1}{2}}\,(x_\eta)_{i-\frac{1}{2},j-\frac{1}{2}}.
\end{aligned}
\tag{2.92}
$$

The formulas needed to compute the $\hat{g}$ coefficient are

$$
\begin{aligned}
(x_\xi)_{i,j} &= \frac{x_{i+1,j} - x_{i-1,j}}{2\,\Delta\xi}, \\
(x_\eta)_{i,j} &= \frac{x_{i,j+1} - x_{i,j-1}}{2\,\Delta\eta}, \\
(y_\xi)_{i,j} &= \frac{y_{i+1,j} - y_{i-1,j}}{2\,\Delta\xi}, \\
(y_\eta)_{i,j} &= \frac{y_{i,j+1} - y_{i,j-1}}{2\,\Delta\eta}, \\
J_{i,j} &= (x_\xi)_{i,j}\,(y_\eta)_{i,j} - (y_\xi)_{i,j}\,(x_\eta)_{i,j}.
\end{aligned}
\tag{2.93}
$$

The difference equation (2.85) must hold for all interior points $1 \le i \le M-1$, $1 \le j \le N-1$. Consequently, there are $(M-1)\times(N-1)$ equations for determining the $(M-1)\times(N-1)$ unknowns $\hat{f}_{i,j} = f_{i,j}$, $1 \le i \le M-1$, $1 \le j \le N-1$.

Exercise 2.4.2 Verify the formulas for $\mathrm{NW}_{i,j}$, $\mathrm{SE}_{i,j}$, and $C_{i,j}$. Write the difference equations for $\hat{f}_{i,j}$ in matrix form and then verify that the coefficient matrix is $(M-1)(N-1)$ by $(M-1)(N-1)$, is made up of $(M-1)(N-1)$ tridiagonal blocks, is symmetric, is banded, and has row-sums zero. Note that not all off-diagonal entries of the matrix are necessarily positive. However, if the entries are all positive, then the row-sums zero condition implies that the matrix is (not strictly) diagonally dominant. §

The next project illustrates the power of the method presented; it is quite difficult to do with other standard methods.

Project 2.4.3 Use the code discussed in Appendix B Section B.3.2 to solve the following boundary value problem. Recall that the partial differential equation can be written (see 2.45)

$$\nabla \cdot (\mathcal{T}\,\nabla f) = g\,, \tag{2.94}$$

where the matrix $\mathcal{T}$ is given by

$$\mathcal{T} = \begin{pmatrix} \alpha & \beta \\ \beta & \gamma \end{pmatrix}. \tag{2.95}$$

The matrix will be defined as the rotation of a diagonal matrix:

$$\mathcal{T} = P^{-1}\,D\,P\,, \tag{2.96}$$

where

$$P = \begin{pmatrix} \cos(t) & \sin(t) \\ -\sin(t) & \cos(t) \end{pmatrix}, \tag{2.97}$$

and

$$D = \begin{pmatrix} d_1 & 0 \\ 0 & d_2 \end{pmatrix}. \tag{2.98}$$

Here t is a parameter and

$$d_1 = 1 + 2\,x^2 + y^2\,, \qquad d_2 = 1 + x^2 + 2\,y^2\,. \tag{2.99}$$

The region Ω is given by the image of the unit square under the transformation

$$x = \xi + \epsilon\,\cos\left(\frac{\pi}{4}(\xi+\eta)\right), \qquad y = \eta + \epsilon\,\sin\left(\frac{\pi}{4}(\xi+\eta)\right), \tag{2.100}$$

where again ϵ is a parameter.

To create a test problem, take $t = \pi/8$ and $\epsilon = 1/2$. Choose

$$f(x, y) = \sin(\pi x)\ \sin(\pi y) \tag{2.101}$$

and then calculate the inhomogeneous term g and the Dirichlet boundary conditions so that f is a solution of (2.45). Note that $\epsilon = 0$ gives a good preliminary test for this code.

If $f(x, y)$ is the exact solution of a problem and $\hat{f}_{i,j}$ is an approximate solution with $0 \le i \le M$, $0 \le j \le N$, then the error in the approximate solution is defined as

$$E_{M,N} = \max |\hat{f}_{i,j} - f(x_{i,j}, y_{i,j})|, \quad 0 \le i \le M, \quad 0 \le j \le N. \tag{2.102}$$

For a second-order method, the convergence rate constant is defined by

$$C = \lim_{M,N \to \infty} M\,N\,E_{M,N}. \tag{2.103}$$

The theory of finite differences requires C to be a finite nonzero number if the method is second-order accurate. Perform a convergence-rate test on the resulting program, that is, compute C. To do this, choose $M = N = 2^K$ for $1 \le K \le 7$ and compute $C_M = M^2\,E_{M,M}$.

A central point is that the region Ω used in the problem can be changed by making a few trivial changes in the code. §

2.5 SUMMARY

Using grid generation for solving partial differential equations consists of, first, two preliminary algebraic steps:

- transform the PDE to general coordinates (Section 2.3 or Section 7.3); and
- discretize the differential equation (Section 2.4 and Steinberg and Roache [194]),

and second, three numerical steps:

- generate a grid (see the rest of this book);
- evaluate the stencil coefficients of the discretized equation;
- solve the discretized equation.

The last step is not discussed in this chapter; the reader is referred to multigrid, preconditioned conjugate-gradient, Newton's, and operator-splitting methods. These methods can be used to solve grid-generation equations as well (see, for example, Camarero and Younis [23]).

Note that the coefficients of the hosted equation, the boundary conditions, and the geometric region can be varied with only a few minor changes in the computer program that implements this method; no changes are necessary in the algorithm. This flexibility is a major benefit of this approach. Moreover, grid-generation techniques can be applied to time-dependent problems (see, for example Thomas and Lombard [204] or Pulliam and Steger [150]), nonlinear problems, and those with moving boundaries (Yeung and Vaidhyanathan [233]).

Each step in the solution process can be pushed to a much higher level. The purpose of this book is to extend the grid-generation step, using differential geometric and variational ideas, to a powerful set of algorithms capable of generating grids on a wide variety of geometric objects. Extensions of the discretization techniques and solution methods for the discretized equations are not pursued further here.

Chapter 3

Grid Generation on the Line

3.1 INTRODUCTION

The goal in this chapter is to present a theory of one-dimensional grid generation on the line to illustrate some of the basic concepts involved in grid-generation algorithms. A major goal of one-dimensional grid-generation algorithms is to create grids with specified distances between the grid points, thus the chapter begins with a simple and intuitive geometric argument to show how to produce second-order differential equations whose solutions are one-dimensional grids with specified segment lengths. This discussion provides a firm foundation for developing *solution-adaptive* grid generators.

There are many ways to generate grids in one dimension; the interest here is in only those methods that generalize to higher dimensions. Two Poisson grid generators are described; the second of these generators is of particular importance since it is widely used, especially in higher dimensions. This is followed by a discussion of how to numerically implement all one-dimensional generators.

One of the objectives of this text is to present as many grid-generation results as possible within the context of classical *calculus of variations*, following and extending the work of Steinberg and Roache [194]. The one-dimensional problem provides an introduction to the geometric basis for choosing variational principles in higher dimensions. A simple discrete minimization problem is discussed first, then extended to the continuum to produce a variational algorithm for one-dimensional grid generation. Basic results in the classical calculus of variations are reviewed. More details and proofs of relevant classical variational results can be found in one of the many texts on this subject, e.g., Gelfand and Fomin [78]. The variational section is completed by presenting a variational derivation of a higher-order grid generator. Variational ideas are extended to higher dimensions in Chapter 6, *Variational Planar Grid Generation.*

Another alternative is to generate grids as solutions to fourth-order differential equations. Unfortunately, such methods are not as general as the weighted variational algorithms.

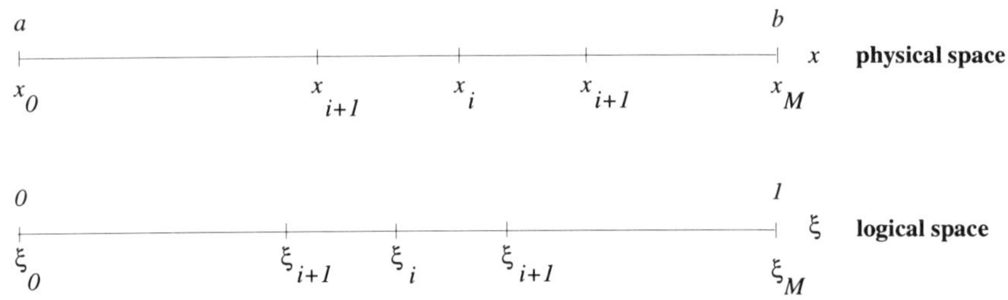

Figure 3.1 *One-dimensional grid*

Alternate approaches to the topic of this chapter have been presented in several other places (e.g., Thompson et al. [219]). The bulk of interest in one-dimensional algorithms stems from their being a model for higher-dimensional grid generation and from their application to solution adaptivity in the numerical solution of partial differential equations (e.g., Thompson et al. [219], Anderson [5, 6, 8], Dwyer et al. [58], Ghia and Ghia [81], and Nakamura [146]). Thus, the chapter closes with a discussion of one-dimensional solution-adaptive methods.

3.2 GENERATORS THAT CONTROL GRID SPACING

The purpose of this section is to derive *second-order differential equations* for generating grids where the length of the segments in the grid are specified by a *weight function*; weight functions that depend on either the logical variable or the physical variable are considered. The derived differential equations are closely related to those obtained by variational methods and, consequently, provide substantial insight into those methods.

Consider a weight function $\phi(\xi)$ that depends on the logical-space variable ξ, for ξ in the interval $[0, 1]$. The lengths of the grid intervals are to be *positive* and *proportional* to ϕ, so it is natural to assume that $\phi = \phi(\xi) > 0$. If M is a positive integer, then an unfolded grid on the interval $[a, b]$ containing $M + 1$ points is given by x_i, $0 \le i \le M$, where $x_0 = a$, $x_M = b$, and $x_i < x_{i+1}$, $0 \le i \le M - 1$. The problem is to generate a grid so that the lengths of the intervals $[x_i, x_{i+1}]$ are proportional to the value of ϕ at the midpoint of the interval. More precisely, find x_i so that

$$x_{i+1} - x_i = K\,\phi\left(\frac{\xi_{i+1} + \xi_i}{2}\right), \tag{3.1}$$

where $0 \le i \le M - 1$ and K is some constant that is to be found.

If the grid is given by a transformation from logical to physical space, then $x_i = x(\xi_i)$ where $\xi_i = i/M = i\Delta\xi$. Observe that if ϕ is continuous, then for $\Delta\xi$ going to zero, the left-hand side of equation (3.1) goes to zero while the right-hand side does not. To fix this, set $K = C\,\Delta\xi$ where C is another constant and divide equation (3.1) by $\Delta\xi$ to obtain

$$\frac{x(\xi_{i+1}) - x(\xi_i)}{\Delta\xi} = C\,\phi\left(\frac{\xi_{i+1} + \xi_i}{2}\right). \tag{3.2}$$

Equation (3.2) can be expressed at a general point $\xi \in [0,1]$ as

$$\frac{x(\xi + \Delta\xi) - x(\xi)}{\Delta\xi} = C\,\phi\left(\xi + \frac{\Delta\xi}{2}\right). \tag{3.3}$$

If x is continuous, then the limit of equation (3.3) as $\Delta\xi \to 0$ yields the ordinary differential equation

$$x_\xi(\xi) = C\,\phi(\xi). \tag{3.4}$$

Dividing (3.4) by ϕ and differentiating with respect to ξ gives

$$\left(\frac{x_\xi}{\phi}\right)_\xi = 0. \tag{3.5}$$

Note that the constant C has been removed from the problem. If ϕ is differentiable and x is twice differentiable, the quotient rule for derivatives and a little algebra gives

$$x_{\xi\xi} - \frac{\phi_\xi}{\phi}\,x_\xi = 0. \tag{3.6}$$

The transformation must satisfy the boundary conditions

$$x(0) = a, \qquad x(1) = b. \tag{3.7}$$

These boundary conditions along with the *linear* differential equation given in equation (3.6) uniquely determine the transformation x. The Jacobian of the map is $J = x_\xi$ and according to (3.4), the resulting transformation has the property that

$$J(\xi) = C\,\phi(\xi), \tag{3.8}$$

which is the continuum analog of (3.2) which says that the segment lengths are *proportional to* ϕ.

The differential equation in (3.5) is a variable coefficient Laplace equation; it has the by now familiar *conservative form* of the differential equation given in equation (2.11) in Chapter 2 where $\alpha = 1/\phi$. The conservative form (3.5) does not require ϕ to be smooth; it therefore has an advantage over the nonconservative form (3.6) since in some applications the weight function is constructed numerically. Either (3.5) or (3.6) can be discretized and the boundary conditions (3.7) applied to obtain a numerical algorithm for one-dimensional grid generation. This is done in Section 3.4.

The constant C, and consequently K, is determined by integrating equation (3.4) between $\xi = 0$ and $\xi = 1$:

$$\frac{1}{C} = \frac{1}{b-a}\int_0^1 \phi(\xi)\,d\xi. \tag{3.9}$$

The constant C does not appear in either (3.5) or (3.6), so it is not needed in the numerical algorithms; however, the expression for C in (3.9) is useful in the following exercises.

Exercise 3.2.1 Show that if $\phi \equiv 1$, then $x(\xi) = (b-a)\,\xi + a$. §

Exercise 3.2.2 Show that the solution to (3.6) and (3.7) is

$$x(\xi) = a + C \int_0^{\xi} \phi(\mu)\, d\mu\,, \tag{3.10}$$

where C is given by (3.9). §

Exercise 3.2.3 Suppose the domain in physical space is translated by a real number s. Find the solution to (3.6) with boundary conditions $x(0) = a + s$ and $x(1) = b + s$; show that C is invariant to the translation and that the interior points of the grid are translated by s. §

Exercise 3.2.4 Let $0 < \xi_0 < 1$ and suppose that

$$\phi(\xi) = \begin{cases} 1, & 0 \le \xi \le \xi_0\,, \\ 2, & \xi_0 < \xi \le 1\,. \end{cases} \tag{3.11}$$

Show that the solution to (3.5) with (3.7) is

$$x(\xi) = \begin{cases} \frac{b-a}{2-\xi_0}\,\xi + a\,, & 0 \le \xi \le \xi_0\,, \\ \frac{2\,(b-a)\,\xi + 2\,a - b\,\xi_0}{2-\xi_0}\,, & \xi_0 < \xi \le 1\,. \end{cases} \tag{3.12}$$

Note that $x(\xi)$ is continuous but $x_\xi(\xi)$ has a jump at ξ_0; x_ξ/ϕ is differentiable. §

The purpose of the weight function is to modify the basic grid in Exercise 3.2.1 by adapting toward or away from particular locations in the physical domain (see Section 3.8 on solution adaptivity). There is a significant limitation to the use of logical-space weight functions in this regard. Suppose, for example, that one desires to change the grid spacing at the point $x = x_0$ in physical space, using the weight in the previous exercise. Then one must determine ξ_0 in (3.11) from the relation $x_0 = x(\xi_0)$, which leads to

$$\xi_0 = \frac{2\,(x_0 - a)}{x_0 + b - 2\,a}\,. \tag{3.13}$$

For more complicated logical-space weight functions, it may be difficult or impossible to explicitly determine the values of parameters such as ξ_0 in the weights to ensure that the stretching of the grid occurs at the desired location in physical space. In higher dimensions this difficulty becomes quite annoying. One way to fix the problem is through the use of physical-space weighting.

A *physical-space weight function* depends on the dependent variable x. Assume that the weight is given by $w = w(x) > 0$ which is defined on $[a, b]$. As before, the problem is to generate a grid so that the lengths of the intervals $[x_i, x_{i+1}]$ are given by the value of w at the midpoint of the interval, that is, find x_i so that

$$x_{i+1} - x_i = K w\left(\frac{x_{i+1} + x_i}{2}\right)\,, \tag{3.14}$$

with $0 \le i \le M-1$ and K is some constant to be found. In the previous discussion, choose $\phi(\xi) = w(x(\xi))$; then (3.4) implies that x satisfies the differential equation

$$\frac{x_\xi(\xi)}{w(x(\xi))} = C. \tag{3.15}$$

Differentiating with respect to ξ gives

$$\left(\frac{x_\xi(\xi)}{w(x(\xi))}\right)_\xi = 0. \tag{3.16}$$

The quotient rule and some algebra gives

$$x_{\xi\xi} - \frac{w_x}{w}\, x_\xi^2 = 0. \tag{3.17}$$

Note that this equation is not quite the same as equation (3.6), but can be put in this form using the chain rule:

$$x_{\xi\xi} - \frac{w_\xi}{w}\, x_\xi = 0. \tag{3.18}$$

As before, $x(0) = a$ and $x(1) = b$, so that there are the correct number of boundary conditions to solve equation (3.17) for a function that generates the desired grid.

The differential equation in (3.16) again has the same form as the differential equation in (2.11) where $\alpha = 1/w$. Now, however, the differential equation in (3.16) is *nonlinear* because the coefficient is a function of the dependent variable (the equation is actually *quasi-linear*). In general, solutions to nonlinear equations need not exist and if they exist they may not be unique. Nonlinear equations are significantly more troublesome to solve numerically than linear equations (see Section 3.4).

The constant C, and consequently K, is determined by rearranging (3.15), giving

$$C \frac{d\xi}{dx} = \frac{1}{w(x)}. \tag{3.19}$$

Integrating this between $x = a$ and $x = b$ gives

$$C = \int_a^b \frac{dx}{w(x)}. \tag{3.20}$$

In contrast to the logical-space weight case, the general solution to (3.16) cannot be explicitly given due to the nonlinearity.

Exercise 3.2.5 Suppose $a < x_0 < b$ and let

$$w(x) = \begin{cases} 1, & a \le x \le x_0, \\ 2, & x_0 < x \le b. \end{cases} \tag{3.21}$$

Show that (3.15), (3.7), and (3.20) give

$$x(\xi) = \begin{cases} a + \frac{x_0 - a}{\xi_0}\,\xi\,, & 0 < \xi < \xi_0\,, \\ \frac{(x_0 - b)\,\xi + (b\,\xi_0 - x_0)}{\xi_0 - 1}\,, & \xi_0 < \xi < 1\,, \end{cases} \tag{3.22}$$

where ξ_0 is chosen so that $x(\xi_0) = x_0$:

$$\xi_0 = \frac{x_0 - a}{C}\,. \tag{3.23}$$

Note that x_ξ has a jump at ξ_0, but that x_ξ / w is differentiable at ξ_0. The advantage of physical-space weighting is that one does not need to compute ξ_0 to determine the weight function. Also show that if the physical domain is translated by s, then the grid is also translated by s, provided $x_0 \to x_0 + s$. §

To summarize, it is possible to generate one-dimensional grids whose local spacing is proportional to a given weight function. The approach makes use of second-order partial differential equations; both conservative and nonconservative forms of the equations are used. The weight functions can be given in terms of either the logical or physical variable. The grid generation equation for a logical-space weight function is linear and translation invariant. The logical-space weight function can be difficult to construct so that the desired effect occurs at a predictable location in physical space. On the other hand, the grid generation equation for a physical-space weight function is nonlinear, but still translation invariant, provided the weight function is modified appropriately. The physical-space weight function is relatively easy to construct and permits adaptation at predictable locations in physical space.

3.3 ONE-DIMENSIONAL POISSON GRID GENERATORS

Poisson equations are commonly used by engineers to generate grids in higher dimensions. One-dimensional Poisson grid generators are studied in this section; basic properties are noted and the generators are compared to those in the previous section. The Poisson generators are one-dimensional forms of the AH (Amsden and Hirt [3]) and the TTM (Thompson, Thames, and Mastin [212]) generators. Poisson grids are determined by assuming that the transformation that generates the grid satisfies a second-order differential equation known as the Poisson equation. The reason for this assumption is that solutions to Poisson equations are smooth (provided there are sufficient restrictions on the smoothness of the inhomogeneous term)—a highly desirable property of finite difference grids.

The *AH generator* assumes that the transformation satisfies the Poisson equation

$$x_{\xi\xi} = P(\xi)\,, \quad x(0) = a\,, \quad x(1) = b\,, \tag{3.24}$$

where P is some given continuous function that serves as a logical-space weight.

This is a *linear* boundary value problem and has the general solution

$$x(\xi) = (b-a)\,\xi + a + \xi \int_0^1 \zeta P(\zeta)\,d\zeta - \xi \int_\xi^1 P(\zeta)\,d\zeta - \int_0^\xi \zeta P(\zeta)\,d\zeta\,, \tag{3.25}$$

as can be easily checked [also, see Equation (2.14)]. Equations (3.24)-(3.25) imply that $x(\xi)$ and its first two derivatives are continuous. In fact, the solution and its first derivative make sense and are continuous if P is only piecewise continuous. Thus, x is considerably smoother than P. This is the main reason that Poisson generators are useful.

The Jacobian of (3.25) is

$$x_\xi = b - a - \int_\xi^1 P(\zeta)\,d\zeta + \int_0^1 \zeta P(\zeta)\,d\zeta\,, \tag{3.26}$$

which shows the main difficulty with this generator; namely, that the relation between the Jacobian (i.e., segment length) and the weight function is nontrivial. For example, even if $P > 0$, one cannot guarantee $J > 0$. It is not clear that a physical-space weight function $P(x)$ would permit more effective control over the grid lengths.

The AH generator (3.24) is the same as one of the generators given by (3.6) or (3.17) provided that one of the equalities

$$P = \frac{\phi_\xi}{\phi} x_\xi\,, \quad P = \frac{w_x}{w} x_\xi^2 \tag{3.27}$$

holds. The latter generators are clearly preferable to AH since the effect of their weight function is unambiguous.

In two dimensions, a natural generalization of the AH generator produces folded grids on rather simple regions (see Section 5.4.1). The two-dimensional generalization of the *TTM generator* to be described next is considerably more robust (see Section 5.4.2). The basic idea is to consider the inverse $\xi = \xi(x)$ of the transformation and assume that this function satisfies the boundary value problem

$$\xi_{xx} = P(\xi)\,, \quad \xi(a) = 0\,, \quad \xi(b) = 1\,, \tag{3.28}$$

where, as before, P is some given continuous function.

To compare the TTM generator with the previous generators, transform the differential equation (3.28) to an equation in logical space. Recall that $x(\xi)$ and $\xi(x)$ are inverses of each other, so that $x(\xi(x)) = x$. Differentiating this with respect to x gives $x_\xi \xi_x = 1$ or $\xi_x = 1/x_\xi$, which is the standard rule for computing derivatives of inverse functions. The chain rule (2.6) gives

$$\xi_{xx} = \frac{1}{x_\xi}\left(\frac{1}{x_\xi}\right)_\xi = -\frac{x_{\xi\xi}}{x_\xi^3}\,. \tag{3.29}$$

This can be used to transform equation (3.28) to the *nonlinear* boundary value problem

$$x_{\xi\xi} + P(\xi)\,x_\xi^3 = 0\,, \quad x(0) = a\,, \quad x(1) = b\,. \tag{3.30}$$

The transformed equation bears a modest resemblance to (3.17), but here one has a logical-space weight. The transformed equation can be integrated to give a relation for the Jacobian

$$x_\xi^{-2} = 2 \int^\xi P(\zeta)\, d\zeta\,. \tag{3.31}$$

Equation (3.31) shows that the relationship between segment lengths and the weight P is not simple and there is no guarantee that $J = x_\xi > 0$. If P is continuous, then (3.31) implies that x_ξ is continuous and, consequently, x is continuous. Equation (3.30) then implies that $x_{\xi\xi}$ is continuous, that is, x is twice continuously differentiable. Thus, the one-dimensional (1-D) TTM generator with continuous weight functions produces smooth grids.

Exercise 3.3.1 Show that, for either Poisson generator, if $P \equiv 0$ then the grid is linear:

$$x(\xi) = (b-a)\,\xi + a\,.\ \S \tag{3.32}$$

The 1-D TTM generator (3.30) is the same as either of the generators given by (3.6) or (3.17) provided that one of the equalities

$$P\,x_\xi^3 = -\frac{\phi_\xi}{\phi}\,x_\xi\,, \quad P\,x_\xi^3 = -\frac{w_x}{w}\,x_\xi^2 \tag{3.33}$$

holds. These equations show how to choose P in the TTM method to obtain grid spacing proportional to some weight. For example, if $x(\xi)$ is generated by ϕ in (3.6), then one can obtain the same transformation by solving (3.30) using

$$P(\xi) = -\frac{1}{C^2}\frac{\phi_\xi}{\phi^3}\,. \tag{3.34}$$

The equivalence between the TTM generator and the generators of the previous section does not extend to higher dimensions.

Exercise 3.3.2 An alternate form of the weighted TTM generator is sometimes used (Thompson, Warsi, Mastin [219])

$$\xi_{xx} = \xi_x^2\, P(\xi)\,. \tag{3.35}$$

Invert this equation and compare the result to (3.6). §

3.4 NUMERICAL IMPLEMENTATION

All of the grid-generation equations presented in this section can be approximated using the techniques described in Section 2.2. For convenience in comparing the methods and understanding the programs used to generate the grids, difference approximations for all generators are now derived. The problem is to compute x_i, $0 \le i \le M$, where M is a given positive integer and $x_0 = a$ and $x_M = b$.

The *finite-difference approximation* to each of the differential equations of the previous sections always results in a system of algebraic equations of the form

$$\frac{l_i}{\Delta\xi^2}x_{i-1}+\frac{c_i}{\Delta\xi^2}x_i+\frac{r_i}{\Delta\xi^2}x_{i+1}=g_i\,,\qquad 1\le i\le M-1\,,\tag{3.36}$$

where

$$c_i=-\,(l_i+r_i)\;.\tag{3.37}$$

Thus, only r_i, l_i, and g_i are needed to specify the problem. Note that the notation here differs from the notation in Section 2.2: $l_i = L_i\,\Delta\xi^2$, etc.

Equation (3.5),

$$\left(\frac{x_\xi(\xi)}{\phi(\xi)}\right)_\xi=0\,,\tag{3.38}$$

is a linear homogeneous symmetric ordinary differential equation. The stencils for its difference approximation are

$$r_i=\frac{1}{\phi_{i+1/2}}\,,\quad l_i=r_{i-1}\,,\quad g_i=0\,.\tag{3.39}$$

System (3.39) is a *symmetric tridiagonal system* that can be solved using a symmetric tridiagonal solver. Special treatment of this system may be necessary at the end points to employ the tridiagonal solver. For example, at $i = 1$, the stencil equation can be expressed as

$$\frac{c_1}{\Delta\xi^2}x_1+\frac{r_1}{\Delta\xi^2}x_2=-\frac{l_1}{\Delta\xi^2}a\,,\qquad 1\le i\le M-1\,,\tag{3.40}$$

Accordingly, the stencil coefficients are modified to read

$$l_1=0\,,\quad g_1=-\frac{r_0}{\Delta\xi^2}a\,.\tag{3.41}$$

A similar modification will be necessary at $i = M - 1$ and in the numerical algorithms to follow. The location of FORTRAN code implementing this algorithm is given in Appendix B.3.1.

Exercise 3.4.1 Use the material in Section 2.2 to derive (3.39). §

Equation (3.6),

$$x_{\xi\xi}-\frac{\phi_\xi}{\phi}x_\xi=0\,,\tag{3.42}$$

is also a linear homogeneous symmetric ordinary differential equation. However it is not written in symmetric form. Centered differencing gives

$$r_i=1-\frac{\phi_{i+1}-\phi_{i-1}}{4\phi_i}\,,\quad l_i=1+\frac{\phi_{i+1}-\phi_{i-1}}{4\phi_i}\,,\quad g_i=0\,.\tag{3.43}$$

This difference approximation is not symmetric; thus (3.39) is preferred. However, the system (3.43) is still tridiagonal and can be handled by a standard tridiagonal solver.

- Begin loop.
- Evaluate l_i, c_i, r_i, and g_i using x_i^{old}.
- Solve the resulting linear system for $x_i = x_i^{new}$ using a tridiagonal solver.
- Compute $\delta = \max_i |x_i^{new} - x_i^{old}|$.
  - If $\delta > tol$ then set $x_i^{old} = x_i^{new}$ and loop.
  - If $\delta \le tol$ then quit.

Figure 3.2 *Nonlinear iteration*

Exercise 3.4.2 Use the material in Section 2.2 to derive (3.43). §

When the weight function depends on the grid, the ordinary differential equations satisfied by the grid functions are nonlinear, but otherwise they are the same as the previous two differential equations. The usual differencing of (3.16),

$$\left(\frac{x_\xi(\xi)}{w(x(\xi))}\right)_\xi = 0\,, \tag{3.44}$$

gives

$$r_i = \frac{1}{w(x_{i+1/2})}\,, \quad l_i = r_{i-1}\,, \quad g_i = 0\,. \tag{3.45}$$

This is a nonlinear system that can be solved using a nonlinear iteration.

Any problem where the l_i, c_i, r_i, or g_i depend on the grid, that is, on the x_i, is nonlinear. Such problems are solved using a *nonlinear iteration* procedure. A small tolerance parameter tol that determines the accuracy of the solution must be given. To start the iteration, an initial grid x_i^{old} must be generated, say using linear interpolation. Then the loop described in Figure 3.2 is executed. FORTRAN code implementing this algorithm is described in Appendix B.3.1.

Exercise 3.4.3 Use the nonlinear iteration described in Figure 3.2 to solve (3.45). Experiment with values of $tol = 10^{-k}$ for $1 \le k \le 8$ to see what impact this has on the resulting grid. §

A similar approach to that used to obtain (3.43) works for the nonlinear Equation (3.18),

$$x_{\xi\xi} - \frac{w_\xi}{w}\, x_\xi = 0\,. \tag{3.46}$$

Instead of repeating that discussion, it is observed that a different algorithm can be applied. The lower order terms can be evaluated at the old grid to determine the right-hand side of the stencil and the second-order terms can be used to calculate the left-hand side stencil coefficients:

$$r_i = l_i = 1\,, \quad g_i = \frac{(w_{i+1} - w_{i-1})\,(x_{i+1} - x_{i-1})}{4\,w_i\,\Delta\xi^2}\,. \tag{3.47}$$

Because l_i, c_i, and r_i are constant, a particularly efficient algorithm can be used to solve this problem.

Exercise 3.4.4 Modify the algorithm given in Figure 3.2 to solve the system of equations determined by (3.47) by taking the tridiagonal solve outside the loop. The general algorithm is so fast that it is not worth programming this algorithm in the one-dimensional case. However, in higher dimensions, this idea has a substantial payoff. §

Project 3.4.5 Write a code (see Appendix B) to generate the logical-space weighted grid using the stencil coefficients (3.39). Use both the weight in Exercise (3.2.4) and the weight

$$\phi(\xi) = \exp(\lambda \mid \xi - \xi_0 \mid), \quad \lambda > 0, \tag{3.48}$$

on $[a, b] = [-1, 1]$ and with $0 \le \xi_0 \le 1$. Plot the grids in the x-ξ plane. To what point in physical space has the point $\xi = \xi_0$ been mapped? To validate the code, numerically compute x_ξ and C [see (3.9)] at each point of the mesh. §

Project 3.4.6 Write a code (see Appendix B) to produce physical-space weighted grids using the stencil coefficients (3.45). Use both the weight in Exercise (3.2.5) and the weight

$$w(x) = \exp(\lambda \mid x - x_0 \mid), \quad \lambda > 0, \tag{3.49}$$

on the interval $[a, b] = [-1, 1]$ with $-1 \le x_0 \le 1$. Plot the grid in the x-ξ plane. To validate the code, numerically compute x_ξ and C [see (3.20)] at each point of the mesh. §

Project 3.4.7 Write a code (see Appendix B) to solve Equation (3.30) differenced as

$$r_i = l_i = 1, \quad g_i = -P_i \left(\frac{x_{i+1} - x_{i-1}}{2\Delta\xi} \right)^3 . \tag{3.50}$$

Solve the difference equation using the nonlinear iteration described for (3.45). Implement the weight function

$$P(\xi) = \begin{cases} -\frac{\lambda}{A_0^2} e^{+2\lambda(\xi - \xi_0)}, & 0 \le \xi < \xi_0 < 1 \\ \\ +\frac{\lambda}{A_0^2} e^{-2\lambda(\xi - \xi_0)}, & 0 \le \xi_0 < \xi < 1 \end{cases} \tag{3.51}$$

where $\lambda > 0$ and $A_0 \ne 0$. Experiment with the values of A_0, λ, and ξ_0. To what point x_0 in physical space does ξ_0 map? The weight (3.51) is the 1-D analog of the weight used in the higher-dimensional TTM solvers. §

3.5 MINIMIZATION PROBLEMS

The one-dimensional grid generators in Section 3.2 can be derived from variational principles. To build intuition, discrete functionals are introduced, then the limit is taken to obtain a continuum functional. The goal is to find functionals whose minimum is a transformation having the property that the grid spacing is proportional to a given weight function $\phi > 0$.

Let the multivariable function $S > 0$ from R^M to R be defined by

$$S = \sum_{i=1}^{M} \frac{(x_i - x_{i-1})^2}{2\,\phi_{i-1/2}}, \tag{3.52}$$

with $\phi_{i-1/2} = \phi(\xi_{i-1/2})$. The problem is to minimize S subject to the constraints $x_0 = a$ and $x_M = b$. At an extremum, the partial derivatives of S with respect to the grid points must be zero:

$$\frac{\partial S}{\partial x_j} = \frac{x_j - x_{j-1}}{\phi_{j-1/2}} - \frac{x_{j+1} - x_j}{\phi_{j+1/2}} = 0, \qquad 1 \le j \le M-1. \tag{3.53}$$

This is a system of $M - 1$ linear equations. Clearly, if

$$x_j - x_{j-1} = \lambda\,\phi_{j-1/2}, \qquad 1 \le j \le M, \tag{3.54}$$

with λ some constant, then the x_j are solutions of (3.53), i.e., if the grid spacing is proportional to the weight then the grid is an extremum of S. To show the converse, note that if (3.53) is summed for $1 \le i \le k$ then (this is a "collapsing" sum)

$$\frac{x_k - x_{k-1}}{\phi_{j-1/2}} = \frac{x_1 - x_0}{\phi_{1/2}}, \qquad 1 \le k \le M-1, \tag{3.55}$$

that is, with

$$\lambda = \frac{x_1 - x_0}{\phi_{1/2}}, \tag{3.56}$$

the grid must satisfy (3.54).

To show that this is, in fact, a minimum, one examines the *Hessian* matrix $H_{i,j}$ for (3.52):

$$H_{i,j} = \frac{\partial^2 S}{\partial x_i \partial x_j} = \begin{cases} -\dfrac{1}{\phi_{j-\frac{1}{2}}}, & \text{if } i = j-1 \\[2ex] \dfrac{1}{\phi_{j-\frac{1}{2}}} + \dfrac{1}{\phi_{j+\frac{1}{2}}}, & \text{if } i = j \\[2ex] -\dfrac{1}{\phi_{j+\frac{1}{2}}}, & \text{if } i = j+1 \\[2ex] 0, & \text{otherwise} \end{cases} \tag{3.57}$$

where $1 \le i,\ j \le M - 1$. One can readily show that the Hessian is symmetric and positive semidefinite, so S is convex (Minoux [140]) and must have minimum. Since the solution to (3.53) is unique, the minimum must be unique.

Exercise 3.5.1 Show that $H_{i,j}$ given in (3.57) is positive semidefinite. §

Exercise 3.5.2 The first k of equations (3.54) can be summed to give

$$x_k = a + \lambda \sum_{j=1}^{k} \phi_{j-1/2} \tag{3.58}$$

with $1 \le k \le M$ and

$$\lambda = \frac{b-a}{\sum_1^M \phi_{j-1/2}}. \tag{3.59}$$

This equation can be thought of as an implied constraint on the minimization problem. Note that the minimum of S is $\lambda\,(b-a)/2$ and consequently the minimum is never zero for $b > a$. §

The function (3.52) is not the only one that can be minimized in order to obtain the property (3.54). For example, consider

$$\hat{S} = \sum_{i=1}^{M-1} \left\{ \left(\frac{x_{i+1}-x_i}{\phi_{i+1/2}} \right)^2 - \left(\frac{x_i - x_{i-1}}{\phi_{i-1/2}} \right)^2 \right\}^2. \tag{3.60}$$

This function is clearly positive or zero; if the grid x_i satisfies (3.54), the function $\hat{S}$ has the value zero, so the desired grid yields a minimum of the function. The derivatives of $\hat{S}$ in (3.60) are far more complicated than the derivatives of S in (3.52). Invoking the principle of Ockham's Razor, it is best to use the simpler function S since its performance leaves little to be desired. The approach to grid generation outlined here is often referred to as the *direct optimization method* (see Section 6.5 for further discussion). One can minimize such functions directly using the methods of optimization (see Kennon and Dulikravich [109] and Castillo [32]) to obtain grids in all dimensions. The present goal, however, is merely to motivate the continuum functionals emphasized in this text.

Divide (3.52) by $\Delta\xi$ to get

$$\frac{S}{\Delta\xi} = \sum_{i=1}^{M} \left(\frac{x_i - x_{i-1}}{\Delta\xi} \right)^2 \frac{\Delta\xi}{2\,\phi_{i-1/2}}. \tag{3.61}$$

If $x(\xi)$ is differentiable, ϕ continuous, and $x_i = x(\xi_i)$, then as $\Delta\xi \to 0$ the right-hand side of the previous equation converges to

$$I[x] = \int_0^1 \frac{x_\xi^2(\xi)}{2\,\phi(\xi)}\,d\xi. \tag{3.62}$$

$I[x]$ is called a *functional* or a functional of the function $x = x(\xi)$. The problem is to minimize $I[x]$ over all admissible x, that is, over all x that are continuously differentiable and satisfy $x(0) = a$ and $x(1) = b$.

To calculate the minimum of $I[x]$ defined in (3.62), introduce a function $c(\xi)$ that satisfies $c(0) = c(1) = 0$. For any ϵ the function $x(\xi) + \epsilon\,c(\xi)$ satisfies the same boundary conditions as the function $x(\xi)$. Consequently, $\epsilon = 0$ must be a minimum of the function $F : R \to R$ defined by

$$F(\epsilon) = \int_0^1 \frac{(x_\xi(\xi) + \epsilon\,c_\xi(\xi))^2}{2\phi(\xi)}\,d\xi. \tag{3.63}$$

If $x(\xi)$ is to minimize the integral in (3.62), then $F'(0) = 0$, that is,

$$\int_0^1 \frac{x_\xi(\xi)}{\phi(\xi)}\,c_\xi(\xi)\,d\xi = 0. \tag{3.64}$$

The boundary conditions on c and an integration by parts gives

$$\int_0^1 \left(\frac{x_\xi(\xi)}{\phi(\xi)} \right)_\xi c(\xi)\, d\xi = 0. \tag{3.65}$$

This is true for arbitrary c, so

$$\left(\frac{x_\xi(\xi)}{\phi(\xi)} \right)_\xi = 0, \tag{3.66}$$

which is the same as equation (3.5). The differential equation (3.66) is called the *Euler-Lagrange equation* for the minimization problem. It is recognized to be the same as (3.5) and thus all the properties discussed there hold for the minimizing transformation.

If the weight function depends on the physical-space variable x, that is, if a weight function $w = w(x) > 0$, defined on $[a, b]$, is considered, then interestingly it is not possible to take $\phi(\xi) = w(x(\xi))$ in the previous functional and get that the spacing is proportional to w. Instead the spacing turns out to be proportional to the square root of w. Thus, to obtain spacing proportional to w, take $\phi = w^2$ in (3.62) to obtain the variational principle for physical weights: minimize $I[x]$ over all continuously differentiable x where

$$I[x] = \int_0^1 \frac{x_\xi^2(\xi)}{2\, w^2(x)}\, d\xi, \tag{3.67}$$

and x is subject to the constraints $x(0) = a$ and $x(1) = b$.

Exercise 3.5.3 Show that the Euler-Lagrange equation for this minimization problem can be written in the form

$$x_{\xi\xi} - \frac{w_x}{w}\, x_\xi^2 = 0, \tag{3.68}$$

which is the same as equation (3.17). Also, note that the chain rule implies that (3.68) can be written in the form

$$x_{\xi\xi} - \frac{w_\xi}{w}\, x_\xi = 0, \tag{3.69}$$

which has the same form as (3.18). §

This section has demonstrated that both the logical and physical grid-spacing generators in Section 3.2 can be based upon variational principles.

3.6 THE CALCULUS OF VARIATIONS

The classical *calculus of variations* provides the theory for computing the Euler-Lagrange equations for minimization problems. Let $G(r, s, t)$ be a smooth function of three real variables. The problem is to find a function $x(\xi)$ defined on $[0, 1]$ that minimizes the functional

$$I[x] = \int_0^1 G(\xi, x(\xi), x_\xi(\xi))\, d\xi, \tag{3.70}$$

subject to the constraints $x(0) = a$ and $x(1) = b$. The Euler-Lagrange equation for the minimizer x of this functional is derived as above. Introduce the function

$$F(\epsilon) = I[x + \epsilon c], \tag{3.71}$$

where $c = c(\xi)$ is continuously differentiable and $c(0) = c(1) = 0$. Then

$$F'(0) = \int_0^1 \left\{ G_s(\xi, x, x_\xi)\, c + G_t(\xi, x, x_\xi)\, c_\xi \right\} d\xi = 0, \tag{3.72}$$

and then an integration by parts (using the constraints, that is, the boundary conditions, on c) gives

$$\int_0^1 \left\{ G_s(\xi, x, x_\xi) - \left(G_t(\xi, x, x_\xi) \right)_\xi \right\} c\, d\xi = 0. \tag{3.73}$$

Because this must hold for all c, it must be that

$$G_s(\xi, x, x_\xi) - \left(G_t(\xi, x, x_\xi) \right)_\xi = 0. \tag{3.74}$$

Applying the chain rule to the ξ derivative of G_t gives

$$G_s - G_{rt} - G_{st}\, x_\xi - G_{tt}\, x_{\xi\xi} = 0. \tag{3.75}$$

This is a second-order differential equation for x. Recall that the constraints provide two boundary conditions for this differential equation.

It is common to write the functional in the form

$$I[x] = \int_0^1 G(\xi, x, x_\xi)\, d\xi, \tag{3.76}$$

and the Euler-Lagrange equation (3.74) in the form

$$\frac{\partial G}{\partial x}(\xi, x, x_\xi) - \frac{d}{d\xi} \frac{\partial G}{\partial x_\xi}(\xi, x, x_\xi) = 0. \tag{3.77}$$

Exercise 3.6.1 Use the techniques in this section to derive the Euler-Lagrange equations for the functionals (3.62) and (3.67). §

Exercise 3.6.2 Show that the Euler-Lagrange equations for the minimum of the functional

$$I[x] = \int_0^1 \left[P(\xi)\, x(\xi) + \frac{x_\xi^2}{2} \right] d\xi \tag{3.78}$$

give (3.24). §

Exercise 3.6.3 An alternate approach by Brackbill and Saltzman [21] poses variational principles in terms of maps from physical to logical space. For example, a one-dimensional principle of this type is to minimize

$$I[\xi] = \int_0^1 G(x, \xi(x), \xi_x(x))\, dx \tag{3.79}$$

with $\xi(a) = 0$ and $\xi(b) = 1$. Use the approach in this section to obtain the Euler-Lagrange equations for this functional. Use them to find a variational principle for (3.28). Can you find one for (3.35)? §

3.7 A FOURTH-ORDER GRID GENERATION EQUATION

Another 1-D grid-generation equation can be obtained by minimizing

$$I[x] = \int_0^1 (x_{\xi\xi})^2 \, d\xi \tag{3.80}$$

subject to boundary conditions to be determined shortly. The intuition is that $x_{\xi\xi}$ measures the derivative of the rate of change of the distance between grid points, so the minimum of $I[x]$ should result in a transformation with the property that the distance between grid points varies smoothly.

Exercise 3.7.1 Use the methods of the previous section to show that the Euler-Lagrange equation that minimizes the functional (3.80) is the *fourth-order Euler-Lagrange equation*

$$x_{\xi\xi\xi\xi} = 0 . \quad \S \tag{3.81}$$

The advantage of a fourth-order equation is that four boundary conditions can now be imposed on the grid. A natural set of conditions are

$$x(0) = a , \quad x(1) = b , \quad x_\xi(0) = \alpha_1 , \quad x_\xi(1) = \alpha_2 \tag{3.82}$$

with $\alpha_1 > 0$ and $\alpha_2 > 0$. The last two boundary conditions permit direct control over the first and last cell "lengths."

The solution to equation (3.81) is the cubic polynomial

$$x(\xi) = c_0 + c_1\xi + c_2\xi^2 + c_3\xi^3 , \tag{3.83}$$

where the coefficients are determined from the boundary conditions (3.82):

$$c_0 = a , \quad c_1 = \alpha_1 , \quad c_2 = 3\,(b - a) - 2\,\alpha_1 - \alpha_2 , \tag{3.84}$$

and

$$c_3 = \alpha_1 + \alpha_2 - 2\,(b - a) . \tag{3.85}$$

Since x_ξ is a quadratic polynomial in ξ, it is possible for the grid to be folded even if the derivatives at the end points, α_1 and α_2, are positive.

Exercise 3.7.2 Show that any of the following conditions on the c's are sufficient to guarantee an unfolded grid: (i) $c_3 \le 0$, (ii) $c_3 > 0$ and $c_2 > 0$, (iii) $c_3 > 0$ and $c_2 + 3\,c_3 < 0$, or (iv) $c_3 > 0$ and $3\,c_1\,c_3 - c_2^2 > 0$. Note that if $\alpha_1 = \alpha_2 = b - a$, then $c_2 = c_3 = 0$, resulting in the linear unweighted grid. §

The fourth-order grid-generation equation provides no additional capability beyond that available in the weighted variational generator because, given the solution grid $x(\xi)$ in (3.83) generated by (3.81), choose $\phi(\xi) = x_\xi = c_1 + 2c_2\xi + 3c_3\xi^2$ and solve (3.66). On the other hand, since the solution grid to (3.81) is always a cubic function, there is no way to choose the four boundary conditions (3.82) to obtain a grid generated by an arbitrary weight $\phi(\xi)$. From this viewpoint (3.81) is more limited than (3.66).

It is possible to replace the last two boundary conditions in equation (3.82) by the conditions

$$x(\xi_1) = x_1 , \quad x(\xi_2) = x_2 , \tag{3.86}$$

with $0 < \xi_1 < \xi_2 < 1$ and $a < x_1 < x_2 < b$ to fix two internal points of the grid. The solution is still of the form (3.83), but (3.84)–(3.85) does not apply. The coefficients of the polynomial are determined by solving the four equations in four unknowns obtained by applying the new boundary conditions (3.86).

Exercise 3.7.3 Suppose that (3.80) is changed to include a weight function $\phi(\xi)$: minimize

$$I[x] = \int_0^1 \frac{(x_{\xi\xi})^2}{\phi}\, d\xi \tag{3.87}$$

with the boundary conditions (3.82). Show that the Euler-Lagrange equation is

$$\left(\frac{x_{\xi\xi}}{\phi}\right)_{\xi\xi} = 0\,. \tag{3.88}$$

The general solution of this equation has the form

$$x_{\xi\xi} = (\mu\,\xi + \nu)\,\phi(\xi) \tag{3.89}$$

for some constants μ and ν. Use (3.24) and (3.25) to show that

$$\begin{aligned} x(\xi) = (b-a)\,\xi + a + \xi \int_0^1 \zeta\,(\mu\,\zeta + \nu)\,\phi(\zeta)\,d\zeta \\ - \xi \int_\xi^1 (\mu\,\zeta + \nu)\,\phi(\zeta)\,d\zeta - \int_0^\xi \zeta\,(\mu\,\zeta + \nu)\,\phi(\zeta)\,d\zeta \end{aligned} \tag{3.90}$$

satisfies (3.89) and the first two boundary conditions in (3.82). Show that if one chooses the Neumann boundary conditions,

$$\alpha_1 = (b-a) + \nu \int_0^1 (\zeta - 1)\,\phi(\zeta)\,d\zeta\,, \tag{3.91}$$

$$\alpha_2 = (b-a) + \nu \int_0^1 \zeta\,\phi(\zeta)\,d\zeta\,, \tag{3.92}$$

where ν is arbitrary, then

$$x_{\xi\xi} = \nu\phi\,, \tag{3.93}$$

i.e., the second derivative of the transformation is proportional to the given weight. Thus, if one chooses the weight function ϕ and, say, α_1, then ν and α_2 are determined. This approach therefore shows how to extend the approach in (3.24) to obtain control over one of the derivatives of the transformation at the end point. §

3.8 FEATURE-ADAPTIVE MOVING GRIDS

The idea in *moving-grid solution-adaptivity* is to solve hosted equations by moving the grid points using a weighted grid generator so as to reduce the overall truncation error in the solution of the hosted equation. All the theory necessary to do this for one-dimensional steady-state hosted equations is now in place except for the determination of the weight function. There are several ways to choose the weight function. The simplest is the idea of *feature-adaptive weights* in which the weight function depends on features of the solution such as the function itself, its gradient, or second derivative. In this case, the weight function is a physical-space weight function of the form $w = w(f, f_x, f_{xx})$.

Consider the one-dimensional steady-state convection-diffusion equation,

$$f_{xx} - \nu f_x = 0, \tag{3.94}$$

for a model-hosted problem. Assume the domain is $[a, b]$ and Dirichlet data are $f(a) = 0$ and $f(b) = 1$ and that ν a parameter representing the ratio of convective to diffusive terms. When ν is large, the solution of this problem will have a *boundary layer* near one of the boundaries. This means that the solution f changes very rapidly near the boundary. Adaptive grids can be used to resolve such boundary layers.

Exercise 3.8.1 Verify that the solution to (3.94) with $\nu \neq 0$ is

$$f(x) = \frac{e^{\nu(x-a)} - 1}{e^{\nu(b-a)} - 1}. \tag{3.95}$$

Plot the solution for large and small values of the parameter, including both negative and positive values. This provides a good model problem in that it is clearly useful to stretch the grid near the boundary. What is the solution when $\nu = 0$? §

To solve the hosted equation using a general transformation $x(\xi)$, one must first transform the hosted equation to general coordinates. For example, the convection-diffusion equation (3.94) transforms to

$$f_{\xi\xi} - \left(\frac{x_{\xi\xi}}{x_\xi} + \nu x_\xi\right) f_\xi = 0 \tag{3.96}$$

with $f(0) = 0$ and $f(1) = 1$. This equation can be solved once the grid is known. This chapter has given several options for the grid generator, but (3.18) is preferred since it has the useful property that grid spacing is proportional to the physical weight.

Before one can proceed, the weight function must be constructed. The most commonly used feature-adaptive weight function is gradient-based

$$w(x) = \frac{1}{\sqrt{1 + \epsilon^2 f_x^2}} \tag{3.97}$$

with ϵ a nonnegative parameter. Note that if the gradient f_x is small, the weight will be close to one, while if the gradient is large then

$$w(x) \approx \frac{1}{\epsilon\,|f_x|}, \tag{3.98}$$

that is, the weight w is proportional to reciprocal of the gradient. Consequently, large gradients will produce small weights and hence relatively small cell sizes.

The weight function can also be transformed to logical coordinates to obtain a computationally more useful form

$$w(x(\xi)) = \frac{|x_\xi|}{\sqrt{x_\xi^2 + \epsilon^2 f_\xi^2}}. \tag{3.99}$$

Project 3.8.2 Write a code to numerically solve the coupled pair of equations (3.96) and (3.18) using a nonlinear iterative scheme. Use the weight (3.99). Compute the truncation error using the exact solution (3.95) and experiment with the value of ϵ to minimize the truncation error. §

There are endless variations upon this discussion. One can vary the weight function, introduce more parameters in the weight, smooth the weight, and so on. The reader is referred to Anderson and Rai [4], Anderson [7], Daripa [45, 47], Dwyer et al. [57, 58], Ghia et al. [81], Godunov and Prokopov [84], Nakamura [146], Rai and Anderson [155], and Zegeling et al. [238].

3.9 SUMMARY

A theory of grid generation in one dimension has been presented. The choice of weights functions is clear and uncontroversial, even with the Poisson equations (unlike the situation in two dimensions). A grid having the property that segment length is proportional to the given weight is easily obtained by minimizing the proper functional. Positive weights guarantee an unfolded continuum grid. The Poisson and fourth-order grid generation equations do not provide any additional capability beyond the weighted system. As will be seen, numerous difficulties are encountered when the one-dimensional theory is extended to two and three dimensions, but at least a firm and intuitive foundation has been laid.

Chapter 4

Vector Calculus and Differential Geometry

Before proceeding to the study of planar grid generation, a brief pause is made in this chapter to summarize elementary results from vector calculus, general coordinate relationships, and differential geometry. The lengths of and angles between tangent and normal vectors are some of the most useful quantities for developing a geometric understanding of grid generation, so the chapter begins with a review of the scalar and vector products and their geometric meaning (Section 4.1). Next, the tangent to and the arc length of a curve are discussed using vector calculus. Tangents and normals to a surface are defined and their relationship to surface area are described. The relationship between the gradient of a function that implicitly defines a surface and the surface normal vector is given. Many grid generators are formulated in terms of the inverse of a general coordinate map, so relationships in one, two, and three dimensions, between a general coordinate map and its inverse are derived in Section 4.2. The entries in the metric tensor are nothing but the length and inner product of tangent vectors to coordinate lines. Thus these entries contain much of the critical information about grids. Consequently the metric tensor and related concepts are defined in Section 4.3 on differential geometry. The chapter closes with an exposition (Section 4.4) on vector calculus relationships that pertain to planar grid generation. Any book on multidimensional calculus (e.g., Corwin and Szczarba [42]) will provide a more extensive discussion of the material on vector calculus, while Struik [203] and Stoker [201] are good references for the section on differential geometry. Ideas from matrix theory are extensively discussed in Horn and Johnson [97].

4.1 VECTOR CALCULUS

Recall that the *scalar product* (or dot product) and *vector product* (or cross product) of two vectors $\mathbf{u} = (u_1, u_2, u_3)$ and $\mathbf{v} = (v_1, v_2, v_3)$ are given by

$$\mathbf{u} \cdot \mathbf{v} = u_1 v_1 + u_2 v_2 + u_3 v_3 , \tag{4.1}$$

$$\mathbf{u} \times \mathbf{v} = (u_2 v_3 - u_3 v_2, u_3 v_1 - u_1 v_3, u_1 v_2 - u_2 v_1) . \tag{4.2}$$

The vector product is often expressed in terms of the determinant of a matrix:

$$\mathbf{u} \times \mathbf{v} = \det \begin{pmatrix} \mathbf{i} & \mathbf{j} & \mathbf{k} \\ u_1 & u_2 & u_3 \\ v_1 & v_2 & v_3 \end{pmatrix} . \tag{4.3}$$

If θ is the angle between two vectors and $\mathbf{n}$ is a unit vector normal to both $\mathbf{u}$ and $\mathbf{v}$, then these products are also given by

$$\mathbf{u} \cdot \mathbf{v} = |\mathbf{u}|\,|\mathbf{v}| \cos(\theta)\,, \qquad \mathbf{u} \times \mathbf{v} = |\mathbf{u}|\,|\mathbf{v}| \sin(\theta)\,\mathbf{n}\,, \tag{4.4}$$

provided $\mathbf{n}$ points in the direction given by the right-hand rule for cross products.

The following exercises review some basic vector calculus that is applied in the subsequent development.

Exercise 4.1.1 Given two vectors $\mathbf{u}$ and $\mathbf{v}$, show that the area of the parallelogram determined by the two vectors is given by $|\mathbf{u} \times \mathbf{v}|$. Begin by making a drawing of the plane that contains $\mathbf{u}$ and $\mathbf{v}$. If θ is the angle between the two vectors, then from equation (4.4)

$$\cos(\theta) = \frac{\mathbf{u} \cdot \mathbf{v}}{|\mathbf{u}|\,|\mathbf{v}|}\,, \qquad |\sin(\theta)| = \frac{|\mathbf{u} \times \mathbf{v}|}{|\mathbf{u}|\,|\mathbf{v}|}\,. \tag{4.5}$$

The area is given by $A = |\mathbf{u}|\,|\mathbf{v}|\,|\sin(\theta)|$. §

Exercise 4.1.2 Given three vectors $\mathbf{u}$, $\mathbf{v}$, and $\mathbf{w}$, show that the volume of the parallelepiped determined by the three vectors is given by the absolute value of the *triple scalar product*, $|\,[\mathbf{u}, \mathbf{v}, \mathbf{w}]\,|$, where

$$[\mathbf{u}, \mathbf{v}, \mathbf{w}] = \mathbf{u} \cdot \mathbf{v} \times \mathbf{w}\,. \tag{4.6}$$

Make a figure and use the results of Exercise 4.1.1 to illustrate the following comments. The unit vector $\mathbf{n}$, normal to $\mathbf{u}$ and $\mathbf{v}$, can be computed using $\mathbf{n} = \mathbf{u} \times \mathbf{v}/|\mathbf{u} \times \mathbf{v}|$ and the angle ϕ between $\mathbf{n}$ and $\mathbf{w}$ is given by $\cos(\phi) = \mathbf{n} \cdot \mathbf{w}/|\mathbf{w}|$. Your drawing should make it clear that the volume is

$$|\mathbf{u}|\,|\mathbf{v}|\,|\mathbf{w}|\,|\cos(\phi)|\,|\sin(\theta)|\,; \tag{4.7}$$

translating this back to scalar and vector products gives the result. Note that the triple scalar product can be written in several ways in terms of scalar and vector products of $\mathbf{u}$, $\mathbf{v}$, and $\mathbf{w}$. §

Exercise 4.1.3 Prove the following identities:

$$\mathbf{u} \cdot \mathbf{v} \times \mathbf{w} = \det \begin{pmatrix} u_1 & u_2 & u_3 \\ v_1 & v_2 & v_3 \\ w_1 & w_2 & w_3 \end{pmatrix}\,; \tag{4.8}$$

$$[\mathbf{u}, \mathbf{v}, \mathbf{w}] = [\mathbf{v}, \mathbf{w}, \mathbf{u}] = [\mathbf{w}, \mathbf{u}, \mathbf{v}]\,; \tag{4.9}$$

$$(\mathbf{u} \times \mathbf{v}) \cdot (\mathbf{w} \times \mathbf{r}) = (\mathbf{u} \cdot \mathbf{w})(\mathbf{v} \cdot \mathbf{r}) - (\mathbf{u} \cdot \mathbf{r})(\mathbf{v} \cdot \mathbf{w})\,; \tag{4.10}$$

$$\mathbf{u} \times (\mathbf{v} \times \mathbf{w}) = (\mathbf{u} \cdot \mathbf{w})\mathbf{v} - (\mathbf{u} \cdot \mathbf{v})\mathbf{w}.\ \S \tag{4.11}$$

The vector product $\mathbf{u} \times \mathbf{v}$ is perpendicular to both $\mathbf{u}$ and $\mathbf{v}$; this follows from (4.9):

$$[\mathbf{u}, \mathbf{u}, \mathbf{v}] = \mathbf{u} \cdot \mathbf{u} \times \mathbf{v} = 0, \quad [\mathbf{v}, \mathbf{u}, \mathbf{v}] = \mathbf{v} \cdot \mathbf{u} \times \mathbf{v} = 0. \tag{4.12}$$

4.1.1 Vector Geometry of Curves

This section describes the tangent vector to a curve and then uses it to define arc length. Coordinates in three dimensions are x, y, and z and are written as a vector $\mathbf{x} = (x, y, z)$. A curve can be given parametrically by three functions $x(\xi)$, $y(\xi)$, and $z(\xi)$ defined for $0 \le \xi \le 1$ [see Figure (4.1)]. The functions $x(\xi)$, $y(\xi)$, and $z(\xi)$ are assumed continuously differentiable. The parametric description of a curve is also given by the vector $\mathbf{x} = \mathbf{x}(\xi)$. For example, a circle of radius a in the plane $z = b$ is defined by

$$x = a\cos(2\pi\xi), \quad y = a\sin(2\pi\xi), \quad z = b, \quad 0 \le \xi \le 1, \tag{4.13}$$

or in vector form

$$\mathbf{x} = \mathbf{x}(\xi) = (a\cos(2\pi\xi), a\sin(2\pi\xi), b), \qquad 0 \le \xi \le 1. \tag{4.14}$$

The parameterization need not be taken on the interval $[0, 1]$, i.e., $0 \le \xi \le 1$. However, given any curve parameterized by, say, τ for $\tau_0 \le \tau \le \tau_1$, where $\tau_0 < \tau_1$ are given numbers, the curve can be reparameterized using $\xi \in [0, 1]$ by setting $\tau = \tau_1\,\xi + \tau_0\,(1 - \xi)$. Parameterizations on $[0, 1]$ are convenient for grid generation. Relationships for curves in the plane can be obtained from the following results on curves in space by eliminating z, or, equivalently, setting $z = 0$.

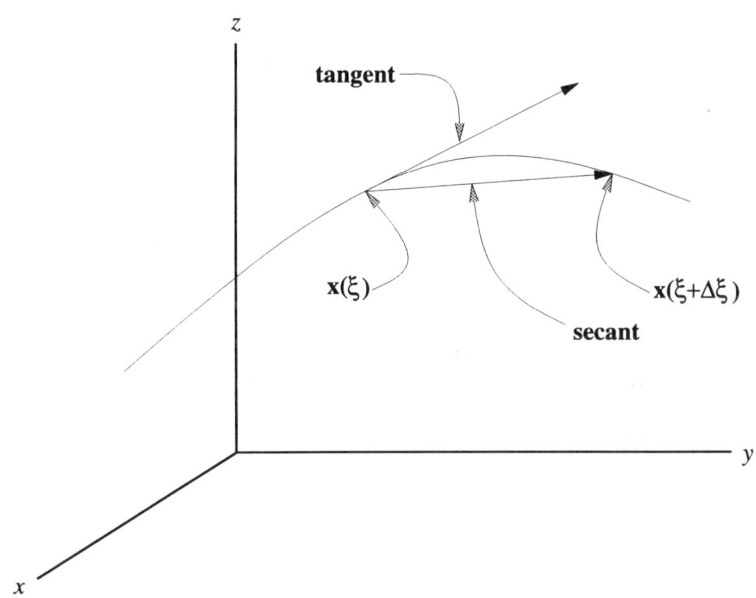

Figure 4.1 *Parametric curve*

Exercise 4.1.4 Write computer programs to plot two- and three-dimensional curves. Such programs are useful as subroutines in codes that display grids. Use your program to plot the circle given in (4.13). Modify the formula for z in (4.13) to read $z = 5\,\xi$ and then plot the resulting spiral. §

The *secant vector* between the points $\mathbf{x}(\xi)$ and $\mathbf{x}(\xi + \Delta\xi)$ on a curve is given by the vector

$$\Delta\mathbf{x}(\xi) = \mathbf{x}(\xi + \Delta\xi) - \mathbf{x}(\xi) \tag{4.15}$$

(see Figure 4.1). The derivative of the vector function $\mathbf{x}(\xi)$ is defined by

$$\mathbf{x}_\xi(\xi) = \frac{d\mathbf{x}}{d\xi}(\xi) = \lim_{\Delta\xi \to 0} \frac{\Delta\mathbf{x}(\xi)}{\Delta\xi}\,. \tag{4.16}$$

From Figure 4.1 it is clear that $\mathbf{x}_\xi$ is a *tangent vector* to the curve. Moreover

$$\Delta\mathbf{x}(\xi) \approx \mathbf{x}_\xi(\xi)\,\Delta\xi\,, \tag{4.17}$$

that is, the secant vector is approximated by the tangent vector times $\Delta\xi$, for $\Delta\xi$ sufficiently small.

The function $\mathbf{x}(\xi)$ maps the interval $[\xi, \xi + \Delta\xi]$ to the arc of the curve between the points $\mathbf{x}(\xi)$ and $\mathbf{x}(\xi + \Delta\xi)$. The length ΔL of this segment is approximately the length of the secant vector

$$\Delta L \approx |\mathbf{x}(\xi + \Delta\xi) - \mathbf{x}(\xi)|\,, \tag{4.18}$$

and by equations (4.15) and (4.17),

$$\Delta L \approx |\mathbf{x}_\xi(\xi)|\,\Delta\xi = \sqrt{x_\xi^2(\xi) + y_\xi^2(\xi) + z_\xi^2(\xi)}\,\Delta\xi\,. \tag{4.19}$$

Consequently, the *arc length* L of the curve is given by

$$L = \int_0^1 |\mathbf{x}_\xi(\xi)|\,d\xi\,. \tag{4.20}$$

Note that $|\mathbf{x}_\xi|\Delta\xi$ is approximately the length of a small piece of a curve in physical space that is the image of a segment of length $\Delta\xi$ in logical space. An important point here is that $|\mathbf{x}_\xi|$ gives the ratio of length in physical space to the length in the parameter space because, from (4.17), $|\mathbf{x}_\xi| \approx |\Delta\mathbf{x}|/\Delta\xi$.

Exercise 4.1.5 Write a computer code to numerically check the approximation given in equation (4.19) for the spiral described in Exercise 4.1.4. §

Exercise 4.1.6 Compute the length of the spiral given in Exercise 4.1.4 using (4.20) and a numerical-integration routine. §

4.1.2 Vector Geometry of Surfaces

At a point on a surface there are two unit normal vectors and an infinity of tangent vectors, i.e., all of the vectors in the plane tangent to the surface. If the surface is described parametrically, then the parameterization defines a unique normal

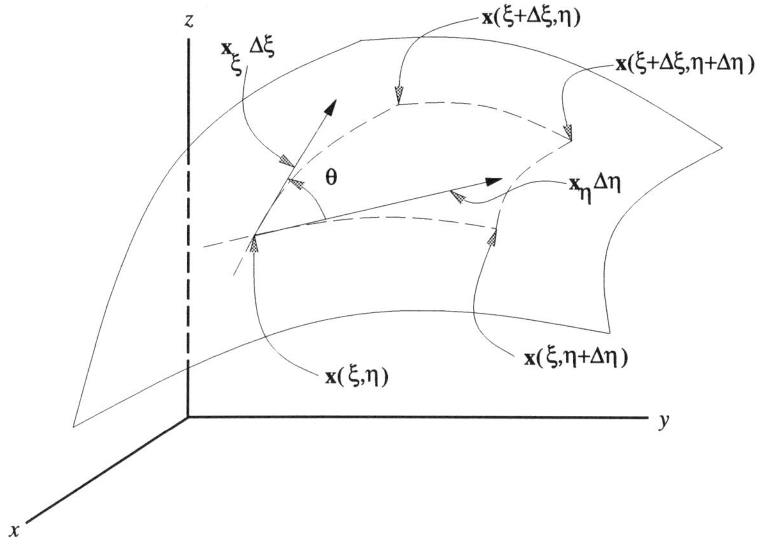

Figure 4.2 *Parametric surface*

vector and two special tangent vectors. These vectors are defined in this section and used to define the concept of surface area. Two parameters are needed to parametrically describe a surface; the vector form of the parametric equation of a surface is

$$\mathbf{x} = \mathbf{x}(\xi, \eta), \quad 0 \le \xi \le 1, \quad 0 \le \eta \le 1, \tag{4.21}$$

while the scalar form of the vector equation is a set of three equations:

$$x = x(\xi, \eta), \quad y = y(\xi, \eta), \quad z = z(\xi, \eta), \tag{4.22}$$

where, as before, $0 \le \xi \le 1$ and $0 \le \eta \le 1$ (see Figure 4.2). Spherical coordinates [see equation (1.40)] can be used to describe the surface of a sphere of radius a:

$$x = a\cos(2\pi\xi)\sin(\pi\eta), \quad y = a\sin(2\pi\xi)\sin(\pi\eta), \quad z = a\cos(\pi\eta), \tag{4.23}$$

or in vector form,

$$\mathbf{x} = (a\cos(2\pi\xi)\sin(\pi\eta), a\sin(2\pi\xi)\sin(\pi\eta), a\cos(\pi\eta)). \tag{4.24}$$

Here $\theta = 2\pi\xi$ is the equatorial angle, while $\phi = \pi\eta$ is the polar angle.

Exercise 4.1.7 Obtain access to graphics software that can be used to plot surfaces. Write a driver routine to plot the sphere given in (4.24) for $a = 1$. Plot the full surface and then use a hidden surface algorithm to plot the surface. §

Exercise 4.1.8 Use a computer code to plot the six logical-space coordinate planes

$$x = \xi\,, \quad y = \eta\,, \quad z = 0\,, \tag{4.25}$$
$$x = \xi\,, \quad y = \eta\,, \quad z = 1\,, \tag{4.26}$$
$$x = 0\,, \quad y = \eta\,, \quad z = \xi\,, \tag{4.27}$$
$$x = 1\,, \quad y = \eta\,, \quad z = \xi\,, \tag{4.28}$$
$$x = \eta\,, \quad y = 0\,, \quad z = \xi\,, \tag{4.29}$$
$$x = \eta\,, \quad y = 1\,, \quad z = \xi\,. \tag{4.30}$$

As always, $0 \le \xi \le 1$ and $0 \le \eta \le 1$. Is there a preferred way of parameterizing these surfaces? For example, surface (4.25) could be parameterized by $x = 1 - \xi$, $y = \eta$, $z = 0$, or $x = \eta$, $y = \xi$, $z = 0$. §

If the value of η is fixed and ξ varies, then a curve on the surface is generated. The tangent T_1 to this curve is given by

$$T_1 = \mathbf{x}_\xi = \frac{\partial \mathbf{x}}{\partial \xi} \tag{4.31}$$

(see Figure 4.2). Similarly, if the value of ξ is fixed and η varies, then a second tangent T_2 is given by

$$T_2 = \mathbf{x}_\eta = \frac{\partial \mathbf{x}}{\partial \eta}\,. \tag{4.32}$$

The curves given by η or ξ fixed are called coordinate curves.

The points

$$\mathbf{x}(\xi,\eta)\,, \quad \mathbf{x}(\xi+\Delta\xi,\eta)\,, \quad \mathbf{x}(\xi,\eta+\Delta\eta)\,, \quad \mathbf{x}(\xi+\Delta\xi,\eta+\Delta\eta) \tag{4.33}$$

on the surface and the coordinate curves joining them define a small patch on the surface (see Figure 4.2). When $\Delta\xi$ and $\Delta\eta$ are sufficiently small, the parallelogram defined by the secant vectors ,

$$\mathbf{x}(\xi+\Delta\xi,\eta) - \mathbf{x}(\xi,\eta)\,, \quad \mathbf{x}(\xi,\eta+\Delta\eta) - \mathbf{x}(\xi,\eta)\,, \tag{4.34}$$

is approximated by the parallelogram defined by the tangent vectors $T_1\,\Delta\xi$ and $T_2\,\Delta\eta$. The area ΔA of the small patch on the surfaces is approximately the area of the parallelogram determined by the tangent vectors. This area is given by the length of the vector product of the two vectors:

$$\Delta A \approx |\mathbf{x}_\xi \times \mathbf{x}_\eta|\,\Delta\xi\,\Delta\eta\,. \tag{4.35}$$

Exercise 4.1.9 Use a computer code to plot some coordinate curves for the surface of the sphere given in equation (4.24). Also plot a few patches, secant, and tangent vectors for this surface, that is, illustrate the discussion in the previous paragraph. §

The vector $\mathbf{x}_\xi \times \mathbf{x}_\eta$ is perpendicular to the two tangent vectors $\mathbf{x}_\xi$ and $\mathbf{x}_\eta$ [as is obvious from (4.12)] and is normal to the surface.

The patch on the surface that is defined by the points in equation (4.33) is the image of this rectangle in logical space given by the points

$$(\xi,\eta)\,,\quad (\xi+\Delta\xi,\eta)\,,\quad (\xi,\eta+\Delta\eta)\,,\quad (\xi+\Delta\xi,\eta+\Delta\eta)\,; \tag{4.36}$$

the rectangle has area $\Delta\xi\,\Delta\eta$. The limit of the ratio of the area of the patch on the surface ΔA [see equation (4.35)] to the area in logical space is

$$\lim_{\max(\Delta\xi,\Delta\eta)\to 0} \frac{\Delta A}{\Delta\xi\,\Delta\eta} = |\mathbf{x}_\xi \times \mathbf{x}_\eta|\,, \tag{4.37}$$

i.e., the length of the vector $\mathbf{x}_\xi \times \mathbf{x}_\eta$ is the ratio of area on the surface to the area in logical space. Consequently, the area A of the surface is given by

$$A = \int_0^1 \int_0^1 |\mathbf{x}_\xi \times \mathbf{x}_\eta|\, d\xi\, d\eta\,. \tag{4.38}$$

Exercise 4.1.10 Show that

$$A = \int_0^1 \int_0^1 \sqrt{J_1^2 + J_2^2 + J_3^2}\, d\xi\, d\eta\,, \tag{4.39}$$

where

$$J_1 = \det\begin{pmatrix} y_\xi & y_\eta \\ z_\xi & z_\eta \end{pmatrix},\quad J_2 = \det\begin{pmatrix} z_\xi & z_\eta \\ x_\xi & x_\eta \end{pmatrix},\quad J_3 = \det\begin{pmatrix} x_\xi & x_\eta \\ y_\xi & y_\eta \end{pmatrix}.\ \S \tag{4.40}$$

Exercise 4.1.11 For a planar curve, $\mathbf{x} = (x(\xi), y(\xi), 0)$, $0 \le \xi \le 1$, show that a normal to the curve is given by $\mathbf{N} = (y_\xi, -x_\xi, 0)$. Show that the normal vector to the surface $x = x(\xi)$, $y = y(\xi)$, $z = \eta$ for $0 \le \xi \le 1$ and $-\infty < \eta < \infty$ is the same as the normal to the curve. §

4.1.3 *Normals and Gradients on Implicit Surfaces and Curves*

Consider a surface given implicitly by the equation $f(x,y,z) = C$, where C is a constant. For example, $x^2 + y^2 + z^2 = 1$ describes the surface of the unit sphere. The *gradient* of f is defined by

$$\nabla_{\mathbf{x}} f = (f_x, f_y, f_z)\,. \tag{4.41}$$

Assume that the curve $\mathbf{x}(\xi)$ lies on the surface, i.e.,

$$f(x(\xi), y(\xi), z(\xi)) = C\,. \tag{4.42}$$

The chain rule gives

$$f_x\, x_\xi + f_y\, y_\xi + f_z\, z_\xi = 0\,, \tag{4.43}$$

i.e.,

$$\nabla_{\mathbf{x}} f \cdot \mathbf{x}_\xi = 0 \tag{4.44}$$

for every curve on the surface. Consequently $\nabla_{\mathbf{x}} f$ must be normal to the surface and therefore be a scalar multiple of $\mathbf{x}_\xi \times \mathbf{x}_\eta$.

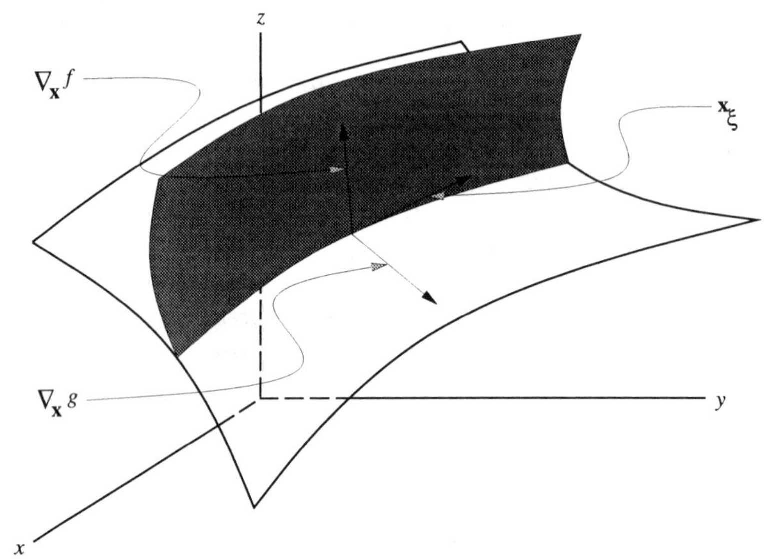

Figure 4.3 *Implicit curve*

Curves can be given implicitly as the intersection of two surfaces:

$$f(x, y, z) = C_1\,, \quad g(x, y, z) = C_2\,, \tag{4.45}$$

where C_1 and C_2 are constants. Because the curve is in each surface, $\nabla_{\mathbf{x}} f$ and $\nabla_{\mathbf{x}} g$ are perpendicular to the curve, and consequently, $\nabla_{\mathbf{x}} f \times \nabla_{\mathbf{x}} g$ is tangent to the curve (see Figure 4.3). If the curve is also given parametrically by $\mathbf{x}(\xi)$, then $\nabla_{\mathbf{x}} f \times \nabla_{\mathbf{x}} g$ is a multiple of $\mathbf{x}_\xi$.

4.2 GENERAL COORDINATE RELATIONSHIPS

General coordinate transformations used in grid generation are often specified in terms of equations involving the inverse map. The relationship between the map and its inverse is expressed in terms of the tangents to the coordinate lines and the normals to the coordinate surfaces, showing that these vectors form a bi-orthogonal set. Formulas for these relationships are derived for one, two, and three dimensions; the curve and surface cases are considered briefly.

4.2.1 Coordinate Relationships in One Dimension

This is a review of the line case of the previous chapter. Let a one-to-one transformation and its inverse be given:

$$x = x(\xi)\,, \quad \xi = \xi(x)\,. \tag{4.46}$$

The fact that these transformations are inverses of each other implies that

$$x = x(\xi(x)), \quad \xi = \xi(x(\xi)). \tag{4.47}$$

The Jacobian matrix $\mathcal{J}$ is the trivial one-by-one matrix,

$$\mathcal{J} = \left(\frac{\partial x}{\partial \xi}\right) = \left(x_\xi\right), \tag{4.48}$$

and the determinant of $\mathcal{J}$ is just the entry in the matrix:

$$J = J(\xi) = x_\xi. \tag{4.49}$$

Although the Jacobian is trivial in this case, J is retained in the formulas to follow, so that they can be compared to the formulas for higher dimensions.

As noted in Chapter 1, *Preliminaries*, the assumption that the transformation is one-to-one implies that $J \neq 0$. If the Jacobian is negative on $[0, 1]$, then the transformation can be replaced by the new transformation $x = x(1 - \xi)$ and the Jacobian of this transformation will be positive. From now on, the Jacobian will be assumed positive.

Exercise 4.2.1 Show that the Jacobians of $x(\xi)$ and $x(1 - \xi)$ have opposite signs. §

Differentiating the first equation in (4.47) with respect to x, and applying the chain rule gives

$$x_\xi\, \xi_x = 1, \tag{4.50}$$

or

$$\xi_x = \frac{1}{x_\xi}. \tag{4.51}$$

Since $J = x_\xi$, (4.50) can be expressed as $JJ^{-1} = 1$. The analog of this formula in higher dimensions plays an important role in grid-generation theory. This is nothing but the standard rule for computing the derivative of the inverse of a function; it is usually written using differentials:

$$\frac{d\xi}{dx} = \frac{1}{dx/d\xi}. \tag{4.52}$$

Exercise 4.2.2 Differentiate the second equation in (4.47) with respect to ξ to re-derive (4.51). §

4.2.2 Coordinate Relationships in Two Dimensions

In Section 1.3 general coordinates for a region $\Omega \subset R^2$ in physical space are given by a map $\mathbf{x} = \mathbf{x}(\boldsymbol{\xi})$ from the unit square $U = U_2$ to Ω. The map can be written using coordinates as

$$x = x(\xi, \eta), \quad y = y(\xi, \eta), \tag{4.53}$$

where $0 \le \xi \le 1$ and $0 \le \eta \le 1$. The map is assumed to be invertible; the inverse is written $\boldsymbol{\xi} = \boldsymbol{\xi}(\mathbf{x})$, or in coordinates

$$\xi = \xi(x, y), \quad \eta = \eta(x, y). \tag{4.54}$$

	Variable	Constant	Tangent
ξ-curve	ξ	η	$\mathbf{x}_\xi$
η-curve	η	ξ	$\mathbf{x}_\eta$

Table 4.1 *Coordinate curves*

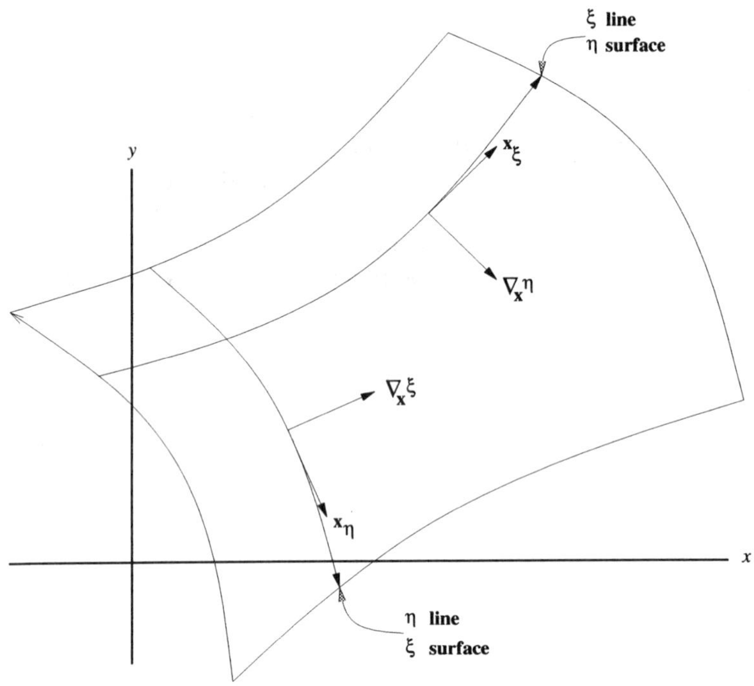

Figure 4.4 *Coordinate lines*

A coordinate curve is given when one of the logical variables is held constant and the other allowed to vary. A ξ-coordinate curve is given when ξ varies and η is constant; a tangent vector to this curve is given by $\mathbf{x}_\xi$. A similar result holds for ξ and η interchanged (see Table 4.1 and Figure 4.4). Holding one of the logical coordinates constant gives an implicit equation for a coordinate curve; for example, $\xi = \xi(x, y)$ with ξ constant is an implicit equation for an η coordinate curve. In three dimensions (see Section 4.2.3 below), holding ξ constant gives a coordinate surface. The two-dimensional case is a bit more confusing than the three-dimensional case, because, in two dimensions, the coordinate surfaces are the same as the coordinate curves. To be consistent with the three-dimensional case, when the coordinate curves are given implicitly, they will be called coordinate surfaces. Thus the ξ-coordinate curve is the same as the η-coordinate surface. A normal to the η-coordinate surface is given by $\nabla_\mathbf{x}\eta$. A similar result holds for an ξ-coordinate surface (see Table 4.2).

	Variable	Constant	Normal
ξ-surface	η	ξ	$\nabla_{\mathbf{x}}\xi$
η-surface	ξ	η	$\nabla_{\mathbf{x}}\eta$

Table 4.2 *Coordinate surfaces*

The fact that the maps (4.53) and (4.54) are inverses implies that $\mathbf{x}(\boldsymbol{\xi}(\mathbf{x})) = \mathbf{x}$ and $\boldsymbol{\xi}(\mathbf{x}(\boldsymbol{\xi})) = \boldsymbol{\xi}$. The first relationship, in coordinates, is

$$x(\xi(x,y),\eta(x,y)) = x\,, \quad y(\xi(x,y),\eta(x,y)) = y\,. \tag{4.55}$$

Using the chain rule to differentiate each of these equations with respect to x and y, produces four equations:

$$x_\xi\,\xi_x + x_\eta\,\eta_x = 1\,, \quad x_\xi\,\xi_y + x_\eta\,\eta_y = 0\,, \tag{4.56}$$

$$y_\xi\,\xi_x + y_\eta\,\eta_x = 0\,, \quad y_\xi\,\xi_y + y_\eta\,\eta_y = 1\,. \tag{4.57}$$

These equations can be written in matrix form:

$$\begin{pmatrix} x_\xi & x_\eta \\ y_\xi & y_\eta \end{pmatrix} \begin{pmatrix} \xi_x & \xi_y \\ \eta_x & \eta_y \end{pmatrix} = \begin{pmatrix} 1 & 0 \\ 0 & 1 \end{pmatrix}. \tag{4.58}$$

Recall that the Jacobian matrix $\mathcal{J}$ for the present case is defined by

$$\mathcal{J} = \begin{pmatrix} x_\xi & x_\eta \\ y_\xi & y_\eta \end{pmatrix}. \tag{4.59}$$

Equation (4.58) implies that $\mathcal{J}$ is invertible, and its inverse $\mathcal{J}^{-1}$ is given by

$$\mathcal{J}^{-1} = \begin{pmatrix} \xi_x & \xi_y \\ \eta_x & \eta_y \end{pmatrix} \tag{4.60}$$

so that (4.58) becomes

$$\mathcal{J}\,\mathcal{J}^{-1} = \mathcal{I}\,, \tag{4.61}$$

where $\mathcal{I}$ is a 2×2 identity matrix. Because $\mathcal{J}$ is invertible, (4.61) implies that $\mathcal{J}^{-1}\,\mathcal{J} = \mathcal{I}$. The next exercise shows how to derive this last relationship from the chain rule.

Exercise 4.2.3 Differentiate (4.54) with respect to the logical variables to show that $\mathcal{J}^{-1}\,\mathcal{J} = \mathcal{I}$. §

The Jacobian J of the map is given by the determinant of the Jacobian matrix:

$$J = \det(\mathcal{J}) = x_\xi\,y_\eta - x_\eta\,y_\xi\,. \tag{4.62}$$

The fact that the determinant of the inverse of a matrix is the inverse of the determinant is expressed by $\det(\mathcal{J}^{-1}) = 1/J$. The inverse of the Jacobian matrix can be computed using cofactors:

$$\mathcal{J}^{-1} = \frac{1}{J}\begin{pmatrix} y_\eta & -x_\eta \\ -y_\xi & x_\xi \end{pmatrix}. \tag{4.63}$$

Exercise 4.2.4 Check formula (4.63). §

Comparing equations (4.60) and (4.63) gives a set of four equations for transforming derivatives of the inverse transformation to derivatives of the transformation

$$\xi_x = \frac{y_\eta}{J}, \quad \xi_y = \frac{-x_\eta}{J}, \tag{4.64}$$

$$\eta_x = \frac{-y_\xi}{J}, \quad \eta_y = \frac{x_\xi}{J}. \tag{4.65}$$

These equations can be solved for the derivatives of the transformation, which gives the inverse relationship:

$$x_\xi = J\,\eta_y; \quad x_\eta = -J\,\xi_y; \tag{4.66}$$

$$y_\xi = -J\,\eta_x; \quad y_\eta = J\,\xi_x. \tag{4.67}$$

Exercise 4.2.5 Show that $\det(\mathcal{J}^{-1}) = \xi_x\eta_y - \xi_y\eta_x = 1/J$ using formulas (4.64)-(4.65). §

Exercise 4.2.6 Verify that the results of Exercise 4.2.3, that is, $\mathcal{J}^{-1}\,\mathcal{J} = \mathcal{I}$, can be written in terms of vectors as

$$\mathbf{x}_\xi \cdot \nabla_\mathbf{x}\xi = 1, \mathbf{x}_\xi \cdot \nabla_\mathbf{x}\eta = 0, \tag{4.68}$$

$$\mathbf{x}_\eta \cdot \nabla_\mathbf{x}\xi = 0, \mathbf{x}_\eta \cdot \nabla_\mathbf{x}\eta = 1. \tag{4.69}$$

Sets of vectors satisfying this type of relationship are called *biorthogonal* (see Figure 4.4). §

4.2.3 *Coordinate Relationships in Three Dimensions*

In Section 1.3 general coordinates for a region $\Omega \subset R^3$ in physical space are given by a map $\mathbf{x} = \mathbf{x}(\boldsymbol{\xi})$ from the unit cube $U = U_3$ to Ω. The map can be written using coordinates as

$$x = x(\xi,\eta,\zeta), \quad y = y(\xi,\eta,\zeta), \quad z = z(\xi,\eta,\zeta), \tag{4.70}$$

where $0 \le \xi \le 1$, $0 \le \eta \le 1$, and $0 \le \zeta \le 1$. The map is assumed to be invertible; the inverse is written $\boldsymbol{\xi} = \boldsymbol{\xi}(\mathbf{x})$, or in coordinates,

$$\xi = \xi(x,y,z), \quad \eta = \eta(x,y,z), \quad \zeta = \zeta(x,y,z). \tag{4.71}$$

A coordinate curve is given when two of the logical variables are held constant and the third is allowed to vary; this produces three types of coordinate curves and their associated tangent vectors (see Table 4.3 and Figure 4.5). A coordinate

	Variable	Constant	Tangent
ξ-curve	ξ	η, ζ	$\mathbf{x}_\xi$
η-curve	η	ξ, ζ	$\mathbf{x}_\eta$
ζ-curve	ζ	ξ, η	$\mathbf{x}_\zeta$

Table 4.3 *Coordinate curves*

	Variable	Constant	Normal
ξ-surface	η, ζ	ξ	$\nabla_\mathbf{x}\xi$
η-surface	ξ, ζ	η	$\nabla_\mathbf{x}\eta$
ζ-surface	ξ, η	ζ	$\nabla_\mathbf{x}\zeta$

Table 4.4 *Coordinate surfaces*

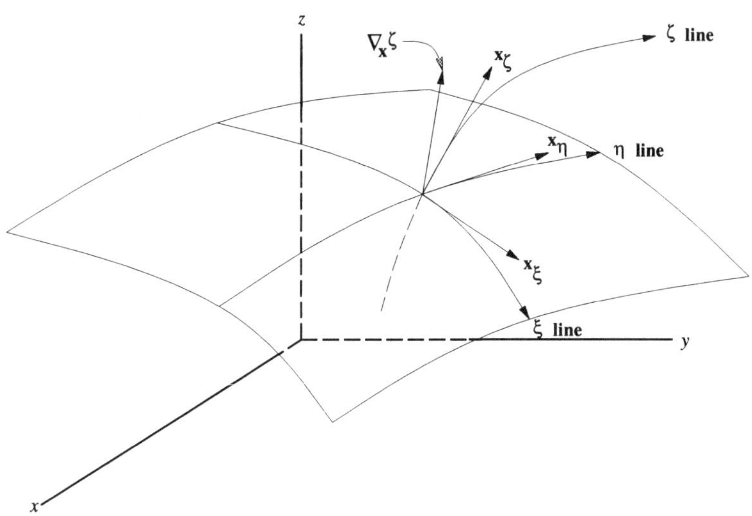

Figure 4.5 *Coordinate curves and surface*

surface is given when one of the logical variables is held constant and the other two are allowed to vary; i.e., the surface is given implicitly. There are three types of coordinate surfaces and their associated normal vectors (see Table 4.4 and Figure 4.5). Because the η and ζ coordinate curves lie in the ξ coordinate surface (and so forth for all cyclic permutations of the logical variables), the normal vector must be proportional to the vector product of the two tangents:

$$\nabla_{\mathbf{x}}\xi \propto \mathbf{x}_\eta \times \mathbf{x}_\zeta\,, \quad \nabla_{\mathbf{x}}\eta \propto \mathbf{x}_\zeta \times \mathbf{x}_\xi\,, \quad \nabla_{\mathbf{x}}\zeta \propto \mathbf{x}_\xi \times \mathbf{x}_\eta\,. \tag{4.72}$$

The fact that the maps (4.70) and (4.71) are inverses implies that $\mathbf{x}(\boldsymbol{\xi}(\mathbf{x})) = \mathbf{x}$ and $\boldsymbol{\xi}(\mathbf{x}(\boldsymbol{\xi})) = \boldsymbol{\xi}$ or, in coordinates, the first relationship gives

$$x(\xi(x,y,z),\eta(x,y,z),\zeta(x,y,z)) = x\,, \tag{4.73}$$
$$y(\xi(x,y,z),\eta(x,y,z),\zeta(x,y,z)) = y\,, \tag{4.74}$$
$$z(\xi(x,y,z),\eta(x,y,z),\zeta(x,y,z)) = z\,. \tag{4.75}$$

Differentiating each of these equations with respect to x, y, and z using the chain rule produces nine equations. These equations can be compactly written using matrices:

$$\begin{pmatrix} x_\xi & x_\eta & x_\zeta \\ y_\xi & y_\eta & y_\zeta \\ z_\xi & z_\eta & z_\zeta \end{pmatrix} \begin{pmatrix} \xi_x & \xi_y & \xi_z \\ \eta_x & \eta_y & \eta_z \\ \zeta_x & \zeta_y & \zeta_z \end{pmatrix} = \begin{pmatrix} 1 & 0 & 0 \\ 0 & 1 & 0 \\ 0 & 0 & 1 \end{pmatrix}. \tag{4.76}$$

Recall that the Jacobian matrix $\mathcal{J}$ for this case is

$$\mathcal{J} = \begin{pmatrix} x_\xi & x_\eta & x_\zeta \\ y_\xi & y_\eta & y_\zeta \\ z_\xi & z_\eta & z_\zeta \end{pmatrix}. \tag{4.77}$$

Equation (4.76) says that $\mathcal{J}$ is invertible and its inverse $\mathcal{J}^{-1}$ is given by

$$\mathcal{J}^{-1} = \begin{pmatrix} \xi_x & \xi_y & \xi_z \\ \eta_x & \eta_y & \eta_z \\ \zeta_x & \zeta_y & \zeta_z \end{pmatrix}. \tag{4.78}$$

As before, (4.76) implies that $\mathcal{J}\,\mathcal{J}^{-1} = \mathcal{I}$, and because $\mathcal{J}$ is square, that $\mathcal{J}^{-1}\,\mathcal{J} = \mathcal{I}$. The next exercise shows how to derive this last fact from from the chain rule.

Exercise 4.2.7 Differentiate $\boldsymbol{\xi}(\mathbf{x}(\boldsymbol{\xi})) = \boldsymbol{\xi}$ with respect to the logical variables to show that $\mathcal{J}^{-1}\,\mathcal{J} = \mathcal{I}$ is obtained. §

The Jacobian J of the map is given by the determinant of the Jacobian matrix

$$\begin{aligned} J = \det(\mathcal{J}) = &+ x_\xi\, y_\eta\, z_\zeta + x_\eta\, y_\zeta\, z_\xi + x_\zeta\, y_\xi\, z_\eta \\ &- x_\xi\, y_\zeta\, z_\eta - x_\eta\, y_\xi\, z_\zeta - x_\zeta\, y_\eta\, z_\xi\,. \end{aligned} \tag{4.79}$$

The fact that the determinant of the inverse of a matrix is the inverse of the determinant gives $\det(\mathcal{J}^{-1}) = 1/J$. The inverse of the Jacobian matrix can be computed using cofactors:

$$\mathcal{J}^{-1} = \frac{1}{J}\begin{pmatrix} y_\eta z_\zeta - y_\zeta z_\eta & x_\zeta z_\eta - x_\eta z_\zeta & x_\eta y_\zeta - x_\zeta y_\eta \\ y_\zeta z_\xi - y_\xi z_\zeta & x_\xi z_\zeta - x_\zeta z_\xi & x_\zeta y_\xi - x_\xi y_\zeta \\ y_\xi z_\eta - y_\eta z_\xi & x_\eta z_\xi - x_\xi z_\eta & x_\xi y_\eta - x_\eta y_\xi \end{pmatrix}. \tag{4.80}$$

Comparing equations (4.78) and (4.80) gives a set of nine equations for transforming derivatives of the inverse transformation to derivatives of the transformation; these equations can be solved for the derivatives of the transformation, which gives the inverse relationship.

Exercise 4.2.8 The Jacobian matrix is given by a matrix of column vectors,

$$\mathcal{J} = \left(\mathbf{x}_\xi, \mathbf{x}_\eta, \mathbf{x}_\zeta\right), \tag{4.81}$$

and the inverse of the Jacobian matrix [equation (4.78)] can be written as the transpose of a matrix of column vectors:

$$\mathcal{J}^{-1} = \left(\nabla_{\mathbf{x}}\xi, \nabla_{\mathbf{x}}\eta, \nabla_{\mathbf{x}}\zeta\right)^T. \tag{4.82}$$

Show that

$$\nabla_{\mathbf{x}}\xi = \frac{1}{J}\mathbf{x}_\eta \times \mathbf{x}_\zeta, \quad \nabla_{\mathbf{x}}\eta = \frac{1}{J}\mathbf{x}_\zeta \times \mathbf{x}_\xi, \quad \nabla_{\mathbf{x}}\zeta = \frac{1}{J}\mathbf{x}_\xi \times \mathbf{x}_\eta. \ \S \tag{4.83}$$

Exercise 4.2.9 Show that the Jacobian and its inverse are given by triple scalar products

$$J = \left[\mathbf{x}_\xi, \mathbf{x}_\eta, \mathbf{x}_\zeta\right], \qquad J^{-1} = \left[\nabla_{\mathbf{x}}\xi, \nabla_{\mathbf{x}}\eta, \nabla_{\mathbf{x}}\zeta\right]. \ \S \tag{4.84}$$

Some of the quantities introduced in this chapter provide important measures of size, that is, lengths, areas, and volumes. These measures are best expressed in terms of ratio of sizes in physical space to sizes in logical space. To summarize:

- $|\mathbf{x}_\xi|$, $|\mathbf{x}_\eta|$, and $|\mathbf{x}_\zeta|$ give ratios of lengths;
- $|\mathbf{x}_\eta \times \mathbf{x}_\zeta|$, $|\mathbf{x}_\zeta \times \mathbf{x}_\xi|$, and $|\mathbf{x}_\xi \times \mathbf{x}_\eta|$ give ratios of areas;
- J gives a ratio of volumes.

Exercise 4.2.10 Verify that the results of Exercise 4.2.7, that is, $\mathcal{J}^{-1}\mathcal{J} = \mathcal{I}$, can be written in terms of vectors as

$$\mathbf{x}_\xi \cdot \nabla_{\mathbf{x}}\xi = 1, \quad \mathbf{x}_\xi \cdot \nabla_{\mathbf{x}}\eta = 0, \quad \mathbf{x}_\xi \cdot \nabla_{\mathbf{x}}\zeta = 0, \tag{4.85}$$

$$\mathbf{x}_\eta \cdot \nabla_{\mathbf{x}}\xi = 0, \quad \mathbf{x}_\eta \cdot \nabla_{\mathbf{x}}\eta = 1, \quad \mathbf{x}_\eta \cdot \nabla_{\mathbf{x}}\zeta = 0, \tag{4.86}$$

$$\mathbf{x}_\zeta \cdot \nabla_{\mathbf{x}}\xi = 0, \quad \mathbf{x}_\zeta \cdot \nabla_{\mathbf{x}}\eta = 0, \quad \mathbf{x}_\zeta \cdot \nabla_{\mathbf{x}}\zeta = 1. \tag{4.87}$$

Sets of vectors satisfying this type of relationship are called *biorthogonal.* §

The vectors introduced are so important that they have special names:

- $\mathbf{x}_\xi$, $\mathbf{x}_\eta$, and $\mathbf{x}_\zeta$ are *covariant* tangent vectors;
- $\nabla_{\mathbf{x}}\xi$, $\nabla_{\mathbf{x}}\eta$, and $\nabla_{\mathbf{x}}\zeta$ are *contravariant* normal vectors.

The covariant and contravariant terminology refers to transformations of physical space to itself; analyzing the situation using linear transformations is sufficient. Assume that a linear transformation

$$\mathbf{w} = A\mathbf{x}, \tag{4.88}$$

that is,

$$w_i = \sum_{j=1}^{3} A_{i,j} x_j , \tag{4.89}$$

of physical space to itself is given. Also, suppose that a transformation $\mathbf{x} = \mathbf{x}(\boldsymbol{\xi})$ of logical space to physical space is given. We want to illustrate how quantities are covariant with respect to the transformation generated by A [not the transformation $\mathbf{x}(\boldsymbol{\xi})$].

Recall that $\mathbf{x}_{\xi_i}$ is a tangent vector to coordinate curve. Differentiating (4.88) with respect to ξ_i gives

$$\mathbf{w}_{\xi_i} = A\mathbf{x}_{\xi_i} , \tag{4.90}$$

that is, the tangent vectors transform in the same way as the coordinates of physical space. This is what is meant by *covariant*.

The inverse of the transformation from logical space to physical space is given by $\boldsymbol{\xi} = \boldsymbol{\xi}(\mathbf{x})$ and a coordinate surface is given by $\xi_i(\mathbf{x}) = C$ where C is a constant. A normal to this surface is given by $\nabla_{\mathbf{x}}\xi_i$. Differentiating $\xi_i(\mathbf{x}) = C$ with respect to $\mathbf{w}$ and using the chain rule with (4.88) gives

$$\nabla_{\mathbf{w}}\xi_i = (A^{-1})^T \nabla_{\mathbf{x}}\xi_i , \tag{4.91}$$

that is,

$$\nabla_{\mathbf{x}}\xi_i = A^T \nabla_{\mathbf{w}}\xi_i . \tag{4.92}$$

This is what is meant by a *contravariant* vector.

4.2.4 Coordinate Relationships on Curves and Surfaces

The Jacobian matrix (1.9) for both curves and surfaces is not square and thus cannot have an proper inverse. Recall that a curve is defined parametrically by three functions of a single variable:

$$x = x(\xi), \quad y = y(\xi), \quad z = z(\xi). \tag{4.93}$$

The Jacobian matrix for a curve is

$$\mathcal{J} = \begin{pmatrix} x_\xi \\ y_\xi \\ z_\xi \end{pmatrix}. \tag{4.94}$$

A surface is defined parametrically by three functions of two variables:

$$x = x(\xi,\eta), \quad y = y(\xi,\eta), \quad z = z(\xi,\eta). \tag{4.95}$$

The Jacobian matrix for a surface is

$$\mathcal{J} = \begin{pmatrix} x_\xi & x_\eta \\ y_\xi & y_\eta \\ z_\xi & z_\eta \end{pmatrix}. \tag{4.96}$$

The fact that the curve and surface Jacobians are not square is a significant inconvenience that can be overcome by introducing the *metric matrix* (or metric tensor), which is well known in differential geometry. Further discussion of the issues raised in this section are reserved for Chapter 10, *Variational Grid Generation on Curves and Surfaces.*

4.3 ELEMENTARY DIFFERENTIAL GEOMETRY

The metric tensor and its applications to grid generation are described in this section. Formulas for the Jacobian matrix can be summarized nicely using index notation. Recall that $\mathbf{x} = (x_1, x_2, \ldots, x_n)$ and $\boldsymbol{\xi} = (\xi_1, \xi_2, \ldots, \xi_k)$ and that the Jacobian matrix $\mathcal{J}$ was defined in equation (1.9) as

$$\mathcal{J}_{ij} = \frac{\partial x_i}{\partial \xi_j}, \quad i = 1, \ldots n, \quad j = 1, \ldots k. \tag{4.97}$$

The *metric matrix* or *metric tensor* $\mathcal{G}$ is one of the important objects introduced in elementary differential geometry. This matrix is important in grid generation because its elements have a geometric interpretation; consequently, many grid-generation algorithms are defined in terms of elements of this matrix, its inverse, and its determinant. The matrix is defined for curves, surfaces, and regions, i.e., all of the objects given in Table 1.2. For a k-dimensional object, that is, when logical space has dimension k, the metric matrix $\mathcal{G}$ is k by k and is defined by

$$\mathcal{G} = \mathcal{J}^T \mathcal{J}. \tag{4.98}$$

For example, the Jacobian matrix for a surface [see equation (4.96)] is

$$\mathcal{J} = \begin{pmatrix} x_\xi & x_\eta \\ y_\xi & y_\eta \\ z_\xi & z_\eta \end{pmatrix}, \tag{4.99}$$

so for a surface, the metric matrix is 2×2 and is given by

$$\mathcal{G} = \begin{pmatrix} x_\xi^2 + y_\xi^2 + z_\xi^2 & x_\xi x_\eta + y_\xi y_\eta + z_\xi z_\eta \\ x_\xi x_\eta + y_\xi y_\eta + z_\xi z_\eta & x_\eta^2 + y_\eta^2 + z_\eta^2 \end{pmatrix}. \tag{4.100}$$

In general, the *metric* g is defined as

$$g = \det(\mathcal{G}) \tag{4.101}$$

and, consequently, $\det(\mathcal{G}^{-1}) = 1/g$. For square matrices, the determinant of a product of matrices is the product of their determinants and the determinant of a transpose is the same as the determinant. For the case $k = n$, $\mathcal{J}$ is square; because equation (4.98) defines $\mathcal{G}$ by $\mathcal{G} = \mathcal{J}^T \mathcal{J}$, g is given by $g = J^2$, where $J = \det(\mathcal{J})$. Thus

$$J = \sqrt{g}\,. \tag{4.102}$$

Based on this identity, the notation J is replaced by $\sqrt{g}$ from here on. It is *convenient* and often *important* to write formulas in terms of $\sqrt{g}$ instead of J.

The formulas for the entries $g_{i,j}$ of $\mathcal{G}$ are easy to write using indexed variables:

$$g_{i,j} = \sum_{l=1}^{n} \frac{\partial x_l}{\partial \xi_i}\frac{\partial x_l}{\partial \xi_j}\,, \qquad 1 \le i,j \le k\,, \tag{4.103}$$

or in vector notation

$$g_{i,j} = \mathbf{x}_{\xi_i} \cdot \mathbf{x}_{\xi_j}\,, \qquad 1 \le i,j \le k\,. \tag{4.104}$$

Interchanging i and j in this definition does not change the formula, so the metric matrix is symmetric [this is also seen from (4.98)]. It is the vector definitions that are the most useful for understanding the geometry of a transformation: $g_{i,j}$ is the dot product of the tangent vector to the ith coordinate curves with the tangent vector to the jth coordinate curve. In particular, a coordinate system is *orthogonal* if and only if the tangents to the coordinate curves are perpendicular:

$$\mathbf{x}_{\xi_i} \cdot \mathbf{x}_{\xi_j} = 0\,, \quad i \ne j\,. \tag{4.105}$$

Thus a coordinate system is orthogonal if and only if

$$g_{i,j} = 0\,, \quad i \ne j\,, \tag{4.106}$$

that is, $\mathcal{G}$ is diagonal.

Another use of the metric matrix is illustrated in the following proposition (see Section 1.3 for the notation and definitions).

PROPOSITION 4.1 For $k \le n$, a map from the logical region U_k to a physical region Ω_k^n is nonsingular at $\boldsymbol{\xi}$ if and only if $g \ne 0$ at $\boldsymbol{\xi}$.

Proof. Recall that $\mathcal{G} = \mathcal{J}^T \mathcal{J}$ (see 4.98). If $k = n$, then $\mathcal{J}$ is square and $\det(\mathcal{G}) = (\det(\mathcal{J}))^2$ and the result follows from the corollary to the Inverse Mapping Theorem 1.1.

For the general case, first assume that $\det(\mathcal{G}) = 0$. Then there is a nonzero vector $\mathbf{v}$ such that $\mathcal{G}\mathbf{v} = \mathbf{0}$. This implies that $\mathcal{J}^T\mathcal{J}\mathbf{v} = \mathbf{0}$ and then that $\mathbf{v} \cdot \mathcal{J}^T\mathcal{J}\mathbf{v} = 0$. However, $\mathbf{v} \cdot \mathcal{J}^T\mathcal{J}\mathbf{v} = \mathcal{J}\mathbf{v} \cdot \mathcal{J}\mathbf{v} = |\mathcal{J}\mathbf{v}|^2$, and consequently $|\mathcal{J}\mathbf{v}| = 0$, which implies that $\mathcal{J}\mathbf{v} = 0$ or that the mapping is singular.

On the other hand, if the mapping is singular, then there is a nonzero vector $\mathbf{v}$ such that $\mathcal{J}\mathbf{v} = 0$ and thus that $\mathcal{G}\mathbf{v} = \mathcal{J}^T\mathcal{J}\mathbf{v} = 0$. This implies that $\det(\mathcal{G}) = 0$. §

When $k = n$, the entries of the inverse of the Jacobian matrix are given by

$$[\mathcal{J}^{-1}]_{i,j} = (\partial \xi_i / \partial x_j)\,. \tag{4.107}$$

Moreover,

$$\mathcal{G}^{-1} = (\mathcal{J}^T \mathcal{J})^{-1} = \mathcal{J}^{-1}\,(\mathcal{J}^T)^{-1} = \mathcal{J}^{-1}\,(\mathcal{J}^{-1})^T\,, \tag{4.108}$$

and consequently the components $g^{i,j}$ of G^{-1} can be written

$$g^{i,j} = \sum_{l=1}^{n} \frac{\partial \xi_i}{\partial x_l} \frac{\partial \xi_j}{\partial x_l}, \qquad 1 \le i,j \le n. \tag{4.109}$$

When $k = n$, the entries in the inverse of the Jacobian matrix can also be written as vector products:

$$g^{i,j} = \nabla_{\mathbf{x}} \xi_i \cdot \nabla_{\mathbf{x}} \xi_j . \tag{4.110}$$

Thus $g^{i,j}$ is the dot product of a normal to the ith coordinate surface with a normal to the jth coordinate surface.

Exercise 4.3.1 In the case $k = n = 3$, use (4.83) and the identity (4.10) to show that

$$g^{i,l} = \frac{1}{g} \left[g_{j,m}\, g_{k,n} - g_{j,n}\, g_{k,m} \right] \tag{4.111}$$

with cyclic permutations on (i,j,k) and (l,m,n) for $i,l = 1,2,3$. Show that this result can also be obtained by directly inverting the metric matrix [i.e., explicitly compute (4.108)]. §

Additional relevance of the metric matrix can be seen from the formula for computing the length of a curve on a surface. Assume that a curve parameterized by τ, $\mathbf{x} = \mathbf{x}(\tau)$, $0 \le \tau \le 1$, is given in physical space. Formula (4.20) gives the length of the curve as

$$L = \int_0^1 \left| \frac{d\mathbf{x}}{d\tau} \right| d\tau , \tag{4.112}$$

where

$$\left| \frac{d\mathbf{x}}{d\tau} \right| = \sqrt{\frac{d\mathbf{x}}{d\tau} \cdot \frac{d\mathbf{x}}{d\tau}} . \tag{4.113}$$

Because physical space is mapped to logical space by $\boldsymbol{\xi} = \boldsymbol{\xi}(x)$, the curve in physical space is mapped to a curve in logical space given by $\boldsymbol{\xi}(\tau) = \boldsymbol{\xi}(x(\tau))$. Assume that logical space has dimension k and is a subset of E^n. Then the chain rule can be used to expand the derivatives:

$$\frac{d\mathbf{x}}{d\tau} = \sum_{i=1}^{k} \frac{\partial \mathbf{x}}{\partial \xi_i} \frac{d\xi_i}{d\tau} \tag{4.114}$$

and consequently

$$\frac{d\mathbf{x}}{d\tau} \cdot \frac{d\mathbf{x}}{d\tau} = \sum_{i,j=1}^{k} \frac{\partial \mathbf{x}}{\partial \xi_i} \cdot \frac{\partial \mathbf{x}}{\partial \xi_j} \frac{d\xi_i}{d\tau} \frac{d\xi_j}{d\tau} = \sum_{i,j=1}^{k} g_{i,j} \frac{d\xi_i}{d\tau} \frac{d\xi_j}{d\tau} . \tag{4.115}$$

The last sum in the previous equation is called the *First Fundamental Form* in differential geometry texts. The derivative of the vector function $\boldsymbol{\xi}$ with respect to τ, $d\boldsymbol{\xi}/d\tau$, is a tangent vector to the logical-space image of the physical-space curve. The last part of equation (4.115) should be thought of as a special dot product, determined by $g_{i,j}$, of the tangent vectors to the curve in logical space. Thus the metric matrix gives the relationship between the length of a curve in physical space to the length of its image in logical space.

4.4 VECTOR CALCULUS AND DIFFERENTIAL GEOMETRY IN THE PLANE

Previous results of this chapter are specialized to the planar case in this section. This material is primarily used in Section 5.6 of the next chapter, *Controlling the Grid Near the Boundary*; it also serves to build geometric intuition involving relationships that hold in the plane. This intuition is useful, for example, in interpreting and suggesting variational grid-generation principles. First-order relationships involving the tangent vectors and their perpendiculars are introduced. The perpendiculars are related to the contravariant normals. Second-order differential relationships begin with the planar *metric identity*, which can be expressed in terms of the *Beltrami operator*. Rates-of-change of the elements of the metric matrix are related to various inner products of the tangent vectors with the second-order vectors and then connected to the Gauss identities. The section ends with a definition of coordinate-line curvature and a discussion of the parallel tangent condition.

4.4.1 First-Order Relationships

Let the point $\mathbf{x} = (x(\xi,\eta), y(\xi,\eta), 0)$ in the physical plane be the image of the point (ξ,η) in logical space. The covariant tangent vectors of the mapping are $\mathbf{x}_\xi = (x_\xi, y_\xi, 0)$ and $\mathbf{x}_\eta = (x_\eta, y_\eta, 0)$; these clearly lie in the plane and are tangents to the coordinate lines of the grid. The unit normal to the plane is $\hat{\mathbf{k}} = (0, 0, 1)$. Recall that the first-order planar metrics are defined in terms of inner products of the covariant tangent vectors

$$g_{ij} = \mathbf{x}_{\xi_i} \cdot \mathbf{x}_{\xi_j} \,. \qquad (4.116)$$

The vector cross product $\mathbf{J} = \mathbf{x}_\xi \times \mathbf{x}_\eta$ is normal to both covariant tangents. Since the tangents lie in the plane, $\mathbf{J}$ is parallel to $\hat{\mathbf{k}}$. The length of the normal is given by the quantity $\sqrt{g}$, i.e., $\mathbf{J} = \sqrt{g}\,\hat{\mathbf{k}}$ where $\sqrt{g}$ is defined by

$$\sqrt{g} = x_\xi\, y_\eta - x_\eta\, y_\xi \,. \qquad (4.117)$$

The square of the length of $\mathbf{J}$ is

$$g = g_{11}\, g_{22} - g_{12}^2 \,. \qquad (4.118)$$

As noted previously, the quantity $\sqrt{g}$ represents the local area of the differential element bounded by the tangents.

Exercise 4.4.1 Verify the identity

$$g = (x_\xi^2 + y_\xi^2)\,(x_\eta^2 + y_\eta^2) - (x_\xi\, x_\eta + y_\xi\, y_\eta)^2. \qquad (4.119)$$

using the definition of $\mathbf{J}$. §

The planar *contravariant normals* were defined in Section 4.2.2, i.e., $\nabla_{\mathbf{x}}\xi = (\xi_x, \xi_y, 0)$ and $\nabla_{\mathbf{x}}\eta = (\eta_x, \eta_y, 0)$. These are the normals to the coordinate surfaces. Both the covariant tangents and contravariant normals play a key role in

$\cdot$	$\mathbf{x}_\xi$	$\mathbf{x}_\eta$	$\mathbf{x}_\xi^\perp$	$\mathbf{x}_\eta^\perp$
$\mathbf{x}_\xi$	g_{11}	g_{12}	0	$-\sqrt{g}$
$\mathbf{x}_\eta$	g_{12}	g_{22}	$\sqrt{g}$	0
$\mathbf{x}_\xi^\perp$	0	$\sqrt{g}$	g_{11}	g_{12}
$\mathbf{x}_\eta^\perp$	$-\sqrt{g}$	0	g_{12}	g_{22}

Table 4.5 *Planar inner product relationships*

the theory of this chapter. However, it is sometimes more convenient to work with the following *tangent perpendiculars* to the covariant tangents

$$\mathbf{x}_\xi^\perp = (-y_\xi, x_\xi, 0) \tag{4.120}$$

$$\mathbf{x}_\eta^\perp = (-y_\eta, x_\eta, 0) \tag{4.121}$$

instead of the contravariant normals. From (4.64) and (4.65), it is clear that

$$\mathbf{x}_\xi^\perp = +\sqrt{g}\,\nabla_{\mathbf{x}}\eta\,, \tag{4.122}$$

$$\mathbf{x}_\eta^\perp = -\sqrt{g}\,\nabla_{\mathbf{x}}\xi\,, \tag{4.123}$$

i.e., the perpendiculars are merely are re-scalings of the lengths of the contravariant normals.

Exercise 4.4.2 Verify that

$$\mathbf{x}_\xi^\perp = \hat{\mathbf{k}} \times \mathbf{x}_\xi\,, \tag{4.124}$$

$$\mathbf{x}_\eta^\perp = \hat{\mathbf{k}} \times \mathbf{x}_\eta\,. \; \S \tag{4.125}$$

Table 4.5 summarizes the inner products of the covariant and contravariant normals. Note, in particular, that $\mathbf{x}_\xi \cdot \mathbf{x}_\xi^\perp = 0$ and $\mathbf{x}_\eta \cdot \mathbf{x}_\eta^\perp = 0$.

Since the tangent perpendiculars lie in the plane, they may be resolved as a linear combination of the covariant tangents (see the following exercise).

Exercise 4.4.3 Use the inner product Table 4.5 to express the perpendiculars $\mathbf{x}_\xi^\perp$ and $\mathbf{x}_\eta^\perp$ in terms of the covariant tangents, obtaining the useful identities:

$$\mathbf{x}_\xi^\perp = \frac{g_{11}\,\mathbf{x}_\eta - g_{12}\,\mathbf{x}_\xi}{\sqrt{g}}\,, \tag{4.126}$$

$$-\mathbf{x}_\eta^\perp = \frac{g_{22}\,\mathbf{x}_\xi - g_{12}\,\mathbf{x}_\eta}{\sqrt{g}}\,. \; \S \tag{4.127}$$

4.4.2 *Second-Order Relationships*

Relations (4.126)-(4.127), may be combined with the easily checked identity

$$-\frac{\partial}{\partial \xi}\mathbf{x}_\eta^\perp + \frac{\partial}{\partial \eta}\mathbf{x}_\xi^\perp = \mathbf{0} \tag{4.128}$$

to obtain the planar *metric identity*

$$\frac{\partial}{\partial \xi}\left(\frac{g_{22}\,\mathbf{x}_\xi - g_{12}\,\mathbf{x}_\eta}{\sqrt{g}}\right) + \frac{\partial}{\partial \eta}\left(\frac{g_{11}\,\mathbf{x}_\eta - g_{12}\,\mathbf{x}_\xi}{\sqrt{g}}\right) = \mathbf{0}. \tag{4.129}$$

This identity holds for any coordinates in the plane. Let $f(\xi, \eta) \in \mathbf{C}^2$ on U_2. The *Beltrami operator* $\Delta_{\mathcal{B}}$ is defined (Struik [203]) by

$$\Delta_{\mathcal{B}} f = \frac{1}{\sqrt{g}}\left[\frac{\partial}{\partial \xi}\left(\frac{g_{22}\,f_\xi - g_{12}\,f_\eta}{\sqrt{g}}\right) + \frac{\partial}{\partial \eta}\left(\frac{g_{11}\,f_\eta - g_{12}\,f_\xi}{\sqrt{g}}\right)\right]. \tag{4.130}$$

Note that the Beltrami operator is the Laplacian if the metric matrix is trivial. The *metric identity* may thus be compactly written $\Delta_{\mathcal{B}}\mathbf{x} = \mathbf{0}$. The metric identity is shown in Section 8.3.2 to be a consequence of a certain variational principle.

Exercise 4.4.4 Compute the Beltrami operator for polar coordinates. §

The second-derivative vectors $\mathbf{x}_{\xi\xi}$, $\mathbf{x}_{\xi\eta}$, and $\mathbf{x}_{\eta\eta}$ lie in the plane. As a direct calculation shows, the following inner product identities relate the second derivative vectors to the rate-of-change of the various first-order metrics:

$$\mathbf{x}_\xi \cdot \mathbf{x}_{\xi\xi} = \frac{1}{2}\,(g_{11})_\xi\,, \tag{4.131}$$

$$\mathbf{x}_\xi \cdot \mathbf{x}_{\xi\eta} = \frac{1}{2}\,(g_{11})_\eta\,, \tag{4.132}$$

$$\mathbf{x}_\xi \cdot \mathbf{x}_{\eta\eta} = (g_{12})_\eta - \frac{1}{2}\,(g_{22})_\xi\,, \tag{4.133}$$

$$\mathbf{x}_\eta \cdot \mathbf{x}_{\xi\xi} = (g_{12})_\xi - \frac{1}{2}\,(g_{11})_\eta\,, \tag{4.134}$$

$$\mathbf{x}_\eta \cdot \mathbf{x}_{\xi\eta} = \frac{1}{2}\,(g_{22})_\xi\,, \tag{4.135}$$

$$\mathbf{x}_\eta \cdot \mathbf{x}_{\eta\eta} = \frac{1}{2}\,(g_{22})_\eta\,. \tag{4.136}$$

Exercise 4.4.5 Derive formula (4.133). §

Because the vectors $\mathbf{x}_{\xi_i\xi_j}$ are in the plane and $\mathbf{x}_\xi$ and $\mathbf{x}_\eta$ are a basis, $\mathbf{x}_{\xi_i\xi_j}$ can be written as a linear combination of $\mathbf{x}_\xi$ and $\mathbf{x}_\eta$. Formulas (4.131) through (4.136) can be used to find the coefficients of the linear combination. This results in the well-known *Gauss identities* which resolve the second-order vectors in terms of the covariant tangents:

$$\mathbf{x}_{\xi\xi} = \Gamma_{11}^1\,\mathbf{x}_\xi + \Gamma_{11}^2\,\mathbf{x}_\eta\,, \tag{4.137}$$

$$\mathbf{x}_{\xi\eta} = \Gamma_{12}^1\,\mathbf{x}_\xi + \Gamma_{12}^2\,\mathbf{x}_\eta\,, \tag{4.138}$$

$$\mathbf{x}_{\eta\eta} = \Gamma_{22}^1\,\mathbf{x}_\xi + \Gamma_{22}^2\,\mathbf{x}_\eta\,, \tag{4.139}$$

where the *Schwarz-Christoffel symbols* Γ are given by

$$2\,g\,\Gamma^1_{11} = g_{22}\,(g_{11})_\xi - g_{12}\,[2\,(g_{12})_\xi - (g_{11})_\eta]\,, \tag{4.140}$$

$$2\,g\,\Gamma^2_{11} = -g_{12}\,(g_{11})_\xi + g_{11}\,[2\,(g_{12})_\xi - (g_{11})_\eta]\,, \tag{4.141}$$

$$2\,g\,\Gamma^1_{12} = g_{22}\,(g_{11})_\eta - g_{12}\,(g_{22})_\xi\,, \tag{4.142}$$

$$2\,g\,\Gamma^2_{12} = g_{11}\,(g_{22})_\xi - g_{12}\,(g_{11})_\eta\,, \tag{4.143}$$

$$2\,g\,\Gamma^1_{22} = -g_{12}\,(g_{22})_\eta + g_{22}\,[2\,(g_{12})_\eta - (g_{22})_\xi]\,, \tag{4.144}$$

$$2\,g\,\Gamma^2_{22} = g_{11}\,(g_{22})_\eta - g_{12}\,[2\,(g_{12})_\eta - (g_{22})_\xi]\,. \tag{4.145}$$

Exercise 4.4.6 Derive one of the formulas (4.140)-(4.145). §

One usually sees these equations in connection with surfaces (planar regions being a special case). In the surface case, terms proportional to the unit surface normal vector must be added to the expressions (4.137)-(4.139).

Exercise 4.4.7 Use the appropriate entries in Table 4.5 to show that

$$(\sqrt{g})_\xi = \mathbf{x}_\xi^\perp \cdot \mathbf{x}_{\xi\eta} - \mathbf{x}_\eta^\perp \cdot \mathbf{x}_{\xi\xi}\,, \tag{4.146}$$

$$(\sqrt{g})_\eta = \mathbf{x}_\xi^\perp \cdot \mathbf{x}_{\eta\eta} - \mathbf{x}_\eta^\perp \cdot \mathbf{x}_{\xi\eta}\,. \text{ §} \tag{4.147}$$

Holding ξ fixed at $\xi = \xi_0$ defines the curve $\mathbf{x}(\xi_0, \eta)$, which is referred to as a constant ξ- line; similarly, the curve $\mathbf{x}(\xi, \eta_0)$ is referred to as a constant η-line. The *curvature* of these coordinate lines is defined by

$$\kappa_1 = \frac{\mathbf{x}_\eta^\perp \cdot \mathbf{x}_{\eta\eta}}{g_{22}^{3/2}}\,, \tag{4.148}$$

$$\kappa_2 = \frac{\mathbf{x}_\xi^\perp \cdot \mathbf{x}_{\xi\xi}}{g_{11}^{3/2}}\,, \tag{4.149}$$

with κ_1 being the curvature of the ξ-line and κ_2 being the curvature of the η-line (further discussion of curvature can be found in Chapter 10).

Exercise 4.4.8 Use the Gauss identities to express the two curvatures in terms of the rate-of-change metrics. §

As an example of the usefulness of the use of the various relations discussed in this section, the condition for parallel tangents (see Figure 4.6) is derived. The condition for $\mathbf{x}_\xi$ to be parallel to itself along constant ξ lines is that

$$\mathbf{x}_\xi(\xi, \eta_1) \propto \mathbf{x}_\xi(\xi, \eta_2) \tag{4.150}$$

for any η_1 and η_2. If the vectors in (4.150) are divided by their lengths, $\sqrt{g_{11}}(\xi, \eta_1)$ and $\sqrt{g_{11}}(\xi, \eta_2)$, the resulting unit vectors have the same direction, that is, $\mathbf{x}_\xi(\xi, \eta)/\sqrt{g_{11}}(\xi, \eta)$ is independent of η. Thus,

$$\frac{\partial}{\partial\eta}\frac{\mathbf{x}_\xi}{\sqrt{g_{11}}} = \mathbf{0}\,, \tag{4.151}$$

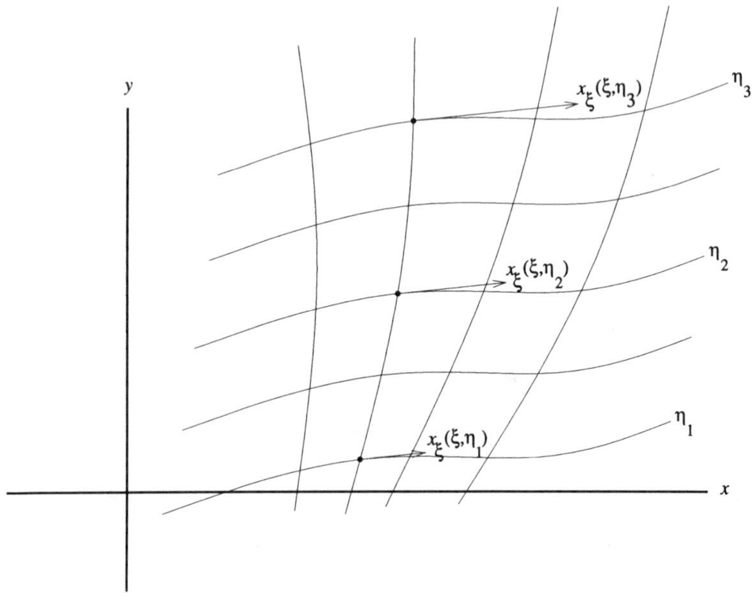

Figure 4.6 *Parallel tangents*

i.e., the rate of change of the direction of the unit tangent along $\mathbf{x}_\xi$ is zero in the η-direction (note that the length of $\mathbf{x}_\xi$ can change). This condition can also be expressed as

$$\mathbf{x}_{\xi\eta} = \frac{1}{2}\frac{(g_{11})_\eta}{g_{11}}\mathbf{x}_\xi \tag{4.152}$$

i.e., $\mathbf{x}_{\xi\eta}$ is parallel to $\mathbf{x}_\xi$ along the constant ξ line.

Exercise 4.4.9 Show that (4.138) and the conditions

$$\Gamma_{12}^2 = 0\,, \tag{4.153}$$

$$\Gamma_{12}^1 = \frac{\partial}{\partial\eta}\log\sqrt{g_{11}} \tag{4.154}$$

are sufficient to guarantee parallel tangents. §

Chapter 5

Classical Planar Grid Generation

5.1 INTRODUCTION

This chapter is devoted to describing some basic planar grid-generation algorithms. The word classical in the title refers to grid generators originally derived from nonvariational methods; these generally appeared in the 1970s to mid-80s. The term should not be regarded as perjorative since these methods form the backbone of most grid-generation software currently in use. This chapter begins with a statement of the basic planar grid-generation problem (Section 5.2). Although the transfinite interpolation method (previously discussed in Section 1.5) is a solution to the basic problem, it has significant limitations. Thus other approaches to the basic problem are often preferred.

The other approaches in this chapter are divided between nonelliptic generators (Section 5.3) and elliptic generators (Section 5.4). This division reflects the authors' bias toward elliptic generators; considerably more space is devoted to the discussion of elliptic grid generation in this book. The justification for this bias is that elliptic grid generators give smooth grids; further, they are quite flexible about the choice of boundary parameterization. The major drawback of the elliptic approach is the relative slowness of the method, as considerably more computational work is generally needed to solve elliptic equations than most of the nonelliptic methods (except the biharmonic approach). There are certainly situations when a nonelliptic generator is advantageous, so a brief survey of such methods is also provided. Nonelliptic methods include algebraic grid generators (Section 5.3.1), conformal and quasiconformal mapping techniques (Section 5.3.2) and orthogonal (Section 5.3.3), hyperbolic (Section 5.3.4), parabolic (Section 5.3.4), and biharmonic (Section 5.3.5) grid generators.

Elliptic methods generate grids by solving boundary value problems for elliptic partial differential equations. For general background information on elliptic partial differential equations see Birkhoff and Lynch [19] for a discussion having a numerical slant or see Epstein [71] for a more mathematical approach. Two elliptic grid generators frequently appear in classical grid generation; these are often referred to as the "Length" (Section 5.4.1) and "Smoothness" (Section 5.4.2) generators. The former is based on Amsden and Hirt [3] and is discussed mainly because its simplicity provides a useful introduction to the subject of elliptic grid generation. The Smoothness generator is usually preferred due to its considerable robustness against folding (but see Section 5.4.3 for an exception). This generator is based on Winslow [235]. It was extended and developed into a powerful tool by Thompson, Thames, and Mastin [212]; thus the method is also referred to as the homogeneous Thompson-Thames-Mastin or TTM generator.

The homogeneous TTM generator can produce reasonable grids on a wide variety of regions, but contains no mechanism for controlling the interior grid other than through the boundary parameterizations. Thompson, Thames, and Mastin [212] added forcing or inhomogeneous terms to the Winslow differential equation to gain control over the interior. Unfortunately, the added terms introduce a variety of arbitrary parameters that a user must specify and whose impact on the grid is imprecise. The net result is that effective use of the inhomogeneous generator is an art instead of an automatic process. A discussion of the inhomogeneous TTM method is found in Section 5.5. Often control over the grid is desired near the boundary of the domain instead of the interior; the most successful method of addressing this need is the Steger-Sorenson method, described in Section 5.6.2. Closely related to the subject of inhomogeneous grid generation is that of solution-adaptive methods (Section 5.7); one classical approach (Adaption with Inhomogeneous TTM, Section 5.7.1) and one nonclassical approach (the Deformation method, Section 5.7.2) are outlined at the end of this chapter.

5.2 THE PLANAR GRID-GENERATION PROBLEM

The basic problem of planar grid generation is: given a simply connected region $\Omega \subset R^2$ in physical space, find a mapping $\mathbf{x} = \mathbf{x}(\xi, \eta)$ from the unit square U_2 in logical space E^2 to the region Ω. The physical region is specified by giving its boundary; this may be done by giving a set of four parametric maps

$$\mathbf{x}_b(s), \quad \mathbf{x}_t(s), \quad 0 \le s \le 1, \tag{5.1}$$

$$\mathbf{x}_l(s), \quad \mathbf{x}_r(s), \quad 0 \le s \le 1. \tag{5.2}$$

[see Figure (1.12)]. The subscripts on $\mathbf{x}$ stand for *bottom*, *top*, *left*, and *right* boundaries of the logical domain. In components, the required boundary maps are

$$x_b(s), \quad x_t(s), \quad x_l(s), \quad x_r(s), \tag{5.3}$$

$$y_b(s), \quad y_t(s), \quad y_l(s), \quad y_r(s), \tag{5.4}$$

while the desired map is written in components as

$$x = x(\xi, \eta), \quad y = y(\xi, \eta). \tag{5.5}$$

It is important that the four consistency checks for the boundary maps given in Table 1.3 be satisfied.

The basic problem has a number of important variations. For example, the given boundary of the region may be reparameterized before extension to the interior is performed. Thus, if $s = s(r)$ with $0 \le r \le 1$, then the composite function $\mathbf{x}_b = \mathbf{x}_b(s(r)) = \hat{\mathbf{x}}_b(r)$ is a reparameterization of the bottom boundary of Ω. Reparameterization is an important technique since the original boundary parameterization is frequently inconvenient.

Exercise 5.2.1 Give a reparameterization of the left and right boundaries of the Modified Horseshoe domain (1.34) that concentrates grid lines toward the bottom boundary of the Horseshoe if transfinite interpolation is used to generate the grid. §

Another modification of the basic planar grid-generation problem is to reduce the number of boundaries that are to be fit. For example, only three boundaries are needed in some approaches to orthogonal grid generation (Eiseman [63]) and only one boundary need be given in hyperbolic grid generation. Of course, the other boundaries are then free and thus beyond user control. The requirement that the region be simply connected may sometimes be relaxed by introducing cuts into the physical domain. For example, an annulus centered at the origin can be cut along the positive x-axis to reduce it to a simply connected domain. Regions having higher degrees of connectivity can sometimes be treated by introducing multiple cuts. Finally, regions not topologically equivalent to a square may require modification of the logical domain. See the first chapter of Thompson, Warsi, and Mastin [219] for an extensive discussion of these latter points.

The transformations are computed using discrete approximations. Grids in physical and logical space are set up as in Section 2.4, that is, let M and N be positive integers and define the logical-grid points (ξ_i, η_j) by

$$\xi_i = \frac{i}{M}, \quad 0 \le i \le M, \quad \eta_j = \frac{j}{N}, \quad 0 \le j \le N. \tag{5.6}$$

Also, let $\Delta\xi = 1/M$ and $\Delta\eta = 1/N$. The transformation $x = x(\xi, \eta)$, $y = y(\xi, \eta)$, carries the logical-space grid to a physical-space grid $\mathbf{x}_{i,j} = (x_{i,j}, y_{i,j})$ where

$$x_{i,j} = x(\xi_i, \eta_j), \quad y_{i,j} = y(\xi_i, \eta_j), \quad 0 \le i \le M, \quad 0 \le j \le N. \tag{5.7}$$

The transfinite interpolation map discussed in Section 1.5 is an example of a map that solves the basic planar grid-generation problem (the reader should review Section 1.5). Transfinite interpolation maps are useful in their own right and as initial guesses for maps that are computed using iterative algorithms. As previously noted, such maps are limited by their lack of smoothness and potential for folding. Many other approaches to the planar grid-generation problem have been proposed; a survey of some of these approaches follows in the next section.

5.3 NONELLIPTIC GRID GENERATORS

Useful nonelliptic grid generators are briefly described in this section. The fact that these appear in this chapter on planar grid generation is merely a matter of convenience; most of the nonelliptic approaches given here have extensions to three dimensions (orthogonal and conformal being the most difficult to extend).

5.3.1 Algebraic Grid Generation

Algebraic grid generation methods have been extensively developed to take advantage of their two main strengths: rapid computation of the grids compared to partial differential equation methods and direct control over grid point locations. These advantages are somewhat offset by the fact that algebraic methods lack any guarantee of grid smoothness; in particular, boundary slope discontinuities may propagate into the interior.

The basic method of algebraic grid generation is *transfinite interpolation* (Gordon and Hall [86]), a special case of which has been described in Section 1.5. Surveys of transfinite interpolation are given in Thompson et al. [219] and Gordon and Thiel [87]. Algebraic methods begin with transfinite interpolation which has its origin in the methods of surface generation in computer-aided design (CAD). The basic building block is unidirectional interpolation along a nondenumerable set of points (also referred to as a "shearing transformation").

Suppose $K+1$ planar curves $\{\ \mathbf{v}_k(\xi),\ k = 0, 1, \dots, K\ \}$ are given and the goal is to interpolate a grid between these curves (see Figure 5.1). This may be accomplished using

$$\mathbf{x}(\xi, \eta) = \sum_{k=0}^{K} \phi_k(\eta)\, \mathbf{v}_k(\xi)\,, \tag{5.8}$$

where the $\phi_k(\eta)$ are interpolating polynomials: $\phi_k(\eta_j) = \delta_{jk}$. Note then that $\mathbf{x}(\xi, \eta_j) = \mathbf{v}_j(\xi)$ for $0 \le j \le K$, that is, the curves have been interpolated.

The simplest choice for the functions ϕ_k are the *Lagrange Interpolating Polynomials* of degree K:

$$\phi_k(\eta) = \prod_{\substack{j=0 \\ j \neq k}}^{K} \frac{\eta - \eta_j}{\eta_k - \eta_j} \tag{5.9}$$

with $\eta_j = j/K$. Recall that this is a Kth degree polynomial that is 1 at $\eta = \eta_k$ and zero at $\eta = \eta_j$, $j \neq k$. The $K+1$ curves can be used to control the grid in the interior of the domain.

Project 5.3.1 Write a code to perform the unidirectional interpolation described above. Use the set of parabolic curves

$$u_k(\xi) = (1 + K - k)\,\xi\,, \tag{5.10}$$

$$v_k(\xi) = k\,(1 + \xi^2)\,. \tag{5.11}$$

Prove analytically that for fixed $\xi = \xi_0$, the coordinate curves are straight between any two consecutive curves $\mathbf{v}_k$ and $\mathbf{v}_{k+1}$; also prove that, in general, the tangent $\mathbf{x}_\eta$ need not be continuous at the $K+1$ curves. §

Other interpolating functions are possible including splines, Hermite polynomials, and Bezier/Bernstein interpolating polynomials. These forms attempt to

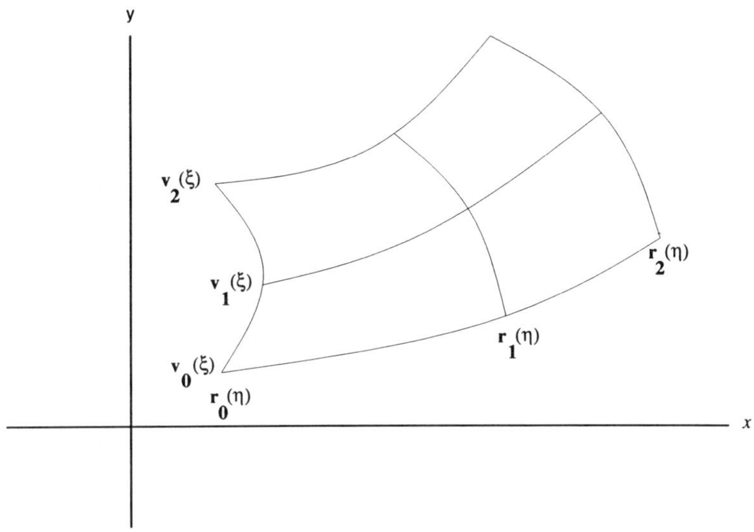

Figure 5.1 *Algebraic grid generation using multiple curves*

smooth the interpolating function so that slopes of grid lines match, resulting in a smoother grid.

Multidirectional interpolation is accomplished by means of *projection operators,* so-called because of their idempotent properties. The right-hand side of (5.8) is an example of a projection operator. To properly interpolate a region, two families of curves must be given (see Figure 5.1):

$$\mathbf{v}_k(\xi), \quad 0 \le k \le K,$$
$$\mathbf{r}_l(\eta), \quad 0 \le l \le L. \tag{5.12}$$

The curves must satisfy the consistency condition

$$\mathbf{r}_l(\eta_k) = \mathbf{v}_k(\xi_l), \quad 0 \le k \le K, \quad 0 \le l \le L. \tag{5.13}$$

The projection operators are defined by

$$\mathbf{P}_\eta = \mathbf{P}_\eta(\xi,\eta) = \sum_{k=0}^{K} \phi_k(\eta)\,\mathbf{v}_k(\xi), \tag{5.14}$$

$$\mathbf{P}_\xi = \mathbf{P}_\xi(\xi,\eta) = \sum_{l=0}^{L} \psi_l(\xi)\,\mathbf{r}_l(\eta), \tag{5.15}$$

where ϕ_k and ψ_l are interpolating functions. The projection $\mathbf{P}_\eta$ is the unidirectional interpolant in the η-direction, while $\mathbf{P}_\xi$ is the unidirectional interpolant in the ξ-direction. The simplest choice for the interpolating functions are again the

Lagrange Interpolating Polynomials: the ϕ_k are given in (5.9), while the ψ_l are given by

$$\psi_l(\xi) = \prod_{\substack{i=0\\ i\neq l}}^{L} \frac{\xi - \xi_i}{\xi_l - \xi_i} \tag{5.16}$$

and $\xi_i = i/L$.

The unidirectional interpolators can be combined to form the Tensor Product Projector:

$$\mathbf{P}_\xi\mathbf{P}_\eta = \mathbf{P}_\xi\mathbf{P}_\eta(\xi,\eta) = \sum_{l=0}^{L}\sum_{k=0}^{K} \phi_k(\eta)\,\psi_l(\xi)\,\mathbf{r}_l(\eta_k) = \sum_{l=0}^{L}\sum_{k=0}^{K} \phi_k(\eta)\,\psi_l(\xi)\,\mathbf{v}_k(\xi_l). \tag{5.17}$$

The definition makes sense because of the consistency condition (5.13).

Exercise 5.3.2 Show that the Tensor Product Projector matches the given curves at the points of their intersection but, in general, nowhere else. This is done by setting $\mathbf{x}(\xi,\eta) = \mathbf{P}_\xi\mathbf{P}_\eta(\xi,\eta)$ and then computing $\mathbf{x}(\xi_i,\eta_j)$. §

To match the given region along all of the given curves, form the Boolean Sum projector

$$\mathbf{P}_\xi \oplus \mathbf{P}_\eta = \mathbf{P}_\xi + \mathbf{P}_\eta - \mathbf{P}_\xi\mathbf{P}_\eta \tag{5.18}$$

and then set the transformation to

$$\mathbf{x}(\xi,\eta) = \mathbf{P}_\xi \oplus \mathbf{P}_\eta(\xi,\eta)\,. \tag{5.19}$$

Exercise 5.3.3 Show that the Boolean Sum projector matches the given curves. Also show that if $K = L = 1$, then the Boolean Sum projector reduces to the transfinite interpolation formula given in Section 1.5. §

In three dimensions, there are 21 distinct projectors using the Tensor product and the Boolean sum. Three of the most useful are

$$\mathbf{P}_\xi\mathbf{P}_\eta\mathbf{P}_\zeta \tag{5.20}$$

which matches the 8 vertices,

$$\mathbf{P}_\xi\mathbf{P}_\eta \oplus \mathbf{P}_\eta\mathbf{P}_\zeta \oplus \mathbf{P}_\zeta\mathbf{P}_\xi \tag{5.21}$$

which matches the 12 edges, and

$$\mathbf{P}_\xi \oplus \mathbf{P}_\eta \oplus \mathbf{P}_\zeta \tag{5.22}$$

which matches the 6 faces of a cube-like region.

Other Algebraic Methods

A generalized unidirectional interpolation method, called "multisurface," is due to Eiseman [60]. In this method, N "control surfaces" are specified. The surfaces do not necessarily correspond to grid surfaces. One then creates a vector field from the surfaces by means of the usual interpolating polynomials. Integrating the resulting expressions lead to the multisurface formulas. The resulting expressions can then be combined as a projector using transfinite interpolation to obtain a grid. The multisurface method reduces to unidirectional interpolation in many cases; e.g., $N = 2$ reduces to the linear Lagrange method, $N = 3$ results in Bezier interpolation, and $N = 4$ becomes the cubic Hermite polynomial interpolation. Eiseman [61, 62] improved the multisurface method by using local interpolant functions in place of the interpolating polynomials to achieve "precise grid control," i.e., grids whose properties can be explicitly specified within a local region independent of the grid elsewhere. This permits local embedding of one grid within another.

A related method of algebraic interpolation is the two-boundary technique of Smith [178]. Smith [179] used only boundary and derivative boundary information, with cubic or linear interpolating polynomials. The method contains terms for controlling orthogonality. Spacing can be controlled by a re-parameterization technique.

This discussion is merely intended as a sketch; the full potential of algebraic grid generation is explored in Gordon [86, 87], Eiseman [59, 60, 61, 62, 69, 70], Shih [172], and Smith [178, 179]. A transfinite interpolation formula for three dimensions is given in Section 9.2.2.

5.3.2 Conformal and Quasiconformal Mapping Techniques

The term *classical* in the chapter title applies best to conformal mapping techniques, which have long served as a means of constructing mappings between planar domains. The theory of conformal mappings is highly developed and the intention here is merely to outline some of the basic results most relevant to grid generation; for more details the reader is referred to Henrici [96] for a discussion of complex variables and conformal mappings with a computational slant and to Moretti [142] for a survey of applications to grid generation.

A *conformal mapping* in the complex plane preserves angles between curves and the "sense" of the angle. Mappings $w = f(z)$ from the complex plane to itself are conformal if f is *analytic*, i.e., df/dz exists and is nonzero. To make the connection with previous notation, set

$$z = \xi + i\eta, \quad w = x + iy. \tag{5.23}$$

Exercise 5.3.4 Let D be the square $1 < x < 2, 0 < y < 1$ in the complex plane and let $w = \exp(z)$. Determine the image of D under this mapping, sketch some of the coordinate lines, and show that angles are preserved, i.e., the mapping is conformal. §

Mappings $w = x(\xi, \eta) + iy(\xi, \eta)$ are conformal if x and y have continuous partial derivatives and satisfy the *Cauchy-Riemann* equations:

$$x_\xi = y_\eta, \quad x_\eta = -y_\xi. \tag{5.24}$$

An immediate consequence of the Cauchy-Riemann equations is that

$$x_\xi x_\eta + y_\xi y_\eta = 0, \tag{5.25}$$

$$x_\xi y_\eta - x_\eta y_\xi = x_\xi^2 + x_\eta^2, \tag{5.26}$$

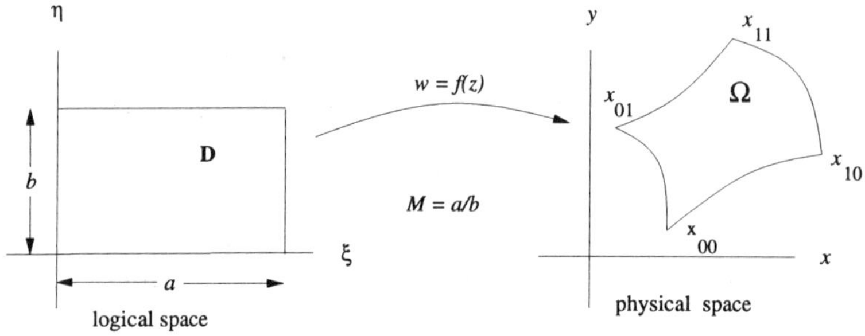

Figure 5.2 *Riemann Mapping Theorem*

i.e., a conformal mapping is orthogonal and its Jacobian is strictly positive (hence, the mapping is locally one-to-one). An important fact about conformal mappings is summarized in the *Riemann Mapping Theorem*:

THEOREM 5.1 Any simply connected domain D with at least two boundary points can be mapped conformally and one-to-one onto the disk $|w| < 1$. The mapping $w = f(z)$ is uniquely determined by setting $f(z_0) = w_0$ and $\arg(f'(z_0)) = 0$ where w_0 is an arbitrary point of the disk and $z_0 \in D$. §

A consequence of the Riemann Mapping Theorem is that it is not possible to map the rectangular domain D in Figure 5.2 to an arbitrary domain Ω having four sides unless the ratio of the length-to-width of D is restricted to a particular constant M known as the *conformal module*. This map can be constructed by solving the pair of partial differential equations

$$M^2 \mathbf{x}_{\xi\xi} + \mathbf{x}_{\eta\eta} = \mathbf{0}. \tag{5.27}$$

The resulting map is then a *quasiconformal* map (see Lehto and Virtaanen [123] and Renelt [156] for a general discussion of quasiconformal maps). The difficulty with this approach is that M is not known a priori since it is domain-dependent; a method of computing M is given in Seidl and Klose [169]. Other approaches to constructing conformal maps by solving partial differential equations can be found in Chakravarthy and Anderson [36], Challis and Burleya [37], and Mastin [132, 134].

Conformal mappings may also be constructed algebraically or by solving integral equations. A method using integral equations to obtain conformal maps in which the logical space is the unit disk is summarized in Thompson et al. [219]. Ives [99] describes several analytic transformation methods including generalizations of the Schwarz-Christoffel Map and the Near-Circle-to-Circle Map which uses the Karman-Trefftz maps (a generalization of the Joukowski Airfoil map). Other references on conformal and quasiconformal mappings include Daripa [46], Fornberg [75], Hayes, Kahaner, and Kellner [94], Menikoff and Zemach [139], and Symm [204].

Conformal mappings permit handling of multiply connected regions with relative ease using multiple branch cuts. The properties of conformal mappings cited in this section are the reason for the continued use of conformal mapping techniques as a means of grid generation even though there are important limitations. Limitations of conformal maps: (i) they are largely restricted to two dimensions, (ii) they result in little control over the interior grid, (iii) finding the map can be very difficult, (iv) multiple valuedness of the functions can lead to difficulties in implementation, (v) arbitrary boundary point distributions are achieved only at the sacrifice of the main advantage, orthogonality, and (vi) conformal mappings are ill-conditioned in the sense that very small changes in the shape of the domain can dramatically alter the position of mapped boundary points.

5.3.3 Orthogonal Grid Generation

Orthogonal grids are defined by having the property $g_{12} = 0$ (see equation 4.105) with $\sqrt{g} \neq 0$. Orthogonality is a desirable property of grids because, as previously mentioned, truncation error is reduced with such grids (Mastin [133]). Additional advantages are that the transformed hosted equation has fewer terms than it would in a nonorthogonal frame and that physical boundary conditions are more easily represented. A major disadvantage of orthogonal coordinate systems is that they only exist for planar domains (Eiseman [63]), and even then they require a particular parameterization of the boundary.

It is difficult to compute the orthogonal grid directly from the relation $g_{12} = 0$ since this is one first-order equation in two unknowns. Haussling and Coleman [91] differentiated this first-order equation with respect to each of the ξ and η variables to obtain two second-order equations. The boundary conditions then uniquely specify the grid. However, for arbitrary boundary conditions, there is no guarantee that the resulting map is orthogonal (in fact, the method can then generate folded grids).

Eiseman [63] proposed the method of orthogonal trajectories wherein one determines the constant ξ-lines from a simple interpolation scheme and the constant η lines are then determined by the requirement of orthogonality. Boundary points can be specified on only three of the four boundaries of the region in this method.

Orthogonal transformations in the plane can be viewed as a quasiconformal map having a real dilation; in this case they are defined by the relation

$$f\,\mathbf{x}_\xi = -\mathbf{x}_\eta^\perp \tag{5.28}$$

with "distortion function" $f = f(\xi, \eta)$. This relation directly implies that $g_{12} = 0$ and that f is the cell-aspect ratio

$$f = \sqrt{\frac{g_{22}}{g_{11}}}. \tag{5.29}$$

Exercise 5.3.5 Derive the relations $g_{12} = 0$ and (5.29) from (5.28). Show that a transformation satisfying (5.28) also satisfies the second-order partial differential equation

$$\frac{\partial}{\partial \xi}(f\,\mathbf{x}_\xi) + \frac{\partial}{\partial \eta}(\mathbf{x}_\eta / f) = \mathbf{0}, \tag{5.30}$$

known as the *Scaled-Laplacian.* §

The most widely studied scheme for generating orthogonal grids involves solving (5.30) (see, for example, Warsi and Thompson [226], Ryskin and Leal [160], Arina [10], Ascoli, Dandy, and Leal [14], and Duraiswami and Prosperetti [54]). In this approach the user generally specifies the cell-aspect ratio f over the domain to control the interior grid (if f is left unspecified, one has the so-called *weak constraint* method). However, solutions to (5.30) are not guaranteed for arbitrary f and, further, the grids may not be orthogonal (one must solve 5.28 directly to guarantee this). A method for constructing f in such a way as to ensure the existence of an orthogonal solution to (5.30) is given in Duraiswami and Prosperetti [54]). The method requires computation of a conformal module for the problem. The solution to (5.30) is, of course, sensitive to the imposed boundary conditions; $g_{12} = 0$ must hold on the boundary in the method of Duraiswami and Propseretti (hence, arbitrary boundary-point distributions cannot be specified in advance). With a given distortion function f, (5.30) is a pair of linear equations, coupled through the boundary condition. The equations are separable and can be efficiently solved by direct methods.

Other work on the subject of orthogonal grid generation may be found in Abrahamsson [1], Tamamidis and Assanis [205], Morice [143], Davies [50], Mobley and Stewart [141], and Potter and Tuttle [152]. Theodoropoulos and Bergeles [206], have proposed an extension of orthogonal grid generation to three dimensions. Allievi and Calisal [2] solves the Scaled-Laplacian equation (5.30) using the finite element approach.

5.3.4 Hyperbolic and Parabolic Grid Generation

This discussion of *hyperbolic* grid generation follows the work of Steger and Chaussee [191], who suggested the following nonlinear, first-order system be used in generating grids about two-dimensional (2-D) *airfoil* surfaces:

$$x_\xi x_\eta + y_\xi y_\eta = 0, \quad x_\xi y_\eta - x_\eta y_\xi = V, \tag{5.31}$$

or in vector notation

$$\mathbf{x}_\xi \cdot \mathbf{x}_\eta = 0, \quad J = V. \tag{5.32}$$

Here $V = V(\xi, \eta)$ is a user-specified control function. The logic behind the system is clear; the first equation enforces grid orthogonality and the second controls the local area of the transformation through the function V, i.e., the system is equivalent to the pair of statements $g_{12} = 0$ and $\sqrt{g} = V$. If the transformation is nonsingular, i.e., $\sqrt{g} \neq 0$, then this system may be expressed vectorially as

$$\mathbf{x}_\eta = \frac{V}{g_{11}} \mathbf{x}_\xi^\perp . \tag{5.33}$$

Note the similarity between equations (5.33) and (5.28); both require the tangent $\mathbf{x}_\eta$ to be proportional to the "perp" of the other tangent. However, if $f = f(\xi, \eta)$, then (5.28) is linear, while (5.33) is nonlinear. More importantly, (5.28) is an *elliptic* system of equations (see Wendland [233] for a definition of *type* for systems of first-order equations). On the other hand, (5.33) is not readily classified due to the nonlinearity. However, as shown in reference [191], if the latter equation is *linearized*, the resulting system of equations has the form

$$\mathcal{A}\mathbf{x}_\xi + \mathcal{B}\mathbf{x}_\eta = \mathbf{v} \tag{5.34}$$

and is *hyperbolic*.

Exercise 5.3.6 Linearize equation (5.33) to find the matrices $\mathcal{A}$ and $\mathcal{B}$ and the vector $\mathbf{v}$ in equation (5.34). Apply the classification scheme in [233] (or some other reference) to determine the type of the linearized system. §

The systems (5.28) and (5.33), as well as the linearized system (5.34), can all be posed as initial-value problems and solved by marching schemes (in which boundary data is given on the inner boundary of an unbounded domain; the grid solution is then obtained by marching the solution outward). However, it is well known that marching schemes for elliptic equations, such as (5.28), are unstable unless certain restrictions are placed upon the cell-aspect ratios of the logical space (see, for example, Roache [157]). On the other hand, there are stable marching schemes for hyperbolic systems (with no restriction on the cell-aspect ratio); thus (5.33) or its linearized version (5.34) are preferred over (5.28) if a marching scheme is to be employed. In the following project, the reader is asked to devise a marching scheme for the nonlinear equation (5.33); details of a marching scheme for the linearized system are given in reference [191].

Project 5.3.7 Use the code described in the Appendix B.5 to solve the following finite difference approximation of equation (5.33):

$$x_{i,j+1} = x_{i,j} - \frac{\Delta\eta}{2\,\Delta\xi}\left(\frac{V}{g_{11}}\right)_{i,j}\left(y_{i+1,j} - y_{i-1,j}\right), \tag{5.35}$$

$$y_{i,j+1} = y_{i,j} + \frac{\Delta\eta}{2\,\Delta\xi}\left(\frac{V}{g_{11}}\right)_{i,j}\left(x_{i+1,j} - x_{i-1,j}\right). \tag{5.36}$$

Let $0 \le \xi \le 1$, $s = \pi\,(1 - 2\xi)$, $a > 0$, $b > 0$, and $V_0 > 0$ and define an ellipse (x, y) by the relations

$$x = a\,\cos(s)\,, \quad y = b\,\sin(s)\,. \tag{5.37}$$

March the solution outward from the ellipse ($\eta = 0$) by applying (5.35)–(5.36). Use the control function $V = V_0\sqrt{g_{11}}\exp\{-\lambda(1-\eta)\}$ with V_0 a scale factor and $\lambda > 0$ a stretching parameter. Experiment with the parameters V_0, a, b, and λ as well as the discretization parameters M and N. Is the discrete grid orthogonal? Compare the cell areas to the control function V. Devise an implicit differencing scheme to solve (5.33) and modify the code accordingly. An example of such a grid is given in Figure 5.3 with $a = 4$, $b = 1$, $V_0 = 4$, and $\lambda = 2$. §

Generation of grids by marching methods is very fast compared to methods that require solving second-order elliptic systems. However, there are limitations inherent in hyperbolic grid generation. The method applies only to unbounded domains, i.e., the outer boundary cannot be specified in advance due to the hyperbolic system of equations. Shock-like behavior of the solution is possible, i.e., nonsmooth grids can be produced. This is especially likely if the initial boundary

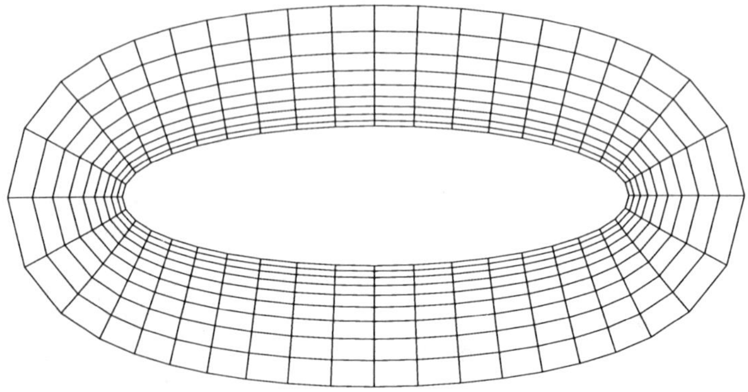

Figure 5.3 *Hyperbolic grid*

curvature causes grid lines to propagate so that the grid lines converge. A numerical dissipation term is generally added to the equations to smooth the grid. The initial boundary curve should be differentiable at all points to ensure the existence of the tangent $\mathbf{x}_\xi$. Since the function V is user-defined, the method is not automatic; experimentation is generally needed to find an effective V for a given application. The method, as differenced in (5.35)-(5.36), is first-order accurate in the marching direction, so large truncation errors are possible, especially if coarse resolutions are used. The method has recently been extended to surface and three-dimensional settings (Steger [192]). A somewhat different approach to hyperbolic grid generation is reported in Starius [189].

To overcome some of the limitations of hyperbolic grid generation, Nakamura, [145], studied a *parabolic* grid-generation scheme in which the speed of the hyperbolic algorithm and the smoothness of elliptic algorithms is claimed. The proposed parabolic grid generation system is

$$x_\eta = A\,x_{\xi\xi} + S_x\,, \tag{5.38}$$

$$y_\eta = A\,y_{\xi\xi} + S_y\,, \tag{5.39}$$

with A a user-specified constant. The source terms S_x and S_y are used to control spacing and orthogonality of the interior mesh. As in hyperbolic grid generation, boundary conditions are specified on the inner boundary of an infinite domain. A marching algorithm progresses toward an outer boundary whose values also influence the source terms, via an interpolation algorithm. Control over orthogonality is indirectly achieved through the source terms. The method is especially useful for O and H-type meshes around *airfoils.* The theory of the source terms does not appear to have been extensively developed, nor has a three-dimensional extension been devised.

5.3.5 Biharmonic Grid Generation

Grid generators based on *fourth-order* partial differential equations permit additional boundary conditions to be imposed. In addition to the usual Dirichlet conditions, for example, one may impose *Neumann* boundary conditions to control boundary orthogonality and spacing. This is especially useful in patching grids on multiple domains together.

The *biharmonic operator* is the only fourth-order generator in use at present. Shubin et al. [174] and Bell et al. [17] propose

$$x_{\xi\xi\xi\xi} + 2\,x_{\xi\xi\eta\eta} + x_{\eta\eta\eta\eta} = 0\,, \tag{5.40}$$

$$y_{\xi\xi\xi\xi} + 2\,y_{\xi\xi\eta\eta} + y_{\eta\eta\eta\eta} = 0\,, \tag{5.41}$$

i.e., $\nabla_{\xi}^4 \mathbf{x} = \mathbf{0}$. Significant penalties are incurred in the biharmonic approach. First, the discretized equations are much harder to solve due to an increase in bandwidth of the matrix equation. Second, although the grids are smooth, there is no direct control over the interior mesh (as is also the case for elliptic generators). Inhomogeneous forms of the biharmonic equations permit indirect control over the interior grid. Third, there is no guarantee against grid folding (refer to the discussion in Section 3.7 for the one-dimensional form of the biharmonic generator).

Sparis [186] applied the biharmonic operator to mappings from physical space to logical space, i.e., $\nabla_{\mathbf{x}}^4 \xi = 0$ and $\nabla_{\mathbf{x}}^4 \eta = 0$, in hope of obtaining a guarantee against folding. Contrary to the claim, however, there is no maximum principle for the biharmonic operator. Therefore, there seems to be little advantage to applying the principle in the physical rather than the logical domain. Extension of the biharmonic grid generator to three dimensions is discussed in Shubin et al. [174].

5.4 ELLIPTIC GRID GENERATION

Elliptic grid generators are based on solving systems of elliptic partial differential equations. A major advantage of this approach is that the interior grid is very smooth, even for nonsmooth boundary data. Grid smoothness is necessary to achieve low truncation error in the solution of the hosted equation by finite difference methods. Another advantage of the elliptic approach is that one may specify all four boundaries of the domain, thus satisfying the basic planar grid-generation problem (Section 5.2). The interior grid is relatively insensitive to the boundary parameterization in the elliptic approach (e.g., see Thomas and Middlecoff [209]; unlike most of the generators mentioned in the previous section, solutions exist for a wide variety of boundary parameterizations and change only gradually as the boundary data is changed. The major penalty incurred in the elliptic approach is the loss of speed relative to the other methods mentioned; it takes considerably more work to solve an elliptic equation (there are large numbers of researchers who search for fast ways to solve elliptic problems). Another significant problem is that it is much harder to control the interior grid when it is generated by solving an elliptic system; this point is discussed in Section 5.5.

5.4.1 The Simplest Elliptic Generator (Length)

The *simplest* elliptic generator, called the AH (Amsden-Hirt) or Length generator, is a two-dimensional generalization of the AH generator (3.24) that was given in

Section 3.3, with $P = 0$. This generator requires each component of the map to satisfy Laplace's equation:

$$\nabla^2 x = \nabla^2_{\xi} x = x_{\xi\xi} + x_{\eta\eta} = 0; \quad \nabla^2 y = \nabla^2_{\xi} y = y_{\xi\xi} + y_{\eta\eta} = 0. \tag{5.42}$$

The four boundary maps given for x in (5.3) provide the boundary data

$$\begin{aligned} x(\xi,0) &= x_b(\xi), \quad & x(\xi,1) &= x_t(\xi), \\ x(0,\eta) &= x_l(\eta), \quad & x(1,\eta) &= x_r(\eta), \end{aligned} \tag{5.43}$$

for the linear partial differential for x in equation (5.42), while the four boundary maps given for y in (5.4) provide similar boundary data for the linear partial differential for y in equation (5.42).

The equations for determining x and y are *uncoupled, linear,* and formulated on a square logical domain. The partial differential equations are *Laplace equations,* the boundary conditions are called *Dirichlet conditions,* and the boundary value problem is called the Dirichlet boundary value problem for Laplace's equation. It is well known (see for example, Birkhoff and Lynch [19], Forsythe and Wasow [76], or Golub and Ortega [85]) that such problems have a unique solution and that solution is infinitely differentiable (in $\mathbf{C}^{\infty}$) in the interior of U_2 provided the boundary maps are continuous. This latter restriction is not important for grid generation since if the boundary maps are not continuous the physical region is not well defined. The only theoretical question left is: are such maps one-to-one and onto, that is, is the Jacobian of the map nonzero? The answer is no; the maps typically fold for nonconvex regions. In the Rogue's gallery, Appendix C, there are many examples of folded grids generated by this method: they are labeled with "Length," see, for example, Figures C.3, C.4, C.5, C.6, C.7, C.9, C.10, C.11, and C.12.

In spite of the lack of a folding guarantee, the AH generator is still attractive in some cases. The most important case is that of a convex domain having one or more boundary points where the boundary has discontinuous slope. In that case, algebraic generators fail to produce a smooth grid, while the AH generator would produce a smooth, unfolded grid. The computation of the AH grid is relatively fast compared to other nonlinear elliptic generators, because the generator equations are linear.

The numerical solution of the AH equations is a fundamental task in numerical partial differential equations. The second derivatives in Laplace's equation are discretized using the standard central differences:

$$\frac{1}{\Delta\xi^2}\left(x_{i-1,j} - 2\,x_{i,j} + x_{i+1,j}\right) + \frac{1}{\Delta\eta^2}\left(x_{i,j-1} - 2\,x_{i,j} + x_{i,j+1}\right) = 0, \tag{5.44}$$

for $1 \le i \le M-1$, $1 \le j \le N-1$. The numerical boundary conditions corresponding to (5.43) are given by

$$\begin{aligned} x_{i,0} &= x_b(\xi_i), \quad & x_{i,N} &= x_t(\xi_i), \quad & 0 \le i \le M, \\ x_{0,j} &= x_l(\eta_j), \quad & x_{M,j} &= x_r(\eta_j), \quad & 0 \le j \le N. \end{aligned} \tag{5.45}$$

There are many computer codes for rapidly computing numerical solutions to such problems (see, for example, Birkhoff and Lynch [19]). A program to solve such problems may be found by sending a request by computer mail to **netlib**.

Exercise 5.4.1 Compute the five-point stencil for the AH generator. What is this stencil when $\Delta\xi = \Delta\eta$? §

Exercise 5.4.2 Obtain a computer code from a software library that solves Laplace's equation for the Dirichlet problem. Use this software to generate grids for the Modified Horseshoe, Swan, and Chevron regions given in Section 1.5. §

More sophisticated elliptic grid generators are needed to handle the folding problem on nonconvex domains. One of the best is the Winslow generator, discussed in the next subsection.

5.4.2 *The Winslow or Smoothness Grid Generator*

The most widely used of all elliptic grid generators is the *Winslow* or *homogeneous Thompson-Thames-Mastin* (TTM) generator. The method is also sometimes referred to as the *smoothness* generator, especially within the context of variational grid generation (see Chapter 6). The generator is a two-dimensional generalization of the generator (3.28) that was discussed in Section 3.3, with $P = 0$. The method arose from the need for an elliptic generator that would produce unfolded grids, i.e., a transformation with positive Jacobian. The method requires the components of the inverse transformation,

$$\xi = \xi(x,y), \quad \eta = \eta(x,y), \tag{5.46}$$

to satisfy Laplace's equation:

$$\nabla^2\xi = \nabla^2_{\mathbf{x}}\xi = \xi_{xx} + \xi_{yy} = 0; \quad \nabla^2\eta = \nabla^2_{\mathbf{x}}\eta = \eta_{xx} + \eta_{yy} = 0 \tag{5.47}$$

on a convex logical domain. Dirichlet boundary conditions are given by the inverses of the boundary maps. By requiring the inverse of the map to be harmonic (instead of the map itself, as in the previous section), it is possible to show that the mapping is one-to-one and onto. Rado's Theorem from the theory of Harmonic Mapping shows that the Winslow mapping is, in fact, a diffeomorphism provided the boundary map is a homeomorphism (Liao [129]). Thus, the continuum solution to the Winslow equations results in an unfolded transformation on the interior of the physical domain.

As with the previous AH elliptic grid generator, if the boundary of Ω and the boundary maps are sufficiently smooth, then the Winslow problem possess a unique solution that is infinitely differentiable in the interior of Ω.

It is difficult to numerically solve the Winslow equations on an arbitrary domain directly. To develop a convenient numerical algorithm for computing the Winslow map, equations (5.47) are transformed to logical space. It must be assumed that $g \neq 0$ so that the mapping is invertible. The result of inverting equations (5.47) is

$$g_{22}\, x_{\xi\xi} - 2\, g_{12}\, x_{\xi\eta} + g_{11}\, x_{\eta\eta} = 0, \quad g_{22}\, y_{\xi\xi} - 2\, g_{12}\, y_{\xi\eta} + g_{11}\, y_{\eta\eta} = 0, \tag{5.48}$$

where (see equation (4.103))

$$\begin{aligned} g_{11} &= x_\xi^2 + y_\xi^2, \\ g_{12} &= x_\xi\, x_\eta + y_\xi\, y_\eta, \\ g_{22} &= x_\eta^2 + y_\eta^2. \end{aligned} \tag{5.49}$$

Recall that g_{11} is the length-squared of a tangent vector to a ξ-coordinate line, g_{22} is the length-squared of a tangent vector to a η-coordinate line, and g_{12} is the inner product of the two tangent vectors. This provides a useful geometric interpretation of the coefficients in the transformed equations and a more compact form for writing the coefficients:

$$g_{i,j} = \mathbf{x}_{\xi_i} \cdot \mathbf{x}_{\xi_j} \,. \tag{5.50}$$

A sophisticated derivation of the inverted equations (5.48) is given in Section 7.3.5. The following exercise uses the ideas in Chapter 2 to find the transformed equations.

Exercise 5.4.3 Use the formulas in Section 2.3 to find the transformed equations (5.48). §

In contrast to (5.42), equations (5.48) are neither linear nor uncoupled; they are coupled through the *metric-matrix* coefficients, which depend on both x and y. These coefficients also make the equations quasilinear instead of linear. It is useful to introduce an operator notation for the TTM equations as this notation is more compact. Thus, let

$$\mathcal{Q}_w = g_{22} \frac{\partial^2}{\partial \xi^2} - 2\, g_{12} \frac{\partial^2}{\partial \xi \partial \eta} + g_{11} \frac{\partial^2}{\partial \eta^2} \,. \tag{5.51}$$

Then the TTM equations (5.48) can be written in vector form as

$$\mathcal{Q}_w \mathbf{x} = \mathbf{0} \tag{5.52}$$

where $\mathcal{Q}_w$ stands for quasilinear Winslow operator.

The finite-difference form of the TTM equations follows directly from the formulas given in Section 2.4. Substantially more work is involved in numerically solving the Winslow system of equations than the AH system due to the coupling and nonlinearity. A code described in Appendix B.6 implements a *Picard* or *successive-substitution* iteration algorithm for computing grids using the TTM generator. This type of algorithm is described in Figure 3.2. Dirichlet boundary conditions for the system of nonlinear partial differential equations (5.48) are given by (5.43).

The algorithm proceeds as follows. An initial grid, $(x_{i,j}^{old}, y_{i,j}^{old})$ on the physical region is generated, typically using transfinite interpolation. The first step in the outer loop is to use the old values (x^{old}, y^{old}) to compute the metric matrix entries g_{11}, g_{12}, and g_{22}, i.e., the metric matrix calculations are lagged. The linearized finite difference equations are then solved using an SOR relaxation, giving new values $(x_{i,j}^{new}, y_{i,j}^{new})$. The outer loop is exited if the difference between the old and new values is less than a given tolerance. If the tolerance is not satisfied, the old values of (x, y) are set equal to the new values and the outer iteration is repeated. Note that the equations for both x and y are the same; this can be used to reduce storage and increase efficiency in numerical algorithms for generating the grids.

It is possible to replace the Picard iteration with a Newton iteration; the Newton iteration typically converges faster than the Picard iteration, but is more trouble to code than Picard iterations. Faster but more complex alternatives are offered by multigrid methods (see Jain [104, 105]).

Exercise 5.4.4 Look up the code described in Appendix B that solves the Winslow equations and verify that the formulas for one of the coefficients is in agreement with the discussion in this section. Describe the numerical algorithm by making a flow chart for the code. Use the homogeneous TTM generator to produce a grid on one of the nonconvex regions given in the Rogue's Gallery, Appendix C. §

5.4.3 Truncation Error in Grid Generation: A Case Study

Grids generated as numerical solutions to partial differential equations are subject to *truncation error* effects. As a result, the discrete grid that satisfies the discretized equations may not share the properties (such as orthogonality, specified area, etc.) possessed by the continuum map, except in the limit. A particularly striking example is encountered in applying the TTM generator to the Modified Horseshoe domain described in Chapter 1. The theorem stated in the previous section implies that for a wide class of domains, folded maps are not solutions to the TTM equations. This is an asymptotic result and thus may not hold for any discrete grid. As demonstrated in this section, truncation error can play an important role in grid generation.

Project 5.4.5 Use the computer code described in Section B.6 of Appendix B to solve the homogeneous TTM equations on the Modified Horseshoe domain, given by equations (1.34) in Chapter 1. Generate and plot $n \times n$ grids for the resolutions $n = 4, 8, 16, 32$, and the parameter values $R = 2$, $R = 4$, and $R = 5$. Are the grids folded? §

The following theorem and project indicate that the results of the previous project are due to truncation error.

THEOREM 5.2 Let $\mathbf{x} = \mathbf{x}(r(\xi,\eta), s(\xi,\eta))$ be the composition of two maps, the first from a unit logical space to a domain in the (r,s) plane and the second from the (r,s) domain to a domain in physical space. If (i) the *primary* map $(r(\xi,\eta), s(\xi,\eta))$ satisfies the TTM equations, and (ii) the *secondary* map $\mathbf{x}(r,s)$ is a *conformal map*, then the composite map satisfies the TTM equations.

Proof. Define the metrics of the primary map as follows:

$$\hat{g}_{11} = \mathbf{x}_r \cdot \mathbf{x}_r \,, \tag{5.53}$$

$$\hat{g}_{12} = \mathbf{x}_r \cdot \mathbf{x}_s \,, \tag{5.54}$$

$$\hat{g}_{22} = \mathbf{x}_s \cdot \mathbf{x}_s \,. \tag{5.55}$$

Transform the TTM grid generator on the composite map

$$\mathcal{Q}_w \mathbf{x} = g_{22}\,\mathbf{x}_{\xi\xi} - 2\,g_{12}\,\mathbf{x}_{\xi\eta} + g_{11}\,\mathbf{x}_{\eta\eta} \tag{5.56}$$

using the chain rule, beginning with

$$g_{11} = \hat{g}_{11}\,r_\xi^2 + 2\,\hat{g}_{12}\,r_\xi\,s_\xi + \hat{g}_{22}\,s_\xi^2 \,, \tag{5.57}$$

$$g_{12} = \hat{g}_{11}\,r_\xi\,r_\eta + \hat{g}_{12}(r_\xi\,s_\eta + r_\eta\,s_\xi) + \hat{g}_{22}\,s_\xi\,s_\eta \,, \tag{5.58}$$

$$g_{22} = \hat{g}_{11}\,r_\eta^2 + 2\,\hat{g}_{12}\,r_\eta\,s_\eta + \hat{g}_{22}\,s_\eta^2 \,. \tag{5.59}$$

Let $\sqrt{\tau} = r_\xi s_\eta - r_\eta s_\xi$. Then the chain rule applied to the TTM equations results in

$$\begin{aligned} \mathcal{Q}_w \mathbf{x} = {} & \sqrt{\tau}\,[\hat{g}_{22}\,\mathbf{x}_{rr} - 2\hat{g}_{12}\,\mathbf{x}_{rs} + \hat{g}_{11}\,\mathbf{x}_{ss}] \\ & + [g_{22}\,r_{\xi\xi} - 2g_{12}\,r_{\xi\eta} + g_{11}\,r_{\eta\eta}]\,\mathbf{x}_r \\ & + [g_{22}\,s_{\xi\xi} - 2g_{12}\,s_{\xi\eta} + g_{11} s_{\eta\eta}]\mathbf{x}_s\,. \end{aligned} \tag{5.60}$$

Assume that the secondary map is conformal, i.e., it satisfies the Cauchy-Riemann conditions

$$x_r = y_s\,, \tag{5.61}$$

$$-x_s = y_r\,. \tag{5.62}$$

Immediate consequences are $\hat{g}_{12} = 0$, $\hat{g}_{22} = \hat{g}_{11}$, and $\mathbf{x}_{rr} + \mathbf{x}_{ss} = \mathbf{0}$. As a result, the first bracketed term in (5.60) is zero. Also, the composite map metrics (5.57)-(5.59) reduce to

$$g_{11} = \hat{g}_{11}\left(r_\xi^2 + s_\xi^2\right), \tag{5.63}$$

$$g_{12} = \hat{g}_{11}\left(r_\xi\, r_\eta + s_\xi\, s_\eta\right), \tag{5.64}$$

$$g_{22} = \hat{g}_{11}\left(r_\eta^2 + s_\eta^2\right), \tag{5.65}$$

so that the second and third bracketed terms of (5.60) are also zero:

$$\hat{g}_{11}\left[(r_\eta^2 + s_\eta^2)\,r_{\xi\xi} - 2\,(r_\xi r_\eta + s_\xi s_\eta)\,r_{\xi\eta} + (r_\xi^2 + s_\xi^2)\,r_{\eta\eta}\right]\mathbf{x}_r = 0\,, \tag{5.66}$$

$$\hat{g}_{11}\left[(r_\eta^2 + s_\eta^2)\,s_{\xi\xi} - 2\,(r_\xi\, r_\eta + s_\xi\, s_\eta)\,s_{\xi\eta} + (r_\xi^2 + s_\xi^2)\,s_{\eta\eta}\right]\mathbf{x}_s = 0\,, \tag{5.67}$$

since the primary mapping from logical space to the (r, s) domain satisfies the TTM equations. Thus, $\mathcal{Q}_w\,\mathbf{x} = \mathbf{0}$ for the composite map. §

In general, the composition of two maps each solving the TTM equations need not solve the composite TTM equations.

Project 5.4.6 Map the horseshoe domain to the (r, s) domain in Figure 5.4 using the map given by

$$r(\xi, \eta) = \xi r_R(\eta), \tag{5.68}$$

$$s(\xi, \eta) = \xi s_R(\eta) + (1 - \xi)\pi\eta, \tag{5.69}$$

where

$$r_R(\eta) = \log(2) + \frac{1}{2}\log\left(\cos^2(\pi\eta) + R^2 \sin^2(\pi\eta)\right), \tag{5.70}$$

$$s_R(\eta) = \arctan(R\,\tan(\pi\eta))\,, \tag{5.71}$$

and $R = \exp(\alpha)$. Find a parameterization of the bottom, top, and left boundaries so that the composite map preserves the boundary parameterization of the Modified Horseshoe. Then solve the TTM equation on this new domain (with $R = 4.0$)

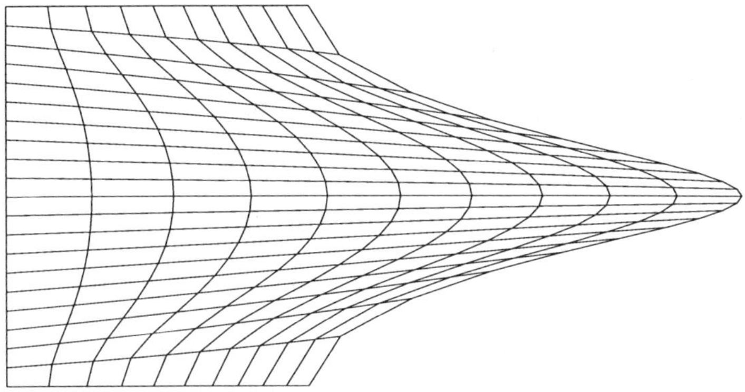

Figure 5.4 *Horseshoe region*

and map the result onto the horseshoe domain via the conformal map

$$-x(r,s) = \exp(s)\,\cos(r)\,, \tag{5.72}$$

$$y(r,s) = \exp(s)\,\sin(r)\,, \tag{5.73}$$

to obtain an unfolded grid. Use this unfolded grid as the initial guess to start the iteration to solve the discrete TTM equations (5.56) to retrieve the folded grid in the previous project. This result strongly suggests that folding of the horseshoe grid in Project 5.4.5 is due to the effects of *truncation error.* §

5.5 THE INHOMOGENEOUS TTM GRID GENERATOR

The purpose of this section is to present the planar version of the *inhomogeneous Thompson-Thames-Mastin* generator and show how it is used to control the grid on the interior of a domain. The generator can be used to attract or repel grid nodes toward specified points in the logical domain and to move coordinate lines in a similar fashion. The original approach is presented in Thompson, Thames, and Mastin [212], who added source terms to the homogeneous partial differential equation (5.47). There are two basic forms of the inhomogeneous grid generator; both require the user to specify two weight functions to be used in the inhomogeneous terms. The original inhomogeneous approach is

$$\nabla_{\mathbf{x}}^2 \xi = P\,, \quad \nabla_{\mathbf{x}}^2 \eta = Q\,, \tag{5.74}$$

while the reformulated approach (Warsi [228])

$$\nabla_{\mathbf{x}}^2 \xi = \frac{g_{22}}{g} P\,, \quad \nabla_{\mathbf{x}}^2 \eta = \frac{g_{11}}{g} Q\,, \tag{5.75}$$

is sometimes preferred because the control functions P and Q for this latter form of the equations are orders of magnitude smaller than for the original form. The reader is cautioned that the weight function Q should not be confused with the grid-generation operator $\mathcal{Q}_w$.

Rado's Theorem (Liao [129]) provides a theoretical basis for the homogeneous TTM or Winslow method because it implies that, in the continuum, the grid is not folded. As shown in the previous section, this does not imply that the discrete grid is unfolded. When inhomogeneous terms are added to the equations, Rado's Theorem no longer applies, i.e., there is no guarantee against folding of the inhomogeneous continuum grid. Moreover, the inhomogeneous continuum grid need not be smooth even though one is using an "elliptic" generator. For instance, if a nonsmooth grid is given (say, by an algebraic generator) then it is possible to compute the grid metrics, tangents, etc., and therefore P and Q using the inhomogeneous equation. The computed P and Q would then generate the nonsmooth grid if used to solve the inhomogeneous grid equations. Restrictions on P and Q are clearly needed if a smooth, unfolded grid is to be obtained.

Inverting the original form (5.74) of the TTM equations gives

$$\mathcal{Q}_w \mathbf{x} = -g\,(P\,\mathbf{x}_\xi + Q\,\mathbf{x}_\eta) \tag{5.76}$$

while inverting the reformulated equations (5.75) gives

$$\mathcal{Q}_w \mathbf{x} = -g_{22}\,P\,\mathbf{x}_\xi - g_{11}\,Q\,\mathbf{x}_\eta\,. \tag{5.77}$$

The latter equations bear a close resemblance to the one-dimensional differential equation obtained from the variational problem with a logical-space weight; compare equation (3.66) of Section 3.5, equation (3.33) of Section 3.3, and equation (3.6) of Section 3.2. Steger and Sorenson [190] use the original TTM approach (5.76) while Anderson [6] and Thomas and Middlecoff [209] follow the newer approach (5.77).

Local control of the grid is obtained by choosing P and Q to have the exponential forms (Thompson, Thames, and Mastin [212]):

$$\begin{aligned} P(\xi,\eta) = &-\sum_{m=1}^{M} a_m \frac{\xi-\xi_m}{|\xi-\xi_m|}\, e^{-c_m|\xi-\xi_m|} \\ &-\sum_{i=1}^{I} b_i \frac{\xi-\xi_i}{|\xi-\xi_i|}\, e^{-d_i[(\xi-\xi_i)^2+(\eta-\eta_i)^2]^{\frac{1}{2}}}\,, \end{aligned} \tag{5.78}$$

$$\begin{aligned} Q(\xi,\eta) = &-\sum_{m=1}^{M} a_m \frac{\eta-\eta_m}{|\eta-\eta_m|}\, e^{-c_m|\eta-\eta_m|} \\ &-\sum_{i=1}^{I} b_i \frac{\eta-\eta_i}{|\eta-\eta_i|}\, e^{-d_i[(\xi-\xi_i)^2+(\eta-\eta_i)^2]^{\frac{1}{2}}}\,, \end{aligned} \tag{5.79}$$

where M is the number of lines and I is the number of points the grid is to be attracted to, and where a_m, b_i, c_m, d_i, ξ_i, and η_i are parameters. Notice that P and Q are logical-space weight functions.

These source terms contain arbitrary parameters that are varied to obtain a desired grid; there does not appear to be any particular methodology to guide the selection of parameter values. The $P - Q$ approach to interior grid control is reasonably effective but lacks automation, i.e., user intervention is required. Determination of the parameters in the source terms requires experience and skill. Control of the grid is imprecise and nonautomatic. As discussed in Section 3.8, physical-space weighting is more useful than logical-space weighting; the former is not directly achievable with this scheme. The EAGLE code manual (Thompson et al. [221]) provides further details concerning the use and capabilities of this algorithm. There are week-long short courses available for those who would like to use the EAGLE code.

Numerical implementation of the inhomogeneous generator requires only minor modification of the homogeneous generation code. A code partially implementing the $P - Q$ weights in the form (5.77) is described in Appendix B, Section B.6.

Project 5.5.1 Use the code described in Appendix B, Section B.6, to attract grid lines to the coordinate line $\eta = \eta_s$. The code assumes

$$P = 0\,, \tag{5.80}$$

$$Q = -Q_0 \,\mathrm{sgn}(r)\, e^{-\lambda |r|}\,, \tag{5.81}$$

where $r = \eta - \eta_s$, $\mathrm{sgn}(r) = r/|r|$, and Q_0 and $\lambda > 0$ are user-supplied parameters. Set $\eta_s = 1/2$ and experiment with the parameters Q_0 and λ on the Trapezoid domain to determine reasonable values. Verify that if $Q_0 > 0$, lines are attracted to $\eta = \eta_s$, while if $Q_0 < 0$, they are repelled. Can you predict the location and degree of attraction in physical space that is obtained by changing Q_0 and λ? What happens to the grid as the parameters become large? Modify the code to run the form (5.76) of the inhomogeneous equations. Which form do you prefer and why? Modify the code to include the general weights P and Q given in (5.78)–(5.79). §

The particular case $Q_0 = 5$, $\lambda = 1$, $\eta_s = 1/2$ is shown in Figure 5.5 for the Parallelogram, Swan, Trapezoid, and Valley regions. Grid lines do appear attracted to $\eta_s = 1/2$, but the location in physical space and the degree of attraction are unpredictable. The solution iteration fails to converge on the Annulus, Horseshoe, and several other domains. Better results might be obtained by further experimenting with the input parameters. Re-parameterization of the boundary would also improve the grids. The grids often fold badly for certain seemingly reasonable choices of the parameter Q_0. In the case $\eta_s = 0$ on the Airfoil, there is a strong sensitivity to Q_0, with the grid folding for values of Q_0 greater than two. Even when unfolded grids are obtained, the grid lines are not parallel to the airfoil surface. This fails to meet an essential requirement of *viscous-flow* calculations.

See Chapter 9 or Thompson [214] for a discussion of the volume or three-dimensional form of the inhomogeneous TTM method. Only limited control over the interior grid generated by an elliptic equation can be achieved by re-parameterization of the boundary. The experience reported by Thomas and Middlecoff [209] illustrates this and is consistent with the theory concerning properties of solutions to elliptic equations. Inhomogeneous grid generators can also be used to alter the grid near the boundary as shown in the next section.

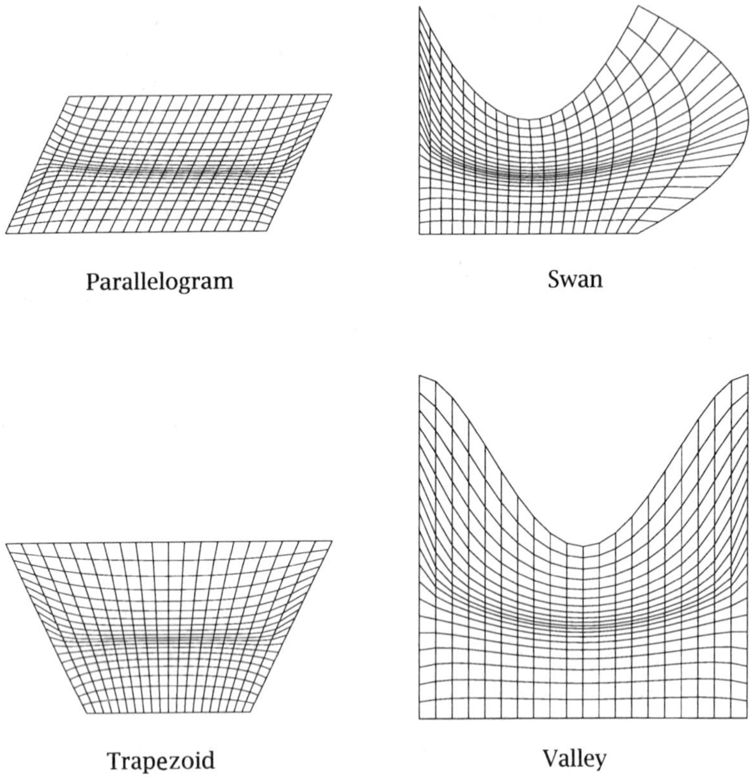

Figure 5.5 *Inhomogeneous TTM grids*

5.6 CONTROLLING THE GRID NEAR THE BOUNDARY

A particularly important requirement for a grid generator is to be able to control the behavior of the transformation near the boundary, since often this is where the behavior of the solution changes most rapidly (e.g., boundary-layer phenomena in fluid dynamics). In such an application, control over grid spacing and orthogonality near the boundary is often required. The most effective means of achieving this is to apply Neumann (gradient) boundary conditions to the elliptic partial differential equations of the grid generator. However, most elliptic grid generators (such as those discussed in this chapter) are second-order partial differential equations, which can support only one set of boundary conditions. If Neumann conditions are applied at the boundary, then the Dirichlet boundary conditions must be dropped. The resulting mapping would then no longer be boundary-conforming.

Both Dirichlet and Neumann boundary conditions can be enforced if the order of the equation is increased. The fourth-order *biharmonic* grid generator (Section 5.3.5) is a likely candidate. However, although the biharmonic approach would work in principle, it is relatively hard to implement and solve a fourth-order PDE compared to a second-order PDE. Furthermore, control over the interior grid is very indirect and there is a tendency for grids generated by the biharmonic to fold.

One alternate approach is proposed in Thomas and Middlecoff [209]. While this method has enjoyed some success, its general use is not advocated because the derivation of the method is not mathematically justified (i.e., the final equations do not guarantee that the resulting transformation will possess the desired boundary properties). There are also practical instances in which the method has failed to perform effectively (Salari [163]).

Another approach to the problem is described in this section. The grid near the boundary is controlled using second-order *inhomogeneous*, elliptic equations with Dirichlet boundary conditions. Neumann conditions are simulated through the inhomogeneous source terms. Geometric consequences of specifying a Neumann boundary condition along a boundary with known Dirichlet data are described in the first part of this section, while the second part uses these results in a general exposition of the widely used Steger and Sorenson [190] method of simulating Neumann boundary conditions. The generality of the discussion permits extension of the original Steger-Sorenson method to other elliptic grid generators beside the original TTM context in which it was proposed.

5.6.1 *The Neumann Boundary Condition*

A surprising amount of information on the boundary tangents, metrics, and rate-of-change metrics of a transformation can be deduced if both Dirichlet and Neumann boundary conditions are simultaneously imposed on the boundary of a domain. The reader is referred to Section 4.4 for a review of the relevant planar differential geometry needed for this section. In general, boundary conditions can be applied to any part or parts of the four boundaries of a planar region. To concretize this discussion, suppose the goal is to control the grid along the entire $\eta = 0$ boundary (the discussion that follows is readily extended to other boundaries). Assume that Dirichlet data $\mathbf{x}(\xi)$ is given on this boundary and that the boundary has sufficient smoothness for the derivatives required in this discussion to exist. Then from the Dirichlet data, one may compute the tangent $\mathbf{x}_\xi$, its perpendicular $\mathbf{x}_\xi^\perp$, the metric g_{11}, and the second-derivative vector $\mathbf{x}_{\xi\xi}$ all along the $\eta = 0$ boundary.

Assume that the following general Neumann boundary condition is also to be imposed on the grid at $\eta = 0$:

$$\mathbf{x}_\eta = \frac{1}{\sqrt{g_{11}}}\left[\sigma_1(\xi)\,\mathbf{x}_\xi + \sigma_2(\xi)\,\mathbf{x}_\xi^\perp\right] \tag{5.82}$$

with σ_1 and σ_2 given differentiable functions on $0 < \xi < 1$. Since $\mathbf{x}_\eta$ is known at $\xi = 0$ and at $\xi = 1$ (from the Dirichlet data on the other boundaries), a pair of consistency conditions on σ_1 and σ_2 are required, namely, that

$$\frac{1}{\sqrt{g_{11}}}\,[\sigma_1(0)\,\mathbf{x}_\xi + \sigma_2(0)\,\mathbf{x}_\xi^\perp] = \mathbf{x}_\eta(0)\,, \tag{5.83}$$

$$\frac{1}{\sqrt{g_{11}}}\,[\sigma_1(1)\,\mathbf{x}_\xi + \sigma_2(1)\,\mathbf{x}_\xi^\perp] = \mathbf{x}_\eta(1)\,. \tag{5.84}$$

Given consistent σ_1 and σ_2, the remaining metric terms are obtained from (5.82) by forming the relevant inner products. On the $\eta = 0$ boundary:

$$g_{12} = \sqrt{g_{11}}\,\sigma_1\,, \tag{5.85}$$

$$g_{22} = \sigma_1^2 + \sigma_2^2\,, \tag{5.86}$$

$$\sqrt{g} = \sigma_2\,\sqrt{g_{11}}\,. \tag{5.87}$$

The expression (5.87) shows that the additional requirement $\sigma_2(\xi) > 0$ should be imposed to ensure a positive Jacobian on the boundary.

Exercise 5.6.1 Verify the relations (5.85)-(5.87). Also show that for arbitrary planar vectors $\mathbf{a}^{\perp\perp} = -\mathbf{a}$ and $(\mathbf{a}+\mathbf{b})^\perp = \mathbf{a}^\perp + \mathbf{b}^\perp$. Apply these to (5.82) to show $\mathbf{x}_\eta^\perp$ on the boundary is

$$\mathbf{x}_\eta^\perp = \frac{1}{\sqrt{g_{11}}}\,[\sigma_1(\xi)\,\mathbf{x}_\xi^\perp - \sigma_2(\xi)\,\mathbf{x}_\xi]\,.\ \S \tag{5.88}$$

So far, then, $\mathbf{x}_\xi$, $\mathbf{x}_\xi^\perp$, $\mathbf{x}_\eta$, $\mathbf{x}_\eta^\perp$, $\mathbf{x}_{\xi\xi}$, and the metrics g_{11}, g_{12}, g_{22}, and $\sqrt{g}$ are known or computable from the given Dirichlet and Neumann data. To use the Steger-Sorenson method, most of the rate-of-change metrics on the boundary are also needed. To begin, the boundary condition (5.82) can be differentiated with respect to ξ to obtain the rate-of-change of $\mathbf{x}_\eta$, i.e., the cross derivative $\mathbf{x}_{\xi\eta}$, in terms of known quantities:

$$\mathbf{x}_{\xi\eta} = \frac{1}{\sqrt{g_{11}}}\left\{\sigma_1\,\mathbf{x}_{\xi\xi} + \sigma_2\,\mathbf{x}_{\xi\xi}^\perp + (\sigma_1)_\xi\,\mathbf{x}_\xi + (\sigma_2)_\xi\,\mathbf{x}_\xi^\perp - \frac{(g_{11})_\xi}{2\sqrt{g_{11}}}\,\mathbf{x}_\eta\right\}. \tag{5.89}$$

Exercise 5.6.2 Use (4.131)-(4.136) and (5.85)-(5.87) to verify the following computable expressions for the rate-of-change metrics:

$$\frac{1}{2}\,(g_{11})_\xi = \mathbf{x}_\xi \cdot \mathbf{x}_{\xi\xi}\,, \tag{5.90}$$

$$(g_{12})_\xi = \sqrt{g_{11}}\,(\sigma_1)_\xi + \frac{\sigma_1}{2}\,\frac{(g_{11})_\xi}{\sqrt{g_{11}}}\,, \tag{5.91}$$

$$\frac{1}{2}\,(g_{22})_\xi = \sigma_1(\sigma_1)_\xi + \sigma_2(\sigma_2)_\xi\,, \tag{5.92}$$

$$(\sqrt{g})_\xi = \frac{\sigma_2}{2}\,\frac{(g_{11})_\xi}{\sqrt{g_{11}}} + \sqrt{g_{11}}\,(\sigma_2)_\xi\,, \tag{5.93}$$

$$\frac{1}{2}\,(g_{11})_\eta = (g_{12})_\xi - \mathbf{x}_\eta \cdot \mathbf{x}_{\xi\xi}\,.\ \S \tag{5.94}$$

The other rate-of-change metrics involving derivatives with respect to η are unknown since $\mathbf{x}_{\eta\eta}$ cannot be computed from (5.82) as the latter holds only along the $\eta = 0$ boundary. In particular, $(g_{12})_\eta$ and $(g_{22})_\eta$ are unknown. The Christoffel symbols Γ^1_{11}, Γ^2_{11}, Γ^1_{12}, and Γ^2_{12} can be computed from the known rate-of-change metrics using (4.140)–(4.143), but Γ^1_{22} and Γ^2_{22} cannot.

The special case $\sigma_1 = 0$, $\sigma_2 = s > 0$ is of particular importance since it is the requirement that the grid be orthogonal on the boundary, with constant spacing s. In this case, (5.82) reduces to

$$\mathbf{x}_\eta = \frac{s}{\sqrt{g_{11}}}\,\mathbf{x}_\xi^\perp\,. \tag{5.95}$$

This is a compact way of stating that orthogonality and a specified spacing s are required on the boundary. From the previous results,

$$g_{12} = 0\,, \tag{5.96}$$

$$g_{22} = s^2\,, \tag{5.97}$$

$$\sqrt{g} = s\,\sqrt{g_{11}}\,, \tag{5.98}$$

$$(g_{12})_\xi = 0\,, \tag{5.99}$$

$$(g_{22})_\xi = 0 \tag{5.100}$$

on the boundary.

Exercise 5.6.3 Find Γ^1_{11}, Γ^2_{11}, Γ^1_{12}, Γ^2_{12} for the special case described above. §

5.6.2 *The Steger-Sorenson Approach*

The Neumann boundary conditions (5.82) can be simulated with an inhomogeneous grid generator using the Steger and Sorenson approach [190]. The method was originally proposed for the Winslow operator, but it is emphasized here that the basic approach can be applied to other second-order operators as well. Let $\mathcal{Q}\mathbf{x}$ be any second-order quasilinear elliptic generator (e.g., Length, AO, or Winslow) and let the form of the inhomogeneous generator be

$$\mathcal{Q}\mathbf{x} = -g\,\{\phi(\xi,\eta)\,\mathbf{x}_\xi + \psi(\xi,\eta)\,\mathbf{x}_\eta\}\,. \tag{5.101}$$

To simulate Neumann boundary conditions on the $\eta = 0$ boundary, the inhomogeneous weight functions are chosen to be

$$\phi(\xi,\eta) = \phi_0(\xi)\,e^{-\lambda\eta}\,, \tag{5.102}$$

$$\psi(\xi,\eta) = \psi_0(\xi)\,e^{-\lambda\eta}\,, \tag{5.103}$$

so that (5.101) becomes

$$\mathcal{Q}\mathbf{x} = -g\,\{\phi_0(\xi)\,\mathbf{x}_\xi + \psi_0(\xi)\,\mathbf{x}_\eta\}\,e^{-\lambda\eta} \tag{5.104}$$

with $\lambda > 0$ a user-supplied parameter that serves to blend the boundary condition into the homogeneous condition in the interior. Computational experience shows that the grid is relatively insensitive to λ.

To determine the weight functions, form the inner products

$$-\mathbf{x}_\xi^\perp \cdot \mathcal{Q}\mathbf{x} = \psi_0\, g^{\frac{3}{2}}\, e^{-\lambda\eta}, \tag{5.105}$$

$$\mathbf{x}_\eta^\perp \cdot \mathcal{Q}\mathbf{x} = \phi_0\, g^{\frac{3}{2}}\, e^{-\lambda\eta}. \tag{5.106}$$

Generally, (5.104) does not hold on the boundary, but only on the interior. If one requires (5.104) to hold on the boundary as well as in the interior, ϕ_0 and ψ_0 can be found by evaluating (5.105)-(5.106) at $\eta = 0$:

$$g^{\frac{3}{2}}\, \phi_0(\xi) = (\mathbf{x}_\eta^\perp \cdot \mathcal{Q}\mathbf{x})_0\,, \tag{5.107}$$

$$-g^{\frac{3}{2}}\, \psi_0(\xi) = (\mathbf{x}_\xi^\perp \cdot \mathcal{Q}\mathbf{x})_0\,. \tag{5.108}$$

Formula (5.87) is used to evaluate $g^{3/2}$. Finding the inhomogeneous weights therefore requires evaluating $\mathcal{Q}\mathbf{x}$ on the boundary. Unfortunately, the elliptic operators contain the unknown vector $\mathbf{x}_{\eta\eta}$, so the evaluation cannot be done a priori, but requires the use of an iterative solution procedure.

Before describing the iteration, expressions for the weights in (5.107)-(5.108) are derived. The various generators $\mathcal{Q}\mathbf{x}$ are resolved into combinations of the covariant tangent vectors using the Gauss identities (4.137)-(4.139). For example, the Winslow operator can be expressed as

$$\begin{aligned}\mathcal{Q}_w\mathbf{x} = {} & (g_{22}\,\Gamma_{11}^1 - 2\,g_{12}\,\Gamma_{12}^1 + g_{11}\,\Gamma_{22}^1)\mathbf{x}_\xi \\ & + (g_{22}\,\Gamma_{11}^2 - 2\,g_{12}\,\Gamma_{12}^2 + g_{11}\,\Gamma_{22}^2)\mathbf{x}_\eta\,.\end{aligned} \tag{5.109}$$

Applying (5.107)-(5.108) to this gives the weights

$$-g\,\phi_0 = g_{22}\,\Gamma_{11}^1 - 2\,g_{12}\,\Gamma_{12}^1 + g_{11}\,\Gamma_{22}^1\,, \tag{5.110}$$

$$-g\,\psi_0 = g_{22}\,\Gamma_{11}^2 - 2\,g_{12}\,\Gamma_{12}^2 + g_{11}\,\Gamma_{22}^2\,. \tag{5.111}$$

All the quantities on the right-hand side of these equations, except Γ_{22}^1 and Γ_{22}^2, can be evaluated at $\eta = 0$ in the manner discussed in the previous section.

Exercise 5.6.4 See equation (6.75) of Section 6.3.5 for a description of the AO operator. Show that the weights for the AO operator are

$$-g\,\phi_0 = g_{22}\,\Gamma_{11}^1 + g_{11}\,\Gamma_{22}^1 + (g_{22})_\xi\,, \tag{5.112}$$

$$-g\,\psi_0 = g_{22}\,\Gamma_{11}^2 + g_{11}\,\Gamma_{22}^2 + (g_{11})_\eta\,.\ \S \tag{5.113}$$

In the Steger-Sorenson approach, the unknown quantities Γ_{22}^1 and Γ_{22}^2 contained in ϕ_0, ψ_0 are lagged during the iterative solution. Strong under-relaxation is required to obtain convergence. The iteration is extremely slow as a result; intuitively, this is because one is lagging second-order quantities. The vector $\mathbf{x}_{\eta\eta}$ is numerically evaluated on the boundary using the Pade approximation,

$$\mathbf{x}_{\eta\eta}\big|_0 = \frac{-7\,\mathbf{x}_1 + 8\,\mathbf{x}_2 - \mathbf{x}_3}{2\,\Delta\eta^2} - 3\,\frac{\mathbf{x}_\eta|_0}{\Delta\eta} \tag{5.114}$$

suggested by Steger and Sorenson [190]. The first author has had success using the Steger-Sorenson approach with the AO operator as well as with Winslow. It has not been found necessary to use forward difference approximations for $\mathbf{x}_\eta$ in (5.104), as was reported in Steger and Sorenson [190]. Small grid spacing is needed to reduce the effect of truncation error. If coarse grids are used, the desired continuum boundary conditions (5.82) may not be closely matched. Others have successfully implemented the Steger-Sorenson method on more than one boundary simultaneously.

Project 5.6.5 Write a computer code to perform the Steger-Sorenson iteration for a general operator $\mathcal{Q}\mathbf{x}$. §

5.7 SOLUTION-ADAPTIVE ALGORITHMS

This section extends the discussion initiated in Section 3.8 to two-dimensional solution-adaptive algorithms for moving grids. The goal is to reduce truncation error in the numerical solution of the hosted equations by moving grid nodes as the solution evolves. Weight functions are introduced into the inhomogeneous grid generator to create movement of grid nodes; the weights are based on physical properties of the solution, such as the gradient. The purpose of this section is merely to introduce the reader to the subject. For a more exhaustive treatment, see the review articles on solution-adaptive grid generation by Thompson [220], Eiseman [68], and Hawken [92, 93].

5.7.1 Grid Adaption with Inhomogeneous TTM

As noted in Section 5.5, the inhomogeneous TTM weights (5.78)-(5.79) are rather clumsy to work with. A more fundamental set of weights was derived in Thomas and Middlecoff [209] by setting $P = g^{11}\,\phi(\xi,\eta)$ and $Q = g^{22}\,\psi(\xi,\eta)$. Inversion of equations (5.74) with these new weights leads to an adaptive form of the inhomogeneous TTM equations

$$\mathcal{Q}_w\mathbf{x} = -g_{22}\,\phi\,\mathbf{x}_\xi - g_{11}\,\psi\,\mathbf{x}_\eta\,. \tag{5.115}$$

To determine the weight functions ϕ and ψ, Anderson [7] showed that the following equations are satisfied:

$$(\sqrt{g_{11}})_\xi + \phi\,\sqrt{g_{11}} = 0\,, \tag{5.116}$$

$$(\sqrt{g_{22}})_\eta + \psi\,\sqrt{g_{22}} = 0\,, \tag{5.117}$$

if it is assumed that the transformation satisfies $g_{12} = 0$, $\kappa_1 = 0$, and $\kappa_2 = 0$, where the latter two parameters are the coordinate line curvatures defined in (4.148)-(4.149).

Exercise 5.7.1 Derive (5.116) by forming the inner product of (5.115) with $\mathbf{x}_\eta^\perp$, then setting $g_{12} = 0$, and $\kappa_1 = 0$. Show that $g_{12} = 0$ requires that $\mathbf{x}_\eta^\perp = -g_{22}\,\mathbf{x}_\xi/\sqrt{g}$ and use this fact to complete the derivation. §

Anderson observes that if the weights ϕ and ψ are replaced by new weights w_1 and w_2 through the relationships

$$\phi = \frac{(w_1)_\xi}{w_1}, \tag{5.118}$$

$$\psi = \frac{(w_2)_\eta}{w_2}, \tag{5.119}$$

the local arc length in each coordinate direction will be proportional to the new weights, i.e., $\sqrt{g_{11}} = \beta_1 w_1$ and $\sqrt{g_{22}} = \beta_2 w_2$ are solutions to (5.116)-(5.117) with β_1 and β_2 constants. Anderson thus proposes to solve (5.115) with weights (5.118)-(5.119) determined from some adaptive weight function [similar to (3.99)]. Contrary to the assumptions in the derivation, the solution to these equations will not, in general, be orthogonal nor will it have zero curvature, so the weight functions do not control the local arc length in as direct a fashion as suggested by (5.116)-(5.117). Nevertheless, this approach to adaptivity has been successful in applications. Extension of the approach to three dimensions is straightforward.

In a variant of this approach, Anderson [8] has proposed to solve

$$\mathcal{Q}_w\mathbf{x} = (g_{22}\,\mathbf{x}_\xi - g_{12}\,\mathbf{x}_\eta)\,\frac{D_\xi}{D} - (g_{12}\,\mathbf{x}_\xi - g_{11}\,\mathbf{x}_\eta)\,\frac{D_\eta}{D} \tag{5.120}$$

to control local cell areas through the weight function D. As is true in the arc-length approach, the area control is not exact since certain properties of the solution are assumed to hold in the derivation when, in fact, they do not hold in general.

Roache, Salari, and Steinberg [159] have suggested a modification of the Anderson approach called the hybrid-adaptive method. The objection is made that when the weighting terms in (5.115) or (5.120) are small, the grid relaxes to the homogeneous Thompson-Thames-Mastin grid (Section 5.4.2), i.e., the Anderson approach adapts grids away from the homogeneous TTM grid. In some applications the homogeneous TTM may not be the most desirable base grid (e.g., the Steger-Sorenson airfoil grids may be preferred). The purpose of the hybrid approach is to permit adaptation away from an arbitrary base grid. This is achieved by replacing the weights in (5.115) by $\phi = \phi_b + \phi_a$ and $\psi = \psi_b + \psi_a$, where the subscript b denotes the base grid from which one wants to adapt away from and the subscript a denotes the original adaptive weight terms in the Anderson method. If the Anderson weights are small, the grid will approach the base grid. The base grid weights are determined by evaluating the left-hand side of (5.115) with a consistent discrete approximation and using the base grid, solving for the weights ϕ_b and ψ_b. The reader is referred to the paper for details of implementation and results.

5.7.2 The Deformation Method

A novel approach to solution adaptivity, known as the deformation method, can be found in recent papers of Liao and Anderson [128] and Liao and Su [126, 125], which are based on the work of Moser and Dacorogna. Although only the planar case is specifically considered in this section, the method holds in arbitrary dimensions. The method controls the local area of the mapping through the use of a weight function. This is accomplished by introducing a vector field whose divergence is the specified weight function. The problem of computing the mapping is reduced to solving this divergence relation and a system of ordinary differential equations, one for each node of the grid. The method is based on the *moving grid identity*, which is now derived.

Suppose the mapping functions are dependent on an additional parameter $0 \leq \tau \leq 1$, so that $\mathbf{x} = \mathbf{x}(\xi, \eta, \tau)$ (this parameter need not represent the time variable). A direct computation using the chain rule on $\mathbf{x}_\tau$ shows that the relation

$$\nabla_{\mathbf{x}} \cdot \mathbf{x}_\tau = \frac{(\sqrt{g})_\tau}{\sqrt{g}} \tag{5.121}$$

holds for arbitrary mappings. Using the fact that $(\sqrt{g})_\tau = \mathbf{x}_\tau \cdot \nabla_{\mathbf{x}}\sqrt{g}$, (5.121) can be expressed as

$$\nabla_{\mathbf{x}} \cdot \frac{\mathbf{x}_\tau}{\sqrt{g}} = 0 \,. \tag{5.122}$$

Either of these relations is referred to as the *moving-grid identity*.

The goal in the deformation method is to produce a C^1 mapping on a domain Ω whose Jacobian agrees with a specified weight function. Let $f = f(x, y)$ be the weight function; f must satisfy $f \in C^1(\Omega)$, $f > 0$ in the domain, $f = 1$ on the boundary of the domain, and finally,

$$\int\!\!\int f \, d\Omega = 1. \tag{5.123}$$

Define

$$D(x, y, \tau) = \tau + (1 - \tau) f \,, \tag{5.124}$$

$h = \sqrt{g}\, D$, and the vector field $\mathbf{v} = D\mathbf{x}_\tau$. If $dh/d\tau = 0$, then h is constant with respect to the parameter τ. In that case $(\sqrt{g}\, D)_{\tau=0} = (\sqrt{g}\, D)_{\tau=1}$, but $D_{\tau=0} = f$ and $D_{\tau=1} = 1$, so that if one assumes $\sqrt{g}_{\tau=0} = 1$, then $\sqrt{g}_{\tau=1} = f$. The function h can be made constant with respect to τ, beginning with the moving-grid identity:

$$\nabla_{\mathbf{x}} \cdot \frac{D\mathbf{x}_\tau}{D\sqrt{g}} = 0 \,,$$

$$\nabla_{\mathbf{x}} \cdot \frac{\mathbf{v}}{h} = 0 \,,$$

$$h \, \nabla_{\mathbf{x}} \cdot \mathbf{v} - \mathbf{v} \cdot \nabla_{\mathbf{x}} h = 0 \,,$$

$$\sqrt{g}\, \nabla_{\mathbf{x}} \cdot \mathbf{v} - \mathbf{x}_\tau \cdot \nabla_{\mathbf{x}} h = 0 \,,$$

$$\frac{\partial h}{\partial \tau} + \sqrt{g}\, \nabla_{\mathbf{x}} \cdot \mathbf{v} - \left[\frac{\partial h}{\partial \tau} + \mathbf{x}_\tau \cdot \nabla_{\mathbf{x}} h \right] = 0 \,,$$

$$\sqrt{g}\, [(1 - f) + \nabla_{\mathbf{x}} \cdot \mathbf{v}] - \frac{dh}{d\tau} = 0 \,.$$

Setting

$$\nabla_{\mathbf{x}} \cdot \mathbf{v} = f - 1 \tag{5.125}$$

results in

$$\frac{dh}{d\tau} = 0. \tag{5.126}$$

The deformation method thus entails solving the divergence relationship (5.125) for the vector field $\mathbf{v}$ and then solving the ordinary differential equations $\mathbf{x}_\tau = \mathbf{v}/D$. The divergence relationship is solved with the boundary condition $\mathbf{v} = \mathbf{0}$ to ensure that points on the boundary do not move. The vector field given by (5.125) is only unique up to the curl of an arbitrary second vector field, so it is possible to obtain many mappings whose Jacobians agree with the specified weight function. Explicit constructions for the vector field are given in Liao and Su [126, 125] when the physical domain is a square or the sector of an annulus. Numerical solution of the divergence equation is greatly facilitated by the introduction of a scalar potential function from which the vector field can be derived by the relation $\mathbf{v} = \nabla\phi$; the scalar potential is then found by solving a Poisson equation with appropriate boundary conditions.

Chapter 6

Variational Planar Grid Generation

6.1 INTRODUCTION

One-dimensional variational grid generation has been discussed in Chapter 3, Sections 3.5 and 3.6. The present chapter extends this to the two-dimensional planar case. Variational methods of grid generation have seen relatively little application in industry, so another goal of this chapter is to convince the reader that variational methods are a worthwhile approach to grid generation. An elementary presentation is made to reach a wide audience. A more advanced treatment of this subject is presented in Chapter 8.

The most widely known theory of variational grid generation is due to Brackbill and Saltzman [21] (see also Brackbill [20], Saltzman and Brackbill [161], and Saltzman [162]). This book favors the approach of Steinberg and Roache [191] due to its more geometric (i.e, intuitive) flavor. In Chapter 8, it is shown that the two theories are basically equivalent. Other references include Roache and Steinberg [155], Castillo, Steinberg, and Roache [27, 29, 30], Jacquotte and Cabello [99], and Jacquotte [101].

The chapter begins by reviewing basic ideas from the Calculus of Variations such as the concept of a functional, its first and second variation, and the Euler-Lagrange equations (Section 6.2). The main section in this chapter (Section 6.3) describes the variational approach to grid generation. Variational principles provide a clear and intuitive means of building grid-generation algorithms. Elementary variational principles provide control of the length of segments in the grid, areas of cells in the grid, and the orthogonality of the angles between grid lines. Subsections describe three elementary functionals named Length, Area, and Orthogonality. Two of the functionals contain weight functions for controlling the grids. Euler-Lagrange equations are derived for all the functionals. Minimization of any one of the elementary unweighted principles does not usually lead to useful grids. Weighted variational principles, leading to weighted grid generators, are often more useful. Combinations of the weighted variational principles have proven effective as well, but these have the defect that the user must choose weight parameters (Section 6.3.4). The AO algorithm (Section 6.3.5) is an effective automatic algorithm that contains no parameters. A numerical algorithm for solving the nonlinear Euler-Lagrange equations of variational grid generation is discussed in Section 6.4, while a discrete approach to variational grid generation, known as the Direct Method (6.5), is briefly surveyed in the last section of this chapter.

6.2 THE CALCULUS OF VARIATIONS

The material on the *calculus of variations* from Chapter 3, Section 3.6, is reviewed and extended to two-dimensional transformations for planar grid generation. The concept of a functional and its first and second variation is introduced and necessary conditions for a minimum are discussed. The Euler-Lagrange equations are derived and shown to be the transformations that minimize the functional. The Euler-Lagrange equations are of interest because they are the partial differential equations (PDEs) that are solved to obtain the grid. Euler-Lagrange equations for constrained, unconstrained, and high-order variational principles are given. Proofs and additional results from the calculus of variations can be found in numerous texts, e.g., Gelfand and Fomin [77].

DEFINITION 6.1 A *functional* is a rule $I[\mathbf{x}]$ that assigns a real number to each vector of functions $\mathbf{x}$ belonging to some specified set of vectors of functions. That is, a functional is a function having a set of vectors of functions for its domain and the real numbers for its range.

A functional is often referred to as a variational principle and the set of vectors of functions, the admissible set. The admissible set in the case of grid generation is generally the set of twice differentiable transformations that satisfy the given Dirichlet boundary data. A major goal in the calculus of variations is to minimize a given functional over an admissible set of vectors of functions. An example of a grid-generation functional has already been encountered in Section 3.6. The example is repeated here and extended to the planar case. Let $G(r,s,t)$ a smooth function of three real variables and $x(\xi)$ a smooth function defined on $[0,1]$ satisfying the boundary conditions $x(0) = a$ and $x(1) = b$, where a and b are some given numbers. Then $I[x]$ defined by

$$I[x] = \int_0^1 G(\xi, x(\xi), x_\xi(\xi))\, d\xi \tag{6.1}$$

is a functional to be minimized over the admissible set satisfying the boundary conditions.

To extend this to planar problems, let $G(r_1, r_2, \ldots, r_8)$ be a given smooth function of eight variables and $x(\xi,\eta)$ and $y(\xi,\eta)$ be smooth functions defined on the unit square U_2 and satisfying the given Dirichlet boundary conditions. If $\mathbf{x} = (x,y)$, then $I[\mathbf{x}] = I[x,y]$ defined by

$$I[\mathbf{x}] = \int_0^1 \int_0^1 G(\xi,\eta,x,y,x_\xi,x_\eta,y_\xi,y_\eta)\, d\xi\, d\eta \tag{6.2}$$

is a functional. The goal is to minimize this functional subject to the boundary data. The functional is frequently written in vector notation in the form

$$I[\mathbf{x}] = \int_{U_2} G(\boldsymbol{\xi}, \mathbf{x}, \mathbf{x}_\xi, \mathbf{x}_\eta)\, d\boldsymbol{\xi}\,. \tag{6.3}$$

Transformations that minimize this functional are referred to as the optimal or minimizing grids for the problem. The goal in variational grid generation, then, is to find grids that best fit the desired grid properties specified by the functional.

To perform the minimization requires some results from the calculus of variations. Recall that in calculus, if a multivariable function has a minimum at a point, then all first directional derivatives of the function are zero at that point and all second directional derivatives of the function are positive or zero at that point. The converse of this result is not true; for the converse to hold it must be assumed that the second derivatives are strictly positive. The analog of these results are fundamental in the calculus of variations. First, the analog of directional derivative must be defined.

DEFINITION 6.2 If $I[\mathbf{x}]$ is a functional defined on a space of vectors of functions and $\mathbf{c}$ is a vector of functions such that $\mathbf{x} + \epsilon\,\mathbf{c}$ is in the given space of vectors of functions for all ϵ, then the *first variation* of the functional I in the direction of $\mathbf{c}$ at the point $\mathbf{x}$ is defined by

$$D_{\mathbf{c}} I[\mathbf{x}] = \frac{d}{d\epsilon} I[\mathbf{x} + \epsilon\,\mathbf{c}]\bigg|_{\epsilon=0}\,. \tag{6.4}$$

THEOREM 6.3 A necessary condition for $I[\mathbf{x}]$ to have an *extremum* at $\hat{\mathbf{x}}$ is that $D_{\mathbf{c}} I[\hat{\mathbf{x}}] = 0$ for all admissible $\mathbf{c}$. §

Exercise 6.2.1 Consult a book on vector calculus that reviews the concept of directional derivative. Verify that if $\mathbf{x}$ and $\mathbf{c}$ are taken to be vectors from a finite-dimensional space, then $D_{\mathbf{c}} I[\mathbf{x}]$ is the directional derivative. §

DEFINITION 6.4 With I, $\mathbf{x}$, and $\mathbf{c}$ as in Theorem 6.3, the *second variation* of a functional I is defined as

$$D_{\mathbf{c}}^2 I[\mathbf{x}] = \frac{d}{d\epsilon} D_{\mathbf{c}} I[\mathbf{x} + \epsilon\,\mathbf{c}]\bigg|_{\epsilon=0}\,. \tag{6.5}$$

THEOREM 6.5 A necessary condition for $I[\mathbf{x}]$ to have a *minimum* at $\hat{\mathbf{x}}$ is that $D_{\mathbf{c}}^2 I[\hat{\mathbf{x}}] \geq 0$ for all $\mathbf{c}$ (for a maximum, $D_{\mathbf{c}}^2 I[\hat{\mathbf{x}}] \leq 0$). §

There are also theorems for sufficient conditions (see Gelfand and Fomin [77]).

When the functional I is given in terms of an integral of a function G applied to a vector of functions $\mathbf{x}$, the necessary condition that the first variation is zero at some point $\mathbf{x}$ results in a boundary value problem for $\mathbf{x}$ (the hat on $\mathbf{x}$ has been dropped). The partial differential equations in the boundary value problem are called the *Euler-Lagrange* equations. The boundary conditions are typically enforced by requiring that $\mathbf{c}$ is zero on the boundary of the underlying space on which the functions $\mathbf{x}$ are defined.

The Euler-Lagrange equations are derived by applying (6.4). As an example, the Euler-Lagrange equation for the functional defined in equation (6.1) is now derived. Suppose x is a smooth function of ξ and a minimum of I. Let $c = c(\xi)$ be a smooth function such that $c(0) = c(1) = 0$. These homogeneous boundary conditions guarantee that if $x = x(\xi)$ satisfies the Dirichlet boundary conditions $x(0) = a$ and $x(1) = b$ then so does $x(\xi) + \epsilon\, c(\xi)$. The first variation of I is given by

$$D_c I[x] = \int_0^1 \left\{ G_x(\xi, x, x_\xi)\, c + G_{x_\xi}(\xi, x, x_\xi)\, c_\xi \right\} d\xi\,, \tag{6.6}$$

where $G_x = G_r$ and $G_{x_\xi} = G_t$ when $G = G(r, s, t)$. Integration by parts (using the boundary conditions on c) gives

$$D_c I[x] = \int_0^1 \left\{ G_x(\xi, x, x_\xi) - \left(G_{x_\xi}(\xi, x, x_\xi) \right)_\xi \right\} c\, d\xi\,. \tag{6.7}$$

For a minimum to occur, $D_c I[x] = 0$ for all c. This can only be true if

$$G_x(\xi, x, x_\xi) - \left(G_{x_\xi}(\xi, x, x_\xi) \right)_\xi = 0\,. \tag{6.8}$$

This is the Euler-Lagrange equation, which is conveniently written in the form

$$\frac{\partial}{\partial x} G(\xi, x, x_\xi) - \frac{d}{d\xi} \frac{\partial}{\partial x_\xi} G(\xi, x, x_\xi) = 0\,, \tag{6.9}$$

where the derivative $d/d\xi$ is called a total derivative. If $G_{x_\xi x_\xi} \neq 0$, then this is a second-order differential equation for x. To see this, apply the chain rule to the total derivative to get

$$G_x - G_{\xi x_\xi} - G_{x x_\xi} x_\xi - G_{x_\xi x_\xi} x_{\xi\xi} = 0\,. \tag{6.10}$$

The set of admissible x provides two boundary conditions for this differential equation.

For the planar functional (6.2) or (6.3), the Euler-Lagrange equations are

$$\frac{\partial G}{\partial x} - \frac{d}{d\xi} \frac{\partial G}{\partial x_\xi} - \frac{d}{d\eta} \frac{\partial G}{\partial x_\eta} = 0\,, \tag{6.11}$$

$$\frac{\partial G}{\partial y} - \frac{d}{d\xi} \frac{\partial G}{\partial y_\xi} - \frac{d}{d\eta} \frac{\partial G}{\partial y_\eta} = 0\,, \tag{6.12}$$

where again $d/d\xi$ and $d/d\eta$ are total derivatives. If the total derivatives are computed using the chain rule, then a system of partial differential equations of the form

$$\mathcal{T}_{11}\, \mathbf{x}_{\xi\xi} + \mathcal{T}_{12}\, \mathbf{x}_{\xi\eta} + \mathcal{T}_{22}\, \mathbf{x}_{\eta\eta} + \mathbf{S} = \mathbf{0} \tag{6.13}$$

results where $\mathcal{T}_{11}$, $\mathcal{T}_{12}$, $\mathcal{T}_{22}$ are 2×2 matrices and $\mathbf{S}$ is a 2×1 vector.

Exercise 6.2.2 Calculate the first variation of (6.2) and verify that the Euler-Lagrange equations (6.11) and (6.12) are correct. Note the role played by the Dirichlet boundary conditions on x and y. Compute the 2×2 matrices in (6.13). §

THEOREM 6.6 If $I[\mathbf{x}]$ is given by a smooth multivariate function G applied to $\mathbf{x}$ and $\mathbf{x}$ is a smooth function and satisfies a Dirichlet boundary condition, then a necessary condition for $I[\mathbf{x}]$ to have an extremum at $\mathbf{x}$ is that $\mathbf{x}$ satisfy the Euler-Lagrange equations associated with I. §

Constrained minimizations are also possible. To minimize the functional $I[\mathbf{x}]$ in (6.2) subject to the constraint $u(\mathbf{x}) = 0$, the Euler-Lagrange equations (6.11)–(6.12) must be modified to read:

$$\frac{\partial G}{\partial x} - \frac{d}{d\xi}\frac{\partial G}{\partial x_\xi} - \frac{d}{d\eta}\frac{\partial G}{\partial x_\eta} - \lambda\frac{\partial u}{\partial x} = 0, \tag{6.14}$$

$$\frac{\partial G}{\partial y} - \frac{d}{d\xi}\frac{\partial G}{\partial y_\xi} - \frac{d}{d\eta}\frac{\partial G}{\partial y_\eta} - \lambda\frac{\partial u}{\partial y} = 0, \tag{6.15}$$

where λ is known as the Lagrange Multiplier (see Gelfand and Fomin [77]).

Functionals having higher order derivatives may also be considered in variational grid generation. For the case involving second-order derivatives,

$$G = G(\boldsymbol{\xi}, \mathbf{x}, \mathbf{x}_\xi, \mathbf{x}_\eta, \mathbf{x}_{\xi\xi}, \mathbf{x}_{\xi\eta}, \mathbf{x}_{\eta\eta}), \tag{6.16}$$

the Euler-Lagrange equations for x and y read

$$\begin{aligned} 0 = {} & \frac{\partial G}{\partial x} - \frac{d}{d\xi}\frac{\partial G}{\partial x_\xi} - \frac{d}{d\eta}\frac{\partial G}{\partial x_\eta} \\ & + \frac{d^2}{d\xi^2}\frac{\partial G}{\partial x_{\xi\xi}} + \frac{d^2}{d\xi d\eta}\frac{\partial G}{\partial x_{\xi\eta}} + \frac{d^2}{d\eta^2}\frac{\partial G}{\partial x_{\eta\eta}}, \end{aligned} \tag{6.17}$$

$$\begin{aligned} 0 = {} & \frac{\partial G}{\partial y} - \frac{d}{d\xi}\frac{\partial G}{\partial y_\xi} - \frac{d}{d\eta}\frac{\partial G}{\partial y_\eta} \\ & + \frac{d^2}{d\xi^2}\frac{\partial G}{\partial y_{\xi\xi}} + \frac{d^2}{d\xi d\eta}\frac{\partial G}{\partial y_{\xi\eta}} + \frac{d^2}{d\eta^2}\frac{\partial G}{\partial y_{\eta\eta}}. \end{aligned} \tag{6.18}$$

A variational principle for the inverse map $\boldsymbol{\xi} = \boldsymbol{\xi}(x, y)$ reads

$$I[\boldsymbol{\xi}] = \int_\Omega \tilde{G}(\mathbf{x}, \boldsymbol{\xi}, \nabla_{\mathbf{x}}\xi, \nabla_{\mathbf{x}}\eta)\, dx\, dy, \tag{6.19}$$

for which the Euler-Lagrange equations are

$$\frac{\partial \tilde{G}}{\partial \xi} - \frac{d}{dx}\frac{\partial \tilde{G}}{\partial \xi_x} - \frac{d}{dy}\frac{\partial \tilde{G}}{\partial \xi_y} = 0, \tag{6.20}$$

$$\frac{\partial \tilde{G}}{\partial \eta} - \frac{d}{dx}\frac{\partial \tilde{G}}{\partial \eta_x} - \frac{d}{dy}\frac{\partial \tilde{G}}{\partial \eta_y} = 0. \tag{6.21}$$

6.2.1 Minimization Theory

For the more mathematically inclined, a short introduction to the *theory* of minimizing functionals is given. To keep this material simple, the variational problem for the hosted equation (2.45),

$$I[f] = \int_\Omega \{(\nabla f) \cdot (\mathcal{T}\, \nabla f) + 2\, g\, f\}\, dx\, dy\,, \tag{6.22}$$

is used as an example. Recall that $f = f(x,y)$ and $g = g(x,y)$ are smooth real-valued functions, with f zero on the boundary of Ω, $\mathcal{T} = \mathcal{T}(x,y)$ a 2×2 real symmetric positive matrix, and $\nabla = \nabla_{\mathbf{x}} = (\partial/\partial x, \partial/\partial y)$.

Exercise 6.2.3 For the functional (6.22), verify that the first variation is given by

$$D_c I[f] = \int_\Omega \{(\nabla c) \cdot (\mathcal{T}\, \nabla f) + (\nabla f) \cdot (\mathcal{T}\, \nabla c) + 2\, g\, c\}\; dx\, dy\,, \tag{6.23}$$

and the second variation is given by

$$D_c^2 I[f] = 2 \int_\Omega \nabla c \cdot \mathcal{T}\, \nabla c\, dx\, dy\,.\ \S \tag{6.24}$$

An integration by parts and the use of symmetry gives

$$D_c I[f] = -2 \int_\Omega (\nabla \cdot \mathcal{T}\, \nabla f - g)\, c\, dx\, dy\,, \tag{6.25}$$

and consequently, the Euler-Lagrange equation for the functional (6.22) is

$$\nabla \cdot \mathcal{T}\, \nabla f - g = 0\,. \tag{6.26}$$

Note that the second variation does not depend on f; this corresponds to a function having a constant second derivative.

The properties of the functional I depend critically on the properties of the matrix $\mathcal{T}$. For the moment, assume that $\mathcal{T}$ is a constant matrix. Next, let $\mathbf{u}$ and $\mathbf{v}$ be two vectors and then define the bilinear form B by

$$B(\mathbf{u}, \mathbf{v}) = \mathbf{u}^T\, \mathcal{T}\, \mathbf{v} \tag{6.27}$$

and the quadratic form Q by

$$Q(\mathbf{u}) = B(\mathbf{u}, \mathbf{u}) = \mathbf{u}^T\, \mathcal{T}\, \mathbf{u}\,. \tag{6.28}$$

Typically, $\mathcal{T}$ is defined to be *positive definite* by requiring that

$$Q(\mathbf{u}) > 0 \quad \text{when} \quad \mathbf{u} \neq 0\,. \tag{6.29}$$

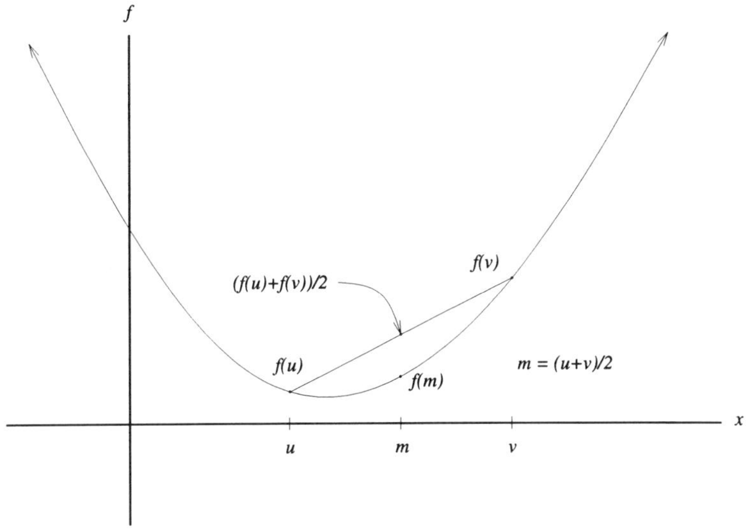

Figure 6.1 *Convex function*

Because $\mathcal{T}$ is symmetric, it can be diagonalized and from this and the positivity of $\mathcal{T}$ it can be seen that

$$c|\mathbf{u}|^2 \le Q(\mathbf{u}) \le C|\mathbf{u}|^2 \tag{6.30}$$

for two positive constants c and C.

In addition, the positivity of $\mathcal{T}$ implies that Q is a *convex function.* Recall that when $f = f(\mathbf{u})$ is a positive real-valued function, then f is (strictly) convex if

$$f\left(\frac{\mathbf{u}+\mathbf{v}}{2}\right) < \frac{f(\mathbf{u}) + f(\mathbf{v})}{2} \quad \text{for} \quad \mathbf{u} \neq \mathbf{v} \tag{6.31}$$

(see Figure 6.1). To see that Q is strictly convex, note that for $\mathbf{u} \neq \mathbf{v}$,

$$0 < Q(\mathbf{u}-\mathbf{v}) = B(\mathbf{u}-\mathbf{v}, \mathbf{u}-\mathbf{v}) = Q(\mathbf{u}) - 2B(\mathbf{u},\mathbf{v}) + Q(\mathbf{v}) \tag{6.32}$$

and consequently

$$2B(\mathbf{u},\mathbf{v}) < Q(\mathbf{u}) + Q(\mathbf{v}) . \tag{6.33}$$

Exercise 6.2.4 Use the previous inequality (6.33) to show that Q is convex. Show that $F(\mathbf{u}) = Q(\mathbf{u}) + L(\mathbf{u}) + C$, where $L(\mathbf{u})$ is linear and C is a constant, is also convex if Q is convex. Also show that both $Q(\mathbf{u})$ and $F(\mathbf{u})$ become infinite as $|\mathbf{u}|$ becomes infinite. Finally, note that F(**u**) is bounded below. §

The results of the previous exercise make it clear that F has at least one minimum. A proof goes as follows: let $m = \inf_{\mathbf{u}} F(\mathbf{u})$. Then there must exist $\mathbf{u}_k$ such that $F(\mathbf{u}_k) \le m + 1/k$, $k > 1$. Because F is large for $\mathbf{u}$ large, the $\mathbf{u}_k$ must be in some bounded set. Bounded sets in finite-dimensional spaces are compact, so there is a subsequence of $\mathbf{u}_k$ that converges to a point $\mathbf{u}$ that is a minimum of F.

Exercise 6.2.5 Use the convexity of F to show that its minimum is unique. §

Most of the previous results can be extended to the functional I. First, if $f = f(x,y)$ and $g = g(x,y)$ are smooth real-valued functions, then let

$$\langle f, g\rangle = \int_\Omega f\,g\,dx\,dy \tag{6.34}$$

be the L^2 inner product and

$$\|f\|^2 = \langle f, f\rangle \tag{6.35}$$

be the L^2 norm. Discussions of inner products and norms can be found in most functional analysis texts and some PDE texts, e.g., Showalter [170]. To study I introduce the bilinear functional

$$\mathcal{B}[f,g] = \int_\Omega B(\nabla f, \nabla g)\,dx\,dy \tag{6.36}$$

and the quadratic form

$$Q[f] = \mathcal{B}[f,f] \tag{6.37}$$

and then note that the functional I given in (6.22) can be written as

$$I[f] = Q[f] + L[f] \tag{6.38}$$

where L is a linear functional. Now, because $\mathcal{T}$ depends on $\mathbf{x}$, it must be assumed that the eigenvalues of $\mathcal{T}$ are bounded below by some positive constant.

Next note that if $\nabla f = 0$ then f is constant. However, only f that are zero on the boundary of Ω are considered here so, in fact, $\nabla f = 0$ implies $f = 0$.

Exercise 6.2.6 Use the previous results to show that I is convex, $I[f]$ goes to infinity as $\|f\|$ goes to infinity, and I is bounded below. §

Unfortunately, the previous argument about the existence of a convergent subsequence is not so elementary in this infinite-dimensional setting. There are two fundamental problems: First, the functional I is not defined for all functions f with finite L^2 norm, and second, bounded sets in infinite-dimensional spaces are not compact in any obvious way.

Exercise 6.2.7 Look up a result (say in Showalter [170]) that implies that I has a minimum and then use the convexity of I to show that the minimum is unique. §

More can be said about I; if $c = c(\mathbf{x})$ is a smooth function that is zero on the boundary of the region Ω, then the Poincaré inequality says that there exists a positive constant k such that

$$k\|c\|^2 \le Q[c] \tag{6.39}$$

(this requires the boundary conditions on c). Use the Poincaré inequality to show that

$$D_c^2 I[f] \ge K\|c\|^2 \tag{6.40}$$

for some positive constant K. As D_c is a directional derivative, it is reasonable to only use c of length one, $\|c\| = 1$. Now it is clear that $D_c^2 I$ is uniformly bounded below, which is just another way of stating that I is strictly convex.

The observation about the second derivative provides another method for showing that a minimum of I is unique.

Exercise 6.2.8 Assume that f and g are minima of I. Let

$$h(\alpha) = I[f + \alpha(g - f)]. \tag{6.41}$$

Show that

$$h'(\alpha) = D_{g-f}I[f + \alpha(g - f)], \tag{6.42}$$

$$h''(\alpha) = D^2_{g-f}I[f + \alpha(g - f)]. \tag{6.43}$$

Then use the facts that $h(0) = h(1)$ and $h''(\alpha) > 0$ and the mean value theorem to show that $f = g$. §

It is important to understand that almost none of the grid-generation functionals are convex, so such an elementary theory as described above is of no help. Liao [124] has studied generalized convexity properties of several functionals. The Liao functional described in Section 8.2.1 is convex but produces poor grids.

6.2.2 Ellipticity

A topic that is closely related to the convexity discussion in the previous section is the notion of *ellipticity* of the Euler-Lagrange equation (6.26). The operator in (6.26) is elliptic provided that

$$\omega^T \mathcal{T} \omega \geq c\,|\omega|^2 \tag{6.44}$$

for all vectors ω and some constant c. This is nothing but the lower half of inequality (6.30). If a partial differential operator P is elliptic then solutions f of $Pf = g$ possess more derivatives than does g (see, for example, Bers, John, and Schecter, p. 135 [18].)

In variational grid generation, it is important to know whether the systems of differential equations comprising the Euler-Lagrange equations are elliptic. A discussion of the theory of elliptic systems is well beyond the scope of this book. Thus, only a method for checking the *ellipticity* of a system of grid-generation equations (defined on a logical space of dimension k) is given. Thus the system of partial differential equations

$$\sum_{i,j=1}^{k} \mathcal{T}_{i,j}\,\mathbf{x}_{\xi_i\xi_j} = F \tag{6.45}$$

(see equation 6.13 for an example) where $\mathcal{T}_{i,j}$ and F depend on $\boldsymbol{\xi}$, $\mathbf{x}$, and $\mathbf{x}_{\boldsymbol{\xi}}$ is elliptic provided that

$$\det\left(\sum_{i,j=1}^{k} \mathcal{T}_{i,j}\,\omega_i\,\omega_j\right) \geq c\,|\omega|^{2k} \tag{6.46}$$

for all $\omega = (\omega_1, \ldots, \omega_k)$, all $\boldsymbol{\xi}$, and all sufficiently smooth $\mathbf{x}$. Later this definition will be used to check the ellipticity of the Euler-Lagrange equations for several functionals.

Exercise 6.2.9 Apply the ellipticity test to the Winslow equations (5.48). §

6.3 VARIATIONAL GRID GENERATION IN THE PLANE

Both the Length and Smoothness grid generators of Chapter 5 are based on the Laplace equation, which has a well-known variational principle. The variational principle for Length is easily seen to be

$$I[\mathbf{x}] = \frac{1}{2}\int_0^1 \int_0^1 (x_\xi^2 + y_\xi^2 + x_\eta^2 + y_\eta^2)\, d\xi\, d\eta; \tag{6.47}$$

this functional has Euler-Lagrange equations in agreement with (5.42). The Smoothness (Winslow or homogeneous TTM) grid generator can be derived from a variational principle in a similar fashion, following the approach of Brackbill and Saltzman. These observations undoubtedly sparked the first thoughts on variational grid generation. The variational approach is appealing because powerful mathematical results have long been available for variational problems in classical mechanics and more recently for problems in nonlinear elasticity. The hope is that by taking the variational approach, grid generation can be given a solid footing and made more powerful; certainly it already has been made more intuitive.

The immediate goal in variational grid generation is not to find the grid, but to find the best grid-generation equation for the stated application. This is particularly important in two and three dimensions, since the question as to what is the best grid generator is not as readily settled as it was for the one-dimensional case discussed in Chapter 3. The variational approach permits the use of one's intuition in selecting meaningful and appropriate variational principles to be minimized. This is done by employing the geometric interpretation of the tangents and metric quantities described in the previous chapters. Recall that (see Table 4.1) $\mathbf{x}_\xi$ is a tangent vector to an ξ coordinate line, $\mathbf{x}_\eta$ is a tangent vector to an η coordinate line, and the elements of the metric matrix (see Section 4.3) can be defined and interpreted as follows: $g_{11} = \mathbf{x}_\xi \cdot \mathbf{x}_\xi$ as the length-squared of a tangent vector to a ξ-coordinate line, $g_{22} = \mathbf{x}_\eta \cdot \mathbf{x}_\eta$ as the length-squared of a tangent vector to a η-coordinate line, and $g_{12} = \mathbf{x}_\xi \cdot \mathbf{x}_\eta$ as the inner product of the two tangent vectors. The determinant of the Jacobian matrix, $\sqrt{g}$, gives the area of the parallelogram whose sides are given by the tangent vectors $\mathbf{x}_\xi$ and $\mathbf{x}_\eta$ (see Exercise 4.1.1). Grid spacing, area, and orthogonality can be controlled by using these metrics in a variational principle. Examples are given in Sections 6.3.1, 6.3.2, and 6.3.3.

Taken individually, none of the Length, Area, or Orthogonality functionals produce quality grids on a wide variety of regions. To overcome this, a functional that is a weighted combination of the individual functionals is introduced in Section 6.3.4. For appropriate choices of the weight parameters, the functional given by the weighted combination produces quality grids on a wider range of regions. Unfortunately, this approach introduces a new problem, namely that of choosing the parameters automatically. This problem is overcome with the AO functional (see Section 6.3.5) which contains no parameters and thus is automatic. To highlight the strengths and weaknesses of each method, it is useful to compare the various unweighted methods described in this section on the domains of the Rogue's Gallery. Since almost any method will work well on a convex domain, most of the test domains are nonconvex.

The *boundary conditions* applied to the variational grid-generation equations are critical. They must be compatible with the functional or the resulting grid should not be expected to perform as designed. Spacing of the grid points on the boundary can be modified by changing the boundary parameterization, but, at this stage, it is not at all clear a priori what boundary conditions are appropriate. For the examples in the Rogue's gallery, no attempt was made to choose a good (or optimal) boundary grid.

The variational procedure produces a solution that is "best" in a *least squares* sense. A good example is that of the Length generator, which can be derived not only from the principle (6.47), but by attempting to satisfy the Cauchy-Riemann equations (5.24) in a least-squares sense by minimizing

$$I[\mathbf{x}] = \frac{1}{2}\int_0^1\int_0^1 (x_\xi - y_\eta)^2 + (x_\eta + y_\xi)^2 \, d\xi \, d\eta \, . \tag{6.48}$$

It is easy to show that the minimizing transformation again satisfies (5.42); thus, solutions to Laplace's equation can be viewed as satisfying the Cauchy-Riemann equations in a least-squares sense. The advantage of the variational approach is that, with arbitrary boundary conditions, existence of solutions to Laplace equation are guaranteed, whereas solutions to the Cauchy-Riemann equations (5.24) are not. The question of specifying appropriate boundary conditions to permit existence of a solution to (5.24) is also avoided. On the other hand, solutions to (5.24) are obtained from (6.48) only if the proper boundary conditions are applied. If the solution to Laplace equations (5.42) are conformal, then the value of the functional is zero. In general, the functional will be positive at the minimum. Incidentally this example also shows that the same transformation can minimize more than one functional.

Given a variational principle, the grid is determined by computing the first variation of the functional, using the formulas in Section 6.2, *The Calculus of Variations.* The resulting Euler-Lagrange equations are the partial differential equations that are solved to determine the grids. In all but one case, the partial differential equations form a coupled system of nonlinear equations (they are, in fact, quasilinear). The boundary conditions for these partial differential equations are the Dirichlet conditions given by requiring that the boundary of the logical region map to the boundary of the physical region. The nonlinearity of the partial differential equations produces many practical and theoretical problems. For arbitrary variational principles, the Euler-Lagrange equations are probably not elliptic, so smoothness of the grids cannot be assumed to be an automatic by-product of this approach.

The results in Chapter 3, Section 3.5, suggest that if one minimizes a functional that is the ratio of the square of a quantity divided by some weight then the solution of the minimization problem will force the quantity to be proportional to the weight. Unfortunately, this useful property generally breaks down in planar grid generation because the desired grid does not exist. Nevertheless, weighted variational principles are useful since they result in weighted grid generators that permit interior control of the grid. To keep the Euler-Lagrange equations simple, this chapter considers only logical-space weight functions. In this case, the functional has no direct dependence on $\mathbf{x}$, so the first terms drop out of the Euler-Lagrange equations (6.11)–(6.12). Discussion of physical-space weight functions is deferred to Chapter 8. The weight functions in this chapter are taken to be $\phi = \phi(\xi, \eta)$ or, in the case of the Length functional, the two weights $\phi(\xi, \eta)$ and $\psi(\xi, \eta)$. Generally, these weights must be positive in order for a minimum of the functional to exist. Section 6.3.6 comments further on the weights.

6.3.1 The Length Functional

The unweighted Length functional has already been stated in (6.47). A weighted version of the Length functional is presented in this subsection, which generalizes the approach given in Chapter 3, Section 3.2. The goal is to find a grid such that the lengths $\sqrt{g_{11}}$ and $\sqrt{g_{22}}$ of the ξ and η coordinate lines are proportional to two given positive logical weight functions $\phi = \phi(\xi, \eta)$ and $\psi = \psi(\xi, \eta)$. Based on the analogy with the one-dimensional case, it is reasonable to minimize the functional

$$I_L[\mathbf{x}] = \frac{1}{2}\int_0^1\int_0^1 \left(\frac{g_{11}}{\phi} + \frac{g_{22}}{\psi}\right) d\xi\, d\eta\,. \tag{6.49}$$

To derive the Euler-Lagrange equations, assume that $\mathbf{c}$ is a smooth vector function that is zero on the boundary of U_2. The first variation of the Length functional (6.49) is

$$D_{\mathbf{c}} I_L[\mathbf{x}] = \int_0^1\int_0^1 \left(\frac{\mathbf{x}_\xi \cdot \mathbf{c}_\xi}{\phi} + \frac{\mathbf{x}_\eta \cdot \mathbf{c}_\eta}{\psi}\right) d\xi\, d\eta. \tag{6.50}$$

Integration by parts leads to

$$D_{\mathbf{c}} I_L[\mathbf{x}] = -\int_0^1\int_0^1 \mathbf{c} \cdot \left\{\left(\frac{\mathbf{x}_\xi}{\phi}\right)_\xi + \left(\frac{\mathbf{x}_\eta}{\psi}\right)_\eta\right\} d\xi\, d\eta. \tag{6.51}$$

The first variation must be zero for all $\mathbf{c}$ if the functional is to be minimized. Consequently,

$$\left(\frac{\mathbf{x}_\xi}{\phi}\right)_\xi + \left(\frac{\mathbf{x}_\eta}{\psi}\right)_\eta = \mathbf{0}\,. \tag{6.52}$$

This is the Euler-Lagrange equation for the weighted Length functional.

The quotient rule for derivatives can be used to put the Euler-Lagrange equations in the form (6.13)

$$\mathcal{T}_{11}\,\mathbf{x}_{\xi\xi} + \mathcal{T}_{12}\,\mathbf{x}_{\xi\eta} + \mathcal{T}_{22}\,\mathbf{x}_{\eta\eta} + \mathbf{S} = \mathbf{0}\,, \tag{6.53}$$

where

$$\begin{aligned} \mathcal{T}_{11} &= \frac{1}{\phi}\mathcal{I} \\ \mathcal{T}_{12} &= 0\,\mathcal{I} \\ \mathcal{T}_{22} &= \frac{1}{\psi}\mathcal{I}\,, \end{aligned} \tag{6.54}$$

where $\mathcal{I}$ is the identity matrix and

$$\mathbf{S} = \begin{bmatrix} -\frac{\phi_\xi}{\phi^2}\,x_\xi - \frac{\psi_\eta}{\psi^2}\,x_\eta \\ -\frac{\phi_\xi}{\phi^2}\,y_\xi - \frac{\psi_\eta}{\psi^2}\,y_\eta \end{bmatrix}. \tag{6.55}$$

Note that for ϕ and ψ being the same constant, this Euler-Lagrange equation reduces to the unweighted Length generator given by Laplace's equations.

Exercise 6.3.1 Verify equations (6.50), (6.51), and (6.53). §

Exercise 6.3.2 Show that for the Length functional, equation (6.53) is elliptic, that is,

$$\det\left(\mathcal{T}_{11}\,\omega_1^2 + \mathcal{T}_{22}\,\omega_2^2\right) \geq c\left(\omega_1^2 + \omega_2^2\right)^2 \tag{6.56}$$

for some positive constant c. §

Observe that the Euler-Lagrange equations for the weighted Length functional are linear and uncoupled; therefore, the solution grid is unique and exists for arbitrary smooth weights and continuous boundary data. Both linearity and the fact that the two equations are uncoupled conspire to make these grid-generation equations particularly easy to solve. Curiously, the second-order part of the differential operator is not the simple Laplacian; the two weight functions have split the operator. Given the assumption of strictly positive weight functions, the weights do not prevent the operator from remaining elliptic; smooth grids can be expected.

One must be careful not to claim too much for this functional. Contrary to expectations, the solution does not necessarily have the property that the grid lengths are proportional to the given weights. For example, if the physical domain and its boundary coincide with the logical domain, then the solution to (6.52) or the system (6.53) is just $x = \xi$, $y = \eta$ for any choice of constant weight functions! The nice property that held in the one-dimensional case simply does not carry over to the planar problem, a fact that poses a significant obstacle for grid generation in two and three dimensions. Of course, the difficulty here is really with the choice of weights as much as with the grid generator. For example, it is fairly clear that the grid implied by the weights $\phi = 2$, $\psi = 1$ does not exist on the unit square. The uniform grid is the least-squares solution to the problem.

Notice that if the solution $\mathbf{x}$ has the properties $\mathbf{x}_\xi = \mathbf{c}_1\,\phi$ and $\mathbf{x}_\eta = \mathbf{c}_2\,\psi$ where $\mathbf{c}_1$ and $\mathbf{c}_2$ are constant vectors, then $\mathbf{x}$ is a solution of the Euler-Lagrange equation. To have the lengths of the grid segments specified by the weights it is only necessary to have $|\mathbf{x}_\xi| = C_1\,\phi$ and $|\mathbf{x}_\eta| = C_2\,\psi$ for some constants C_1 and C_2. However, the condition on the lengths of the tangent vectors is not sufficient to guarantee that the mapping is a solution of the Euler-Lagrange equations. In general, then, it is not expected that solutions of the Euler-Lagrange equations will have the specified lengths.

The performance of the unweighted Length functional is shown in Appendix C. Grids produced by minimizing the Length functional are quite satisfactory on convex domains such as the Square, Trapezoid, and Dome. The solution on the Trapezoid is the same as that obtained by applying the isoparametric map (1.14). Without exception, the unweighted Length functional produces folded grids on the nonconvex domains. Folding on nonconvex domains is the principle drawback of the unweighted Length equations and the main reason Length is not often considered as a standalone grid generator despite its simplicity.

Project 6.3.3 The purpose of this project is to show that the weight functions ϕ and ψ can be chosen so that the weighted Length equations generate an unfolded grid on a nonconvex domain. For example, the unweighted Length functional produces a folded grid on the Annulus region of the Rogue's Gallery. Modify the computer code described in Appendix B.7.1 that solves the unweighted Length equations so that it solves the weighted system (6.53). Generate grids on the Annulus by choosing the ratio of ϕ and ψ to be approximately equal to the ratio of the "length" of the physical region to the "width" of the physical region. See Castillo, Steinberg, and Roache [30] for more information. §

6.3.2 The Area Functional

A more satisfactory generalization of the one-dimensional case is the *Area functional*. The goal here is to find a grid such that the area of the cells in the grid are proportional to some given weight function $\phi = \phi(\xi, \eta) > 0$. Recall that $g = J^2$ and that J is proportional the area of a cell. Based on analogy with the one-dimensional case, it is reasonable to minimize

$$I_A[\mathbf{x}] = \frac{1}{2}\int_0^1\int_0^1 \frac{g}{\phi}\,d\xi\,d\eta = \frac{1}{2}\int_0^1\int_0^1 \frac{J^2}{\phi}\,d\xi\,d\eta\,. \tag{6.57}$$

The Euler-Lagrange equation for the Area functional I_A has many forms; one of the more convenient is

$$\left(\frac{J\,\mathbf{x}_\eta}{\phi}\right)_\xi - \left(\frac{J\,\mathbf{x}_\xi}{\phi}\right)_\eta = \mathbf{0}\,. \tag{6.58}$$

Using the product rule, this equation can be rewritten as

$$\left(\frac{J}{\phi}\right)_\xi \mathbf{x}_\eta - \left(\frac{J}{\phi}\right)_\eta \mathbf{x}_\xi = \mathbf{0} \tag{6.59}$$

and expanded to obtain

$$J_\xi\,\mathbf{x}_\eta - J_\eta\,\mathbf{x}_\xi - J\,\frac{\phi_\xi}{\phi}\,\mathbf{x}_\eta + J\,\frac{\phi_\eta}{\phi}\,\mathbf{x}_\xi = \mathbf{0}\,. \tag{6.60}$$

Carrying out the derivatives, rearranging terms, and interchanging the equations puts the Euler-Lagrange equations in the expanded form (6.13):

$$\mathcal{T}_{11}\,\mathbf{x}_{\xi\xi} + \mathcal{T}_{12}\,\mathbf{x}_{\xi\eta} + \mathcal{T}_{22}\,\mathbf{x}_{\eta\eta} + \mathbf{S} = \mathbf{0}\,, \tag{6.61}$$

where

$$\mathcal{T}_{11} = \begin{bmatrix} y_\eta^2 & -x_\eta\,y_\eta \\ -x_\eta\,y_\eta & x_\eta^2 \end{bmatrix},$$

$$\mathcal{T}_{12} = \begin{bmatrix} -2\,y_\xi\,y_\eta & +(x_\xi\,y_\eta + x_\eta\,y_\xi) \\ +(x_\xi\,y_\eta + x_\eta\,y_\xi) & -2\,x_\xi\,x_\eta \end{bmatrix},$$

$$\mathcal{T}_{22} = \begin{bmatrix} +y_\xi^2 & -x_\xi\,y_\xi \\ -x_\xi\,y_\xi & +x_\xi^2 \end{bmatrix},$$

$$\mathbf{S} = -\frac{J}{\phi}\begin{bmatrix} y_\eta & -y_\xi \\ -x_\eta & x_\xi \end{bmatrix}\begin{bmatrix} \phi_\xi \\ \phi_\eta \end{bmatrix}. \tag{6.62}$$

Notice that, in contrast to the Length functional, the matrices $\mathcal{T}_{ij}$ for the area functional are independent of the weight functions.

Exercise 6.3.4 Check equations (6.61). §

Exercise 6.3.5 Show that the Area-functional Euler-Lagrange equations (6.61) are not elliptic. To do this choose $x(\xi, \eta) = \xi$ and $y(\xi, \eta) = \eta$ and then evaluate $\mathcal{T}_{11}$, $\mathcal{T}_{12}$, and $\mathcal{T}_{22}$. For these special values of these matrices, show that

$$\det\left(\mathcal{T}_{11}\,\omega_1^2 + \mathcal{T}_{12}\,\omega_1\,\omega_2 + \mathcal{T}_{22}\,\omega_2^2\right) = 0 \tag{6.63}$$

and thus the Euler-Lagrange equations cannot be elliptic. §

In contrast to the Length equations, the Area equations are quasilinear, coupled, and nonelliptic. Solutions to the equations may not exist for arbitrary weight functions or arbitrary boundary data. There are, at present, no theorems to tell us if solutions exist or if they are unique. Computational experience with the unweighted Area functional suggests that indeed it is possible to encounter problems having no solution, but if there is a solution, it appears to be unique. The fact that the equations are coupled merely makes the numerical solution algorithm more difficult to implement (see Section 6.4). Lack of ellipticity results in the strong possibility that the Area grids are not smooth.

If the grid satisfying the original goal that area is proportional to the given weight function (i.e., $J = \beta\phi$, with β a constant) exists, it is a solution to the Area equations, boundary conditions permitting. This is most obvious from (6.59). If the weight ϕ is constant, then the areas of the grid cells should be equidistributed.

The grids in the Rogue's Gallery provide insight into the performance of the unweighted Area functional. This generator performs reasonably well on the Trapezoid, Annulus, Horseshoe, Chevron, Dome, and Valley; however, the grids on some of these domains are not smooth. Solution grids were not obtained for almost half of the test domains, including the Swan, Airfoil, Backstep, Plow, and "C." In these cases, the iteration procedure did not converge, presumably because the equal-area solution does not exist on these domains, at least not for the given boundary data.

6.3.3 *The Orthogonality Functional*

The *Orthogonality functional* is based on a different concept than the Length and Area functionals. The latter functionals were derived by trying to make two quantities proportional. The Orthogonality functional is based on trying to make a quantity zero. Recall that

$$g_{12} = |\mathbf{x}_\xi|\,|\mathbf{x}_\eta|\cos(\theta) \tag{6.64}$$

where θ is the angle between the two tangent vectors to the grid lines. If the grid lines are orthogonal; then $g_{12} = 0$. On the other hand, if $g_{12} = 0$ then either the grid lines are orthogonal or the length of at least one of the tangent vectors is zero. If one of the tangent vectors is zero, then the Jacobian is zero, so the assumption that the transformation is a diffeomorphism implies that the transformation is orthogonal if and only if $g_{12} = 0$. To satisfy the orthogonality condition in a least-squares sense, the Orthogonality functional

$$I_O[\mathbf{x}] = \frac{1}{2}\int_0^1\int_0^1 g_{12}^2\,d\xi\,d\eta \tag{6.65}$$

is minimized. If the minimum is zero, then the transformation is orthogonal, otherwise it is the transformation closest to orthogonal in the least-squares sense. No weight function is needed in the Orthogonality principle.

The Euler-Lagrange Equation for this functional is

$$(g_{12}\mathbf{x}_\eta)_\xi + \left(g_{12}\mathbf{x}_\xi\right)_\eta = \mathbf{0}\,. \tag{6.66}$$

Full differentiation of this expression gives the Euler-Lagrange equations in the form (6.13):

$$\mathcal{T}_{11}\,\mathbf{x}_{\xi\xi} + \mathcal{T}_{12}\,\mathbf{x}_{\xi\eta} + \mathcal{T}_{22}\,\mathbf{x}_{\eta\eta} + \mathbf{S} = \mathbf{0}\,. \tag{6.67}$$

where

$$\mathcal{T}_{11} = \begin{bmatrix} x_\eta^2 & x_\eta\,y_\eta \\ x_\eta\,y_\eta & y_\eta^2 \end{bmatrix},$$

$$\mathcal{T}_{12} = \begin{bmatrix} (4\,x_\xi\,x_\eta + 2\,y_\xi\,y_\eta) & (x_\xi\,y_\eta + x_\eta\,y_\xi) \\ (x_\xi\,y_\eta + x_\eta\,y_\xi) & (4\,y_\xi\,y_\eta + 2\,x_\xi\,x_\eta) \end{bmatrix},$$

$$\mathcal{T}_{22} = \begin{bmatrix} x_\xi^2 & x_\xi\,y_\xi \\ x_\xi\,y_\xi & y_\xi^2 \end{bmatrix}, \tag{6.68}$$

and $\mathbf{S} = \mathbf{0}$. The Orthogonality equations are quasilinear, coupled, and nonelliptic.

Orthogonal grids derived as solutions to (6.67) are not given for most of the domains in the Rogue's Gallery. The iterative procedure often failed to converge, presumably because orthogonal solutions do not exist, at least for the given boundary data. In general, orthogonal transformations matching the boundary conditions seldom exist on severely distorted domains such as the ones in the Rogue's Gallery. Grids were obtained only for the Square, the Parallelogram, the Annulus (which gives polar coordinates), and the Valley. In view of these results, the use of Orthogonality as a standalone method for automatically generating grids is not recommended.

The grid generated on the parallelogram is another example of the least-squares nature of the solutions in variational grid generation. The grid generated by the Orthogonality equations in this example is clearly not orthogonal, yet it minimizes the Orthogonality functional by equidistributing the metric g_{12}.

6.3.4 Combinations of Functionals

To overcome the limitations of the individual functionals, it is natural to consider minimizing *combinations* of the Length, Area, and Orthogonality functionals to achieve a compromise between the properties controlled by these functionals. Steinberg and Roache [191] introduced the functional

$$I_W[\mathbf{x}] = w_S I_S + w_A I_A + w_O I_O \,. \tag{6.69}$$

The weight parameters w_S, w_A, and w_O are assumed to be nonnegative constants whose sum is unity. They are used to experimentally adjust the grid by selecting the right compromise between the individual functionals. These weighting parameters should not be confused with the weight functions ϕ and ψ within the individual weighted functionals. The weighted combination of Length and Area, I_{LA}, given by $w_O = 0$ is studied in Castillo et al. [29] with $w_S = 0.1$ and $w_A = 0.9$ recommended as the best values for the weights. This combination often produces smooth, unfolded grids, as demonstrated for the Unit Square, Trapezoid, Annulus, Modified Horseshoe, Chevron, and Dome domains of the Rogue's Gallery given in Appendix C. The Area-Smoothness combination overcomes two important limitations of the individual functionals, namely lack of smoothness for Area and chronic folding for the Length functional. However, the solution to the Area-Smoothness equations fails to exist on the Airfoil, Backstep, Plow, and "C" domains; the grids on the Swan and Valley domains have highly skewed cells. Further experiments with the values of the weight parameters would undoubtedly improve the grids on some of these domains, but it is just such trial and error procedures that prevent robust automatic grid generation.

Exercise 6.3.6 Derive the Euler-Lagrange equations for the weighted combination and show that in general they are quasilinear, coupled, and nonelliptic. §

6.3.5 The AO Functional

A robust automatic generator is obtained by choosing the weight parameters in equation (6.69) to be $w_S = 0$, $w_A = 1/2$, and $w_O = 1/2$. This combination yields the Area-Orthogonality or *AO functional* (Knupp [112]). The name AO derives from the fact that the functional is "halfway" between the equal Area and Orthogonality functionals. The weighted variational principle to be minimized is

$$I_{AO}[\mathbf{x}] = \frac{1}{2}\int_0^1\int_0^1 \frac{g + g_{12}^2}{\phi}\, d\xi\, d\eta = \frac{1}{2}\int_0^1\int_0^1 \frac{g_{11}g_{22}}{\phi}\, d\xi\, d\eta \,. \tag{6.70}$$

The Euler-Lagrange equations take the particularly simple form:

$$\left(\frac{g_{22}\mathbf{x}_\xi}{\phi}\right)_\xi + \left(\frac{g_{11}\mathbf{x}_\eta}{\phi}\right)_\eta = \mathbf{0}\,, \tag{6.71}$$

where g_{11} and g_{22} are defined in (5.49). The expanded equations are as in (6.13):

$$\mathcal{T}_{11}\,\mathbf{x}_{\xi\xi} + \mathcal{T}_{12}\,\mathbf{x}_{\xi\eta} + \mathcal{T}_{22}\,\mathbf{x}_{\eta\eta} + \mathbf{S} = \mathbf{0}\,, \tag{6.72}$$

where

$$\mathcal{T}_{11} = \begin{bmatrix} x_\eta^2 + y_\eta^2 & 0 \\ 0 & x_\eta^2 + y_\eta^2 \end{bmatrix},$$

$$\mathcal{T}_{12} = \begin{bmatrix} 4\, x_\xi\, x_\eta & 2\,(x_\xi\, y_\eta + x_\eta\, y_\xi) \\ 2\,(x_\xi\, y_\eta + x_\eta\, y_\xi) & 4\, y_\xi\, y_\eta \end{bmatrix},$$

$$\mathcal{T}_{22} = \begin{bmatrix} x_\xi^2 + y_\xi^2 & 0 \\ 0 & x_\xi^2 + y_\xi^2 \end{bmatrix},$$

$$\mathbf{S} = -\frac{1}{\phi} \begin{bmatrix} x_\xi & x_\eta \\ y_\xi & y_\eta \end{bmatrix} \begin{bmatrix} g_{22}\, \phi_\xi \\ g_{11}\, \phi_\eta \end{bmatrix}. \tag{6.73}$$

The AO equations are quasilinear and coupled.

Exercise 6.3.7 Use the transformation $x = \xi + \eta$, $y = \epsilon\,(\xi - \eta)$ for sufficiently small ϵ to show that the Euler-Lagrange equations (6.72) are not elliptic. §

Exercise 6.3.8 Show that the sum of the Euler-Lagrange equations for the equal-Area functional (6.61 with $\phi = 1$) and the Orthogonality functional (6.67) give the unweighted AO Euler-Lagrange equations (6.72). §

Exercise 6.3.9 Show that the weighted AO equations can be written in the operator form

$$\mathcal{Q}_{AO}\mathbf{x} = \frac{\phi_\xi}{\phi} g_{22}\mathbf{x}_\xi + \frac{\phi_\eta}{\phi} g_{11}\mathbf{x}_\eta \tag{6.74}$$

where the operator is defined by

$$\mathcal{Q}_{AO}\mathbf{x} = g_{22}\mathbf{x}_{\xi\xi} + 2(\mathbf{x}_\eta \cdot \mathbf{x}_{\xi\eta})\mathbf{x}_\xi + 2(\mathbf{x}_\xi \cdot \mathbf{x}_{\xi\eta})\mathbf{x}_\eta + g_{11}\mathbf{x}_{\eta\eta}\,.\ \S \tag{6.75}$$

As seen in the Rogue's Gallery, unweighted AO grids are smooth, generally nonfolded, have near-uniform areas, and are nearly orthogonal. The latter two properties are not surprising since the functional averages the equal-Area and and Orthogonality functionals. More surprising is the smoothness of the grids since the Length functional is not involved and the operator is nonelliptic.

Unweighted AO grids are not always completely satisfactory; the tendency toward area-uniformity causes cells near the leading edge of the airfoil domain to be nearly the same size as elsewhere in the grid. This is not a useful property of the grid in boundary layer calculations. In principle, this can be repaired using the weighted form of the AO grid generator. Alternatively, the generalized version of the Steger-Sorenson algorithm (Section 5.6.2) can be used with the AO generator.

AO grids can be sensitive to the boundary parameterization (Knupp [112]). This is not too surprising given the close relationship of the AO functional to the Orthogonality principle and corresponding lack of ellipticity on some domains. Grids on the Modified Horseshoe domain (with the same aspect ratios that caused the severe truncation error effects reported in Section 5.4.3) are not folded if generated with AO.

6.3.6 The Reference Grid and the Replication Idea

An effective means of constructing the weight functions ϕ and ψ in the previous variational principles is through the concept of the *reference grid* (Steinberg and Roache [191]). Suppose it is desired to construct a grid on a physical domain Ω such as the one shown in Figure 6.2, and none of the unweighted grid generators produces the desired grid. A weighted form is needed. In the reference grid approach (see Figure 6.2), weight functions are constructed by first choosing a reference domain (in reference space) that closely resembles the physical domain, but on which it is relatively easy to construct a grid having the desired properties. The reference grid is given by a transformation $\mathbf{u} = (u(\xi,\eta), v(\xi,\eta))$ from logical space to reference space. The metrics of the reference grid can be computed from the coordinates: $\tau_{ij} = \mathbf{u}_{\xi_i} \cdot \mathbf{u}_{\xi_j}$ and $\sqrt{\tau} = u_\xi v_\eta - u_\eta v_\xi$. If, for example, the Area functional is used, the weight ϕ is constructed from the reference grid (following the one-dimensional idea of proportionality) by computing the local areas on the reference domain, i.e., one sets $\phi = \sqrt{\tau}$. If the Length functional is used, one uses $\phi = \sqrt{\tau_{11}}$ and $\psi = \sqrt{\tau_{22}}$. For weighted AO, $\phi = \sqrt{\tau_{11}\tau_{22}}$.

Project 6.3.10 Modify the code created in Project 6.3.3 to solve the weighted Length equations (6.53) to obtain a code that constructs weights based on a user-specified reference domain. §

The reference grid approach is explored in Castillo, Steinberg, and Roache [27, 30], and is applied by Yeung and Vaidhyanathan [233]. Although quite powerful, attainment of satisfactory grids by the reference grid approach often requires considerable experimentation on the part of the user. One problem is that of determining a satisfactory parameterization of the boundary of the reference

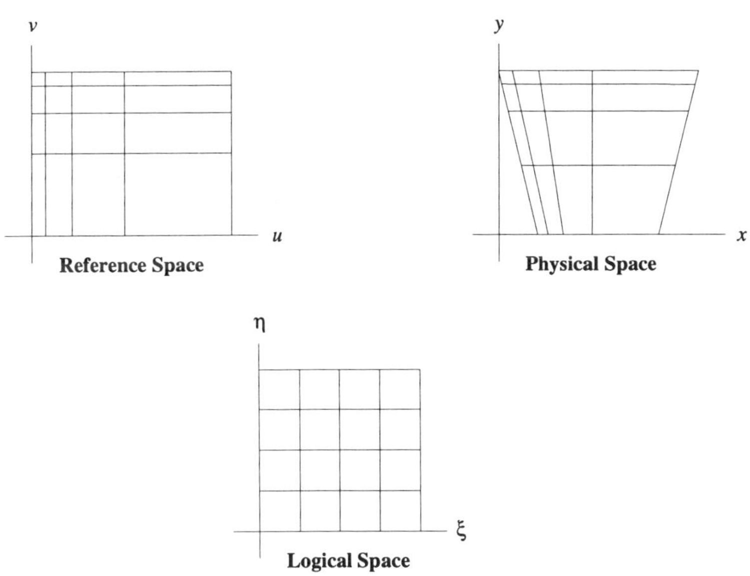

Figure 6.2 *Reference space*

domain. Another is finding reference domains that are close to the physical domain, but for which the desired reference grid can be easily computed.

Closely related to the reference grid concept is the idea of *grid replication.* If the boundary parameterizations of the reference domain and the physical domain coincide, one might reasonably expect the interior grid on the physical domain to be the same as the interior grid on the reference domain. In fact, this is true only for the Area functional. If $\phi = \sqrt{\tau}$ in the Area functional, and the reference and the physical domains coincide, then the Area equations are satisfied by both the physical and reference grids. The physical grid satisfies

$$\left(\frac{J}{\sqrt{\tau}}\right)_\xi \mathbf{x}_\eta - \left(\frac{J}{\sqrt{\tau}}\right)_\eta \mathbf{x}_\xi = \mathbf{0}. \tag{6.76}$$

No matter how the reference grid is constructed, it has $J = \sqrt{\tau}$, so that the Area equations (with weight $\sqrt{\tau}$) are automatically satisfied by the reference grid.

In general, however, replication of the reference grid by the physical grid requires not only coincidence of the domains, but also coincidence of the grid-generation equations, including the weight functions. For example, consider the weighted Length functional, with weights generated using the reference grid approach. The grid on the physical domain satisfies

$$\frac{\mathbf{x}_{\xi\xi}}{\phi} + \frac{\mathbf{x}_{\eta\eta}}{\psi} = \frac{\phi_\xi}{\phi^2}\mathbf{x}_\xi + \frac{\psi_\eta}{\psi^2}\mathbf{x}_\eta \tag{6.77}$$

with $\phi = \sqrt{\tau_{11}}$ and $\psi = \sqrt{\tau_{22}}$. In general, the reference grid need not be generated using the Length equations and thus it will not satisfy equation (6.77) with $\mathbf{x}$ replaced with $\mathbf{u}$. The replication idea is only valid for the Area functional. Thus, even when the reference domain is "close" to the physical domain, the physical grid may not be close to the reference grid unless both grids were generated by similar means.

6.4 NUMERICAL ALGORITHMS FOR VARIATIONAL GENERATORS

To *discretize* the Euler-Lagrange equations for any of the variational generators in the previous section, observe that the expanded forms of the weighted Euler-Lagrange equations (6.53), (6.61), (6.67), and (6.72) can all be written in matrix-vector form as

$$\begin{bmatrix} A_{11} & B_{11} \\ B_{11} & C_{11} \end{bmatrix} \begin{bmatrix} x_{\xi\xi} \\ y_{\xi\xi} \end{bmatrix} +$$

$$\begin{bmatrix} A_{12} & B_{12} \\ B_{12} & C_{12} \end{bmatrix} \begin{bmatrix} x_{\xi\eta} \\ y_{\xi\eta} \end{bmatrix} +$$

$$\begin{bmatrix} A_{22} & B_{22} \\ B_{22} & C_{22} \end{bmatrix} \begin{bmatrix} x_{\eta\eta} \\ y_{\eta\eta} \end{bmatrix} =$$

$$\begin{bmatrix} S_1 \\ S_2 \end{bmatrix} \tag{6.78}$$

where the three coefficient matrices are symmetric and the coefficients $A_{r,s}$, $B_{r,s}$, $C_{r,s}$, and S_r, $1 \le r,\ s \le 2$, generally depend on the first derivatives of the transformation and on the weight function(s) and their derivatives. The equations are nonlinear and coupled, so reasonable choices for a solution algorithm are Picard or Newton iteration. Both methods depend on a starting point for the iteration, that is, it is necessary to already have an initial grid on the region before the iterative algorithm can be applied. Initial grids can be generated using transfinite interpolation or other simple generators. The initial grid need not be of good quality; hopefully, the more sophisticated grid generator will improve the grid quality.

The discretization of the second derivatives is exactly the same as in Section 2.4:

$$\begin{aligned} x_{\xi\xi} &\approx \frac{x_{i-1,j} - 2\,x_{i,j} + x_{i+1,j}}{\Delta\xi^2}\,, \\ x_{\eta\eta} &\approx \frac{x_{i,j-1} - 2\,x_{i,j} + x_{i,j+1}}{\Delta\xi^2}\,, \\ x_{\xi\eta} &\approx \frac{x_{i+1,j+1} - x_{i-1,j+1} - x_{i+1,j-1} + x_{i-1,j-1}}{4\,\Delta\xi\,\Delta\eta}\,, \end{aligned} \tag{6.79}$$

with similar formulas for the y-derivatives. These are the constant coefficient cases of formulas (2.69), (2.73), and (2.79).

The second-derivative discretizations are used to approximate the partial differential equation by a block system of difference equations of the form

$$\begin{aligned} \sum_{|\alpha|,|\beta|\le 1} T_{ij}^{\alpha\beta} x_{i+\alpha,j+\beta} + \sum_{|\alpha|,|\beta|\le 1} U_{ij}^{\alpha\beta} y_{i+\alpha,j+\beta} &= F_{ij}\,, \\ \sum_{|\alpha|,|\beta|\le 1} U_{ij}^{\alpha\beta} x_{i+\alpha,j+\beta} + \sum_{|\alpha|,|\beta|\le 1} V_{ij}^{\alpha\beta} y_{i+\alpha,j+\beta} &= G_{ij}\,. \end{aligned} \tag{6.80}$$

Formulas for the T stencil are

$$\begin{aligned} T_{ij}^{1,0} &= \frac{A_{11}(i,j)}{\Delta\xi^2}\,, \\ T_{ij}^{-1,0} &= \frac{A_{11}(i,j)}{\Delta\xi^2}\,, \\ T_{ij}^{0,1} &= \frac{A_{22}(i,j)}{\Delta\eta^2}\,, \\ T_{ij}^{0,-1} &= \frac{A_{22}(i,j)}{\Delta\eta^2}\,, \\ T_{ij}^{1,1} &= \frac{A_{12}(i,j)}{4\,\Delta\xi\,\Delta\eta}\,, \\ T_{ij}^{1,-1} &= -\frac{A_{12}(i,j)}{4\,\Delta\xi\,\Delta\eta}\,, \\ T_{ij}^{-1,1} &= -\frac{A_{12}(i,j)}{4\,\Delta\xi\,\Delta\eta}\,, \end{aligned}$$

$$T_{ij}^{-1,-1} = \frac{A_{12}(i,j)}{4\,\Delta\xi\,\Delta\eta},$$

$$T_{ij}^{0,0} = -2\left(\frac{A_{11}(i,j)}{\Delta\xi^2} + \frac{A_{22}(i,j)}{\Delta\eta^2}\right). \tag{6.81}$$

Similar formulas hold for the U and V stencils, with the U stencils depending on the $B_{r,s}$ and the V stencils depending on the $C_{r,s}$ coefficients. Also, $F_{ij} = S_1(i,j)$ and $G_{ij} = S_2(i,j)$. Formulas (6.80) and (6.81) hold for all interior points, $1 \le i \le N-1$, $1 \le j \le M-1$; they comprise $2\,(N-1)\,(M-1)$ equations for determining $x_{i,j}$ and $y_{i,j}$. The boundary conditions specify the values of $x_{i,j}$ and $y_{i,j}$ at the boundary points. Consequently, there are $2\,(N-1)\,(M-1)$ unknowns in this problem. This formulation does not need derivatives of the grid coordinates at boundary points of the region.

Variational methods are frequently criticized for their relatively large storage requirements. For the most general planar functionals, there can be as many as twenty-seven stencil arrays and two right-hand side arrays, which must be stored or computed during the iteration. For comparison, the inhomogeneous TTM equation requires eighteen arrays plus the two right-hand side arrays.

The situation is much simpler for the Length functional; only two stencil arrays are needed and the equations are uncoupled. For the AO equations, $C_{11} = A_{11}$, $C_{22} = A_{22}$, and $B_{11} = B_{22} = 0$, so only seventeen stencils need to be stored. The Area and Orthogonality functionals require the full twenty-seven stencils, but the right-hand side arrays are zero for orthogonality.

It is more efficient, however, to store the coefficients $A_{r,s}$, $B_{r,s}$, and $C_{r,s}$ instead of the stencil arrays. Then at most nine arrays need be kept, as opposed to twenty-seven. With this approach, Length again requires two arrays, AO just five, while Area and Orthogonality require nine. The stencils can then be computed from the coefficient arrays as needed, using (6.81).

If the cross-derivative terms in the variational equations are lagged (i.e., included in the right-hand side arrays), additional savings in storage can be obtained. By this means, AO requirements can be reduced to just two coefficient arrays plus the right-hand sides. This is the same storage requirement as in a lagged TTM scheme.

6.5 THE DIRECT OPTIMIZATION METHOD

The *Direct Optimization* method was briefly introduced in Section 3.5. For completeness, a brief overview of the method is made in this chapter on variational methods. The primary authorities on this subject are Castillo [26, 28, 31, 34], Kennon and Dulikravich [107], Pardhanani and Carey [147], Kumar and Kumar [118], Knupp [108, 111], and Barrera et al. [16].

In the continuum variational approach described in the first part of this chapter, a functional is minimized to determine a transformation that is specified by a partial differential equation. It is also possible to minimize functionals by direct methods, i.e, by working with the variational principle itself rather than with the Euler-Lagrange equations (see Dacorogna [42]). The direct approach has not been successfully applied to the continuum variational grid-generation problem as yet, but it has been applied to cases in which grid generation is treated as a discrete optimization problem. In the discrete optimization approach, one minimizes a multivariable function instead of a functional. This results in a discrete optimization problem, as opposed to a problem in the calculus of variations. The Direct Optimization method is sometimes inaccurately referred to as the Direct Variational method.

For an M by N grid, there are $(M-1)(N-1)$ nodes and $2(M-1)(N-1)$ unknowns to be found. Suppose

$$S(\mathbf{x}) = S(\ldots, x_{i,j}, y_{i,j}, \ldots), \quad 1 \le i \le M-1, \quad 1 \le j \le N-1, \tag{6.82}$$

is a positive function from $R^{2(M-1)(N-1)}$ to R. The grid-generation problem is stated as a nonlinear optimization problem (e.g., Fletcher [73] and Minoux [137]): let $(x_{i,j}, y_{i,j})$ be arbitrary points in the plane. Minimize S subject to the constraint that the boundary data is satisfied. The solution grid $\hat{\mathbf{x}}$ must satisfy $S(\hat{\mathbf{x}}) \le S(\mathbf{x})$ for all other grids $\mathbf{x}$.

There are three discrete quantities that are used to make up the grid function S: node-to-node lengths, cell areas or half-areas, and cell angles. Castillo [34] has devised functions named Length, Area, and Orthogonality in analogy to the continuum functionals presented earlier in this chapter. Kennon and Dulikravitch [107] presented an unusual but effective area-like function that appears to have no continuum analog, while Pardhanani and Carey [147] propose alternate Length and Area functions.

An arbitrary discrete functional may not always have a continuum limit. Discrete functions that have a continuum limit are generally preferred, i.e., as the number of nodes is increased, the grid will approach some continuum transformation. Because continuum variational principles may be discretized in different ways, there can be different discrete functions that have the same continuum limit. In the direct method the solution grid satisfies the discrete gradient equations to within a specified tolerance, that is, the solution is essentially exact. However, the resulting discrete grid still only approximates the limiting continuum transformation (assuming there is one) to within some discretization, or truncation, error. As in the continuum variational approach, the grids generally satisfy the underlying geometric conditions in a least-squares, not exact, sense.

Widely studied optimization techniques such as conjugate gradient (Shanno [168]) or the truncated Newton method (Dembo and Steihaug [50]) can be applied to rapidly compute the minimum of S. Such methods require computation of the *gradient*

$$\nabla S = \left(\ldots, \frac{\partial S}{\partial x_{i,j}}, \frac{\partial S}{\partial y_{i,j}}, \ldots\right) \tag{6.83}$$

and often the *Hessian matrix*

$$\mathcal{H} = \begin{pmatrix} \dfrac{\partial^2 S}{\partial x_m \partial x_n} & \dfrac{\partial^2 S}{\partial x_m \partial y_n} \\ \dfrac{\partial^2 S}{\partial y_m \partial x_n} & \dfrac{\partial^2 S}{\partial y_m \partial y_n} \end{pmatrix}. \tag{6.84}$$

For the grid-generation functions, the Hessian is a sparse matrix, so storage requirements are not that severe.

Of course, one can also set the gradient of S to zero (to satisfy the necessary condition for a minimum) and solve the resulting equations by some iterative technique, if the equations are nonlinear, or directly if they are linear. There is little direct evidence of which approach—optimization or solving the linear equations—is best. We suspect that if the fastest optimization techniques are compared to the fastest linear-equation solvers, the contest would be a draw. For those used to numerically solving partial differential equations, the direct method may seem difficult to implement as it makes use of unfamiliar optimization techniques, while those used to solving optimization problems will likely favor the direct approach over the continuum variational method. The best method for a particular user is probably the one with which they are most familiar.

A present advantage of the direct method over the continuum variational formulation is the relative ease with which the properties of the grid on the boundary can be incorporated into the optimization principle (see, for example, the paper on orthogonality by Castillo [28]). Difficulties can occur if the function is not strictly convex, i.e., has local minimae in addition to a global minimum. This actually occurs in the direct method of grid generation when applied to the curve and surface case; in such instances, the problem becomes far more delicate than the planar case (Knupp [111]). The direct method has been extended to weighted and adaptive planar grid generation (Castillo [31]). There are obvious extensions of the method to three dimensions, but no serious work in this direction has yet been performed.

Chapter 7

Tensor Analysis and Transformation Relationships

7.1 INTRODUCTION

Concepts from tensor analysis are introduced in this chapter to assist in transforming hosted equations and set the stage for the advanced treatment of variational grid generation presented in the next three chapters. The reader is assumed to be familiar with classical concepts such as the gradient of a scalar function, the divergence and curl of a vector, and the Laplacian operator. These are extended in Section 7.2 to the gradient of a vector, the divergence and trace of a tensor, the gradient of a tensor, and the tensor product of vectors. These extensions prove quite useful in expressing the transformed operators of hosted equations in compact form. The gradient, divergence, curl, and Laplacian operators are given in general coordinates in terms of transformation matrices such as the Jacobian and its inverse (Section 7.3). As an additional benefit, the transformation relations help tie up a few loose ends left in the previous chapters. Section 7.3.5 derives the Winslow grid-generation equations by inverting the map from physical to logical space, thus completing the discussion in Section 5.4.2. Section 7.3.6 returns to the topic of type invariance of the hosted equation to show that the type of the equation is invariant to a change of coordinates. Section 7.3.7 derives the moving-grid identity mentioned in Section 5.7.2 on the Deformation method and uses it to obtain conservative formulations of time-dependent hosted equations in general coordinates.

7.2 TENSOR ANALYSIS

The basic objects in a grid generator such as scalar, vector, and tensor functions are related through the familiar gradient, divergence, and curl operators and through some nonclassical extensions thereof. Both classical and nonclassical operators are defined in this section; the nonclassical definitions are based on Gurtin [87]. Although this section is basically a catalog of definitions and

Operator	Domain	Range
Trace	Tensor	Scalar
Gradient	Scalar	Vector
Gradient	Vector	Second-Order Tensor
Gradient	Second-Order Tensor	Third-Order Tensor
Divergence	Vector	Scalar
Divergence	Second-Order Tensor	Vector
Curl	Vector	Vector
Laplacian	Scalar	Scalar
Laplacian	Vector	Vector
Tensor Product	Vectors	Tensor

Table 7.1 *Summary of an operator's domain and range*

relationships, some motivation is provided in examples and exercises to relate this section to grid generation. The definitions are also relevant to the transformation of the hosted equations. A large number of relationships can be derived from the basic definitions; an attempt has been made to give only the most useful ones.

Although the definitions and relationships of this chapter hold in quite general settings, only the planar and volume cases are specifically considered since they are relevant to grid generation. The tangent and other vectors reside in an associated vector space $\mathcal{V}$. The term second-order *tensor* is used interchangeably with the term *matrix* to mean any linear transformation $\mathcal{T} : \mathcal{V} \to \mathcal{V}$. Only second- and third-order tensors are needed in this book. For clarification, Table 7.1 summarizes the range and domain of the operators defined in this section.

7.2.1 The Trace of a Tensor

The *trace* of a second-order tensor S is the *scalar*

$$\mathrm{tr}(S) = \sum_j S_{jj}. \tag{7.1}$$

Exercise 7.2.1 Find $\mathrm{tr}(\mathcal{J})$, $\mathrm{tr}(\mathcal{J}^T)$, $\mathrm{tr}(\mathcal{J}^{-1})$, $\mathrm{tr}(\mathcal{G})$, $\mathrm{tr}(\mathcal{G}^{-1})$. §

7.2.2 The Gradient of a Vector

Recall that the gradient ∇f of a scalar $f = f(\xi, \eta)$ is $\nabla_{\boldsymbol{\xi}} f = (f_\xi, f_\eta)$. If $f = f(x(\xi,\eta), y(\xi,\eta))$, then f_ξ and f_η are the covariant components of the gradient vector. The contravariant components are $\nabla_{\mathbf{x}} f = (f_x, f_y)$; the transformation rule between the two sets of components is derived in Section 7.3.1.

The *gradient* can be generalized to operate on vectors. Let $\mathbf{v} \in R^2$ or R^3 be a vector function, then the gradient of this vector is the second-order tensor:

$$[\nabla_{\boldsymbol{\xi}} \mathbf{v}]_{ij} = \frac{\partial v_i}{\partial \xi_j}. \tag{7.2}$$

One of the important applications of this concept in grid generation is $\nabla_\xi \mathbf{x}$, which is readily shown from definition (7.2) to be the Jacobian matrix, i.e., $\mathcal{J} = \nabla_\xi \mathbf{x}$. Only operators with logical-space derivatives will be defined in the rest of this section; because vectors and tensors are independent of the coordinate system, all of the definitions in this section apply with physical-space derivatives as well.

Exercise 7.2.2 Compute $\nabla_\xi \mathbf{x}_\xi$ and $\nabla_\xi \mathbf{x}_\eta$. Show $\mathcal{J}^{-1} = \nabla_\mathbf{x} \xi$. §

7.2.3 The Divergence of a Tensor

The *divergence* of a vector $\mathbf{v} \in R^2$ or R^3 is

$$\text{div}_\xi \cdot \mathbf{v} = \sum_i \frac{\partial v_i}{\partial \xi_i}. \tag{7.3}$$

Exercise 7.2.3 Show that for a vector $\mathbf{v}$,

$$\text{div}_\xi \cdot \mathbf{v} = \text{tr}(\nabla_\xi \mathbf{v}).\ \S \tag{7.4}$$

Exercise 7.2.4 Let $f(\xi, \eta)$ be a scalar function and $\mathbf{v}(\xi, \eta)$ a vector function. Verify the identity

$$\text{div}_\xi \cdot f\,\mathbf{v} = f\,\text{div}_\xi \cdot \mathbf{v} + \mathbf{v} \cdot \nabla_\xi f.\ \S \tag{7.5}$$

The divergence can be generalized to operate on tensors. Let $S = [S]_{ij}$ be a second-order tensor with ℓ rows and m columns ($m = 1, 2$, or, 3), then divergence of this tensor is the *vector* whose ith component ($i = 1, \ldots, \ell$) is

$$(\text{div}_\xi S)_i = \sum_{j=1}^{m} \frac{\partial S_{ij}}{\partial \xi_j}. \tag{7.6}$$

Exercise 7.2.5 Let $f(\xi, \eta)$ be a scalar function. Verify the product rule

$$\text{div}_\xi(fS) = f\text{div}_\xi S + S\nabla_\xi f.\ \S \tag{7.7}$$

The following *divergence identity* is used in the next chapter to derive covariant projections of the Euler-Lagrange equations. For any tensor S and any vector $\mathbf{v}$,

$$(\text{div}_\xi S) \cdot \mathbf{v} = \text{div}_\xi \cdot (S^T \mathbf{v}) - \text{tr}(S^T \nabla_\xi \mathbf{v}). \tag{7.8}$$

Exercise 7.2.6 Prove this identity. §

Two auxiliary matrices

$$C = \sqrt{g}\,(\mathcal{J}^{-1})^T, \qquad C^{-1} = \frac{\mathcal{J}^T}{\sqrt{g}}, \tag{7.9}$$

are introduced to give an example of the use of the divergence of a tensor. In the planar case, these matrices are

$$C = \begin{pmatrix} y_\eta & -y_\xi \\ -x_\eta & x_\xi \end{pmatrix} \tag{7.10}$$

and

$$C^{-1} = \begin{pmatrix} \eta_y & -\eta_x \\ -\xi_y & \xi_x \end{pmatrix}. \tag{7.11}$$

In the surface and volume cases, the auxiliary matrices are more complicated. To show that these auxiliary matrices have some importance of their own, observe that one has directly from (7.10) and (7.11)

$$\mathrm{div}_{\boldsymbol{\xi}} C = \mathbf{0}, \tag{7.12}$$

$$\mathrm{div}_{\mathbf{x}} C^{-1} = \mathbf{0}. \tag{7.13}$$

By observing that $C = \sqrt{g}\mathcal{J}\mathcal{G}^{-1}$, (7.12) is seen to be a restatement of the *metric identity* (4.129):

$$\mathrm{div}_{\boldsymbol{\xi}}(\sqrt{g}\,\mathcal{J}\,\mathcal{G}^{-1}) = \mathbf{0}. \tag{7.14}$$

Exercise 7.2.7 Verify also that

$$\mathrm{div}_{\mathbf{x}}\left(\frac{\mathcal{G}\mathcal{J}^{-1}}{\sqrt{g}}\right) = \mathbf{0}. \tag{7.15}$$

Write this equation explicitly in terms of the physical-space derivatives of the functions $\xi(x, y)$ and $\eta(x, y)$. §

Exercise 7.2.8 Use (7.8) and (7.12)–(7.13) to derive the identities

$$\sqrt{g} = \frac{1}{2}\,\mathrm{div}_{\boldsymbol{\xi}} \cdot C^T\mathbf{x}, \tag{7.16}$$

$$\frac{1}{\sqrt{g}} = \frac{1}{2}\,\mathrm{div}_{\mathbf{x}} \cdot (C^{-1})^T\boldsymbol{\xi}, \tag{7.17}$$

i.e., the area metrics can be expressed as divergences of certain vectors. §

7.2.4 *The Gradient of a Tensor*

Let S be a second order ℓ by m tensor. Then define its *gradient* to be the third-order tensor

$$[\nabla_{\boldsymbol{\xi}} S]^k_{ij} = \frac{\partial S_{ij}}{\partial \xi_k} \tag{7.18}$$

with $k = 1, \ldots, m$. For example, $[\nabla_{\boldsymbol{\xi}}\mathcal{J}]^2_{1,2} = x_{\eta\eta}$. Let $\mathcal{T}$ be an m by m second-order tensor. Then the *contraction* of the gradient of S with $\mathcal{T}$ is the *vector* whose ith component ($i = 1, 2, \ldots, \ell$) is

$$([\nabla_{\boldsymbol{\xi}} S]\mathcal{T})_i = \sum_{j=1}^{m}\sum_{k=1}^{m} \frac{\partial S_{ij}}{\partial \xi_k} T_{jk}. \tag{7.19}$$

Exercise 7.2.9 Show that the Winslow grid operator (5.48) may be expressed as the contraction

$$\mathcal{Q}_w \mathbf{x} = g\,[\nabla_{\boldsymbol{\xi}} \mathcal{J}]\,\mathcal{G}^{-1}\,.\ \S \tag{7.20}$$

Exercise 7.2.10 Let $\mathcal{T} = [T]_{ij}$ be a 2×2 matrix and $\mathcal{J}$ the planar Jacobian matrix. Show that

$$[\nabla_{\boldsymbol{\xi}} \mathcal{J}]\mathcal{T} = T_{11}\,\mathbf{x}_{\xi\xi} + (T_{12} + T_{21})\,\mathbf{x}_{\xi\eta} + T_{22}\,\mathbf{x}_{\eta\eta}\,.\ \S \tag{7.21}$$

Exercise 7.2.11 Verify the product rule

$$\mathrm{div}_{\boldsymbol{\xi}}(S\,\mathcal{T}) = S\,\mathrm{div}_{\boldsymbol{\xi}}\mathcal{T} + [\nabla_{\boldsymbol{\xi}} S]\,\mathcal{T}\,.\ \S \tag{7.22}$$

Exercise 7.2.12 Let $\mathcal{R}$ be an n by ℓ tensor. Verify the vector identity:

$$[\nabla_{\boldsymbol{\xi}}(\mathcal{R}\,S)]\,\mathcal{T} = \mathcal{R}\,([\nabla_{\boldsymbol{\xi}} S]\,\mathcal{T}) + [\nabla_{\boldsymbol{\xi}} R](S\,\mathcal{T}). \tag{7.23}$$

How many components do the vectors in this identity have? §

Exercise 7.2.13 Recall that in the planar case,

$$\frac{1}{\sqrt{g}} = \xi_x\,\eta_y - \xi_y\,\eta_x\,. \tag{7.24}$$

Use (7.22) to verify the contractions

$$\nabla_{\boldsymbol{\xi}}\sqrt{g} = [\nabla_{\boldsymbol{\xi}} \mathcal{J}^T]C\,, \tag{7.25}$$

$$\nabla_{\mathbf{x}}\frac{1}{\sqrt{g}} = [\nabla_{\mathbf{x}}(\mathcal{J}^{-1})^T]C^{-1}\,.\ \S \tag{7.26}$$

Exercise 7.2.14 Use (7.22), (7.12), and (7.13) to show

$$[\nabla_{\boldsymbol{\xi}} C^{-1}]C = \mathbf{0}\,, \tag{7.27}$$

$$[\nabla_{\mathbf{x}} C]C^{-1} = \mathbf{0}\,.\ \S \tag{7.28}$$

7.2.5 *Curl and Laplacian*

The classical definitions of curl and Laplacian are all that are needed in this book. The curl of a two-component vector $\mathbf{v} = (u, v)$ is the vector $\mathrm{curl}_{\boldsymbol{\xi}}\mathbf{v} = (v_\xi - u_\eta)\,\hat{\mathbf{k}}$, while the *curl* of a three-component vector $\mathbf{v} = (u, v, w)$ is the vector

$$\mathrm{curl}_{\boldsymbol{\xi}}\mathbf{v} = (w_\eta - v_\zeta,\ u_\zeta - w_\xi,\ v_\xi - u_\eta)\,. \tag{7.29}$$

These components are obtained from the definition $\mathrm{curl}_{\boldsymbol{\xi}}\mathbf{v} = \nabla_{\boldsymbol{\xi}} \times \mathbf{v}$, provided that in two dimensions $\mathbf{v} = (u, v, 0)$ and

$$\nabla_{\boldsymbol{\xi}} = \left(\frac{\partial}{\partial \xi}, \frac{\partial}{\partial \eta}, 0\right)\,. \tag{7.30}$$

Exercise 7.2.15 Show that for a three-component vector, $\text{div}_{\boldsymbol{\xi}} \cdot (\text{curl}_{\boldsymbol{\xi}} \mathbf{v}) = 0$. Show that $\text{curl}_{\boldsymbol{\xi}}(\nabla_{\boldsymbol{\xi}} f) = \mathbf{0}$ holds in both planar and three-dimensional cases. §

The familiar scalar *Laplacian* or Laplace operator is

$$\nabla_{\boldsymbol{\xi}}^2 f = \text{div}_{\boldsymbol{\xi}} \cdot \nabla_{\boldsymbol{\xi}} f = f_{\xi\xi} + f_{\eta\eta}. \tag{7.31}$$

This is most easily generalized to the *vector* Laplacian using the definition of the divergence of a tensor:

$$\nabla_{\boldsymbol{\xi}}^2 \mathbf{v} = \text{div}_{\boldsymbol{\xi}}(\nabla_{\boldsymbol{\xi}} \mathbf{v}) = \mathbf{v}_{\xi\xi} + \mathbf{v}_{\eta\eta}. \tag{7.32}$$

Exercise 7.2.16 Show that the divergence of the 2×2 Jacobian matrix is the Laplacian of $\mathbf{x}$. §

7.2.6 The Tensor Product

Given two vectors $\mathbf{a} \in R^M$ and $\mathbf{b} \in R^N$, the *tensor product* $(\mathbf{a} \otimes \mathbf{b})$ is the tensor

$$[\mathbf{a} \otimes \mathbf{b}]_{ij} = a_i b_j. \tag{7.33}$$

It is clear from this definition that $(\mathbf{b} \otimes \mathbf{a}) = (\mathbf{a} \otimes \mathbf{b})^T$. Some important examples of the tensor product that occur in planar variational grid generation (see Chapter 8) are given in the next exercise.

Exercise 7.2.17 Verify the following tensor products:

$$\mathbf{x}_\xi \otimes \mathbf{x}_\xi = \begin{pmatrix} x_\xi^2 & x_\xi y_\xi \\ x_\xi y_\xi & y_\xi^2 \end{pmatrix}, \tag{7.34}$$

$$(\mathbf{x}_\xi \otimes \mathbf{x}_\eta) + (\mathbf{x}_\eta \otimes \mathbf{x}_\xi) = \begin{pmatrix} 2x_\xi x_\eta & x_\xi y_\eta + x_\eta y_\xi \\ x_\xi y_\eta + x_\eta y_\xi & 2y_\xi y_\eta \end{pmatrix}, \tag{7.35}$$

$$\mathbf{x}_\eta \otimes \mathbf{x}_\eta = \begin{pmatrix} x_\eta^2 & x_\eta y_\eta \\ x_\eta y_\eta & y_\eta^2 \end{pmatrix}. \ \S \tag{7.36}$$

Exercise 7.2.18 Show that the rank of $\mathbf{a} \otimes \mathbf{b}$ is one. Show that for any vector $\mathbf{v} \in R^N$,

$$(\mathbf{a} \otimes \mathbf{b})\, \mathbf{v} = (\mathbf{b} \cdot \mathbf{v})\, \mathbf{a}. \tag{7.37}$$

Use this property to verify that $\mathbf{x}_\xi$ is an eigenvector of $\mathbf{x}_\xi \otimes \mathbf{x}_\xi$. Verify some of the eigenvectors and other entries given in Table 7.2. §

Exercise 7.2.19 Use (4.126), (4.127), (4.146), and (7.37) to prove the following identity:

$$(\sqrt{g})_\xi \mathbf{x}_\xi = \frac{1}{\sqrt{g}}[g_{11}(\mathbf{x}_\xi \otimes \mathbf{x}_\eta) - g_{12}(\mathbf{x}_\xi \otimes \mathbf{x}_\xi)]\, \mathbf{x}_{\xi\eta}$$

$$+ \frac{1}{\sqrt{g}}[g_{22}(\mathbf{x}_\xi \otimes \mathbf{x}_\xi) - g_{12}(\mathbf{x}_\xi \otimes \mathbf{x}_\eta)]\, \mathbf{x}_{\xi\xi}. \tag{7.38}$$

Tensor	Det.	Tr.	Eigenvector	Eigenvalue	Null Vector
$\mathbf{x}_\xi \otimes \mathbf{x}_\xi$	0	g_{11}	$\mathbf{x}_\xi$	g_{11}	$\mathbf{x}_\xi^\perp$
$\mathbf{x}_\xi \otimes \mathbf{x}_\eta$	0	g_{12}	$\mathbf{x}_\xi$	g_{12}	$\mathbf{x}_\eta^\perp$
$\mathbf{x}_\eta \otimes \mathbf{x}_\xi$	0	g_{12}	$\mathbf{x}_\eta$	g_{12}	$\mathbf{x}_\xi^\perp$
$\mathbf{x}_\eta \otimes \mathbf{x}_\eta$	0	g_{22}	$\mathbf{x}_\eta$	g_{22}	$\mathbf{x}_\eta^\perp$
$(\mathbf{x}_\xi \otimes \mathbf{x}_\eta)+$ $(\mathbf{x}_\eta \otimes \mathbf{x}_\xi)$	$-g$	$2g_{12}$	$\sqrt{g_{22}}\,\mathbf{x}_\xi \pm$ $\sqrt{g_{11}}\,\mathbf{x}_\eta$	$g_{12} \pm$ $\sqrt{g_{11}\,g_{22}}$	$\mathbf{0}$

Table 7.2 *Covariant tangent tensor product properties*

Note how the tensor product property (7.37) can be used to ferret out the second derivatives hidden in the rate-of-change metrics. This identity and ones involving $(\sqrt{g})_\xi\, \mathbf{x}_\eta$, $(\sqrt{g})_\eta\, \mathbf{x}_\xi$, $(\sqrt{g})_\eta\, \mathbf{x}_\eta$ are used in the derivation in Section 8.3.5. §

Exercise 7.2.20 Let f be a scalar function and $\mathbf{v}$ a vector function. Verify the product identity:

$$\nabla_{\boldsymbol{\xi}}(f\,\mathbf{v}) = f\,\nabla_{\boldsymbol{\xi}}\mathbf{v} + \mathbf{v} \otimes \nabla_{\boldsymbol{\xi}} f\,. \text{ §} \tag{7.39}$$

Exercise 7.2.21 Use (7.37) to show that for any four vectors $\mathbf{a}, \mathbf{b}, \mathbf{p}, \mathbf{q} \in R^N$,

$$\mathbf{a} \cdot \{(\mathbf{p} \otimes \mathbf{q})\,\mathbf{b}\} = (\mathbf{a} \cdot \mathbf{p})\,(\mathbf{q} \cdot \mathbf{b})\,, \tag{7.40}$$

$$(\mathbf{a} + \mathbf{b}) \otimes (\mathbf{p} + \mathbf{q}) = (\mathbf{a} \otimes \mathbf{p}) + (\mathbf{a} \otimes \mathbf{q}) + (\mathbf{b} \otimes \mathbf{p}) + (\mathbf{b} \otimes \mathbf{q})\,, \tag{7.41}$$

$$[(\mathbf{a} \otimes \mathbf{p})\,\mathbf{q}] \otimes \mathbf{b} = (\mathbf{a} \otimes \mathbf{p})\,(\mathbf{q} \otimes \mathbf{b})\,. \text{ §} \tag{7.42}$$

Exercise 7.2.22 Recall that $\mathcal{G} = \mathcal{J}^T\mathcal{J}$ and $\mathcal{G}^{-1} = \mathcal{J}^{-1}(\mathcal{J}^{-1})^T$. Verify that the product of the Jacobian matrices in opposite order can be expressed in terms of the following tensor products:

$$g\,(\mathcal{J}^{-1})^T\,\mathcal{J}^{-1} = (\mathbf{x}_\xi^\perp \otimes \mathbf{x}_\xi^\perp) + (\mathbf{x}_\eta^\perp \otimes \mathbf{x}_\eta^\perp)\,, \tag{7.43}$$

$$\mathcal{J}\,\mathcal{J}^T = (\mathbf{x}_\xi \otimes \mathbf{x}_\xi) + (\mathbf{x}_\eta \otimes \mathbf{x}_\eta)\,. \tag{7.44}$$

These equations also show that $\{(\mathbf{x}_\xi \otimes \mathbf{x}_\xi) + (\mathbf{x}_\eta \otimes \mathbf{x}_\eta)\}/\sqrt{g}$ is the inverse of $\{(\mathbf{x}_\xi^\perp \otimes \mathbf{x}_\xi^\perp) + (\mathbf{x}_\eta^\perp \otimes \mathbf{x}_\eta^\perp)\}/\sqrt{g}$. §

Exercise 7.2.23 Prove the first of the following identities:

$$\begin{aligned} g\,(\mathbf{x}_\xi \otimes \mathbf{x}_\xi) &= g_{11}^2\,(\mathbf{x}_\eta^\perp \otimes \mathbf{x}_\eta^\perp) + g_{12}^2\,(\mathbf{x}_\xi^\perp \otimes \mathbf{x}_\xi^\perp) \\ &\quad - g_{12}\,g_{11}\,[(\mathbf{x}_\xi^\perp \otimes \mathbf{x}_\eta^\perp) + (\mathbf{x}_\eta^\perp \otimes \mathbf{x}_\xi^\perp)]\,, \end{aligned} \tag{7.45}$$

$$\begin{aligned} g\,[(\mathbf{x}_\xi \otimes \mathbf{x}_\eta) + (\mathbf{x}_\eta \otimes \mathbf{x}_\xi)] &= 2\,g_{11}\,g_{12}\,(\mathbf{x}_\eta^\perp \otimes \mathbf{x}_\eta^\perp) + 2\,g_{22}\,g_{12}\,(\mathbf{x}_\xi^\perp \otimes \mathbf{x}_\xi^\perp) \\ &\quad - (g_{11}\,g_{22} + g_{12}^2)[(\mathbf{x}_\xi^\perp \otimes \mathbf{x}_\eta^\perp) + (\mathbf{x}_\eta^\perp \otimes \mathbf{x}_\xi^\perp)]\,, \end{aligned} \tag{7.46}$$

$$g\,(\mathbf{x}_\eta \otimes \mathbf{x}_\eta) = g_{22}^2\,(\mathbf{x}_\xi^\perp \otimes \mathbf{x}_\xi^\perp) + g_{12}^2\,(\mathbf{x}_\eta^\perp \otimes \mathbf{x}_\eta^\perp) - g_{12}\,g_{22}\,[(\mathbf{x}_\xi^\perp \otimes \mathbf{x}_\eta^\perp) + (\mathbf{x}_\eta^\perp \otimes \mathbf{x}_\xi^\perp)]\,, \tag{7.47}$$

$$g\,\mathcal{I} = g_{11}\,(\mathbf{x}_\eta^\perp \otimes \mathbf{x}_\eta^\perp) + g_{22}\,(\mathbf{x}_\xi^\perp \otimes \mathbf{x}_\xi^\perp) - g_{12}\,[(\mathbf{x}_\xi^\perp \otimes \mathbf{x}_\eta^\perp) + (\mathbf{x}_\eta^\perp \otimes \mathbf{x}_\xi^\perp)]\,, \tag{7.48}$$

by applying the relation

$$\sqrt{g}\,\mathbf{x}_\xi = g_{12}\,\mathbf{x}_\xi^\perp - g_{11}\,\mathbf{x}_\eta^\perp \tag{7.49}$$

together with (7.41). The symbol $\mathcal{I}$ is used in this book to denote the identity matrix. These identities convert tensor products of covariant vectors to tensor products of contravariant vectors. Identities between mixed covariant and contravariant forms also exist. §

7.3 TRANSFORMATION RELATIONS

Tensor analysis is the natural language for stating the gradient, divergence, curl, and Laplacian operators that occur in the hosted equations in general coordinates. Detailed derivations of these relationships are too lengthy to include here, but readers should verify a few of them on their own. Conservative and nonconservative forms are given for each of the transformed operators. Relationships can be given either in terms of $\mathcal{J}$ or the auxiliary matrix C; the latter is generally preferred due to the identity (7.12), but both forms are useful.

7.3.1 Gradient

Gradient of a Scalar Function

Expressions for the gradient in general coordinates were given in Chapter 2, equations (2.50); these were derived using the chain rule. The relations can be compactly expressed in terms of the auxiliary matrices defined in the previous section:

$$\nabla_{\mathbf{x}} f = \frac{1}{\sqrt{g}}\,C\,\nabla_{\boldsymbol{\xi}} f \tag{7.50}$$

or, inverting,

$$\nabla_{\boldsymbol{\xi}} f = \sqrt{g}\,C^{-1}\,\nabla_{\mathbf{x}} f\,. \tag{7.51}$$

These are the *nonconservative* forms of the gradient transformation.

To obtain the conservative forms (2.51), apply the identities (7.12) and (7.13) along with the product rule (7.7) to find

$$\mathrm{div}_{\boldsymbol{\xi}} C\,f = C\nabla_{\boldsymbol{\xi}} f\,, \tag{7.52}$$

$$\mathrm{div}_{\mathbf{x}} C^{-1}\,f = C^{-1}\nabla_{\mathbf{x}} f\,. \tag{7.53}$$

Thus, the gradient of $f(x,y)$ in *conservative* form is the divergence of a tensor:

$$\nabla_{\mathbf{x}} f = \frac{1}{\sqrt{g}} \operatorname{div}_{\boldsymbol{\xi}}(Cf). \tag{7.54}$$

Inverting:

$$\nabla_{\boldsymbol{\xi}} f = \sqrt{g} \operatorname{div}_{\mathbf{x}}(C^{-1}f). \tag{7.55}$$

The Normal Derivative

Hosted equations, such as (2.45), are often solved with Neumann boundary conditions, stated in terms of the normal derivative. By definition, the normal derivative of $f(\mathbf{x})$ is

$$\frac{\partial f}{\partial \mathbf{n}} = \mathbf{n} \cdot \nabla_{\mathbf{x}} f, \tag{7.56}$$

where $\mathbf{n}$ is a unit normal to a given surface in physical space. In general coordinates, this is

$$\frac{\partial f}{\partial \mathbf{n}} = \frac{\mathbf{n}}{\sqrt{g}} \cdot C \, \nabla_{\boldsymbol{\xi}} f. \tag{7.57}$$

If the surface forms a boundary of the domain, then one of the logical space coordinates is constant on the surface. In two dimensions, for example,

$$\mathbf{n} = \begin{cases} -\dfrac{\mathbf{x}_\eta^\perp}{\sqrt{g_{22}}}, & \xi \text{ constant} \\ \dfrac{\mathbf{x}_\xi^\perp}{\sqrt{g_{11}}}, & \eta \text{ constant} \end{cases} \tag{7.58}$$

One may derive from these the following expressions for the normal derivative in general coordinates in two dimensions:

$$\left.\frac{\partial f}{\partial \mathbf{n}}\right|_{\xi=c} = \frac{1}{\sqrt{g\, g_{22}}} \{g_{22}\, f_\xi - g_{12}\, f_\eta\}, \tag{7.59}$$

$$\left.\frac{\partial f}{\partial \mathbf{n}}\right|_{\eta=c} = \frac{1}{\sqrt{g\, g_{11}}} \{g_{11}\, f_\eta - g_{12}\, f_\xi\}. \tag{7.60}$$

Gradient of a Vector Function

Let $\mathbf{v}$ be a vector function. Then the transformed gradient of the vector function has the nonconservative form:

$$\nabla_{\mathbf{x}} \mathbf{v} = (\nabla_{\boldsymbol{\xi}} \mathbf{v})(\nabla_{\mathbf{x}} \boldsymbol{\xi}) = \frac{1}{\sqrt{g}} (\nabla_{\boldsymbol{\xi}} \mathbf{v}) \, C^T. \tag{7.61}$$

Inverting:

$$\nabla_{\boldsymbol{\xi}} \mathbf{v} = (\nabla_x \mathbf{v})(\nabla_{\boldsymbol{\xi}} \mathbf{x}) = \sqrt{g} \, (\nabla_{\mathbf{x}} \mathbf{v}) \, (C^{-1})^T. \tag{7.62}$$

The conservative form of these transformation rules requires third-order tensors.

Exercise 7.3.1 Define a matrix-vector outer product and the divergence of a third-order tensor so that the following conservative transformation rules for the gradient of a vector make sense:

$$\nabla_{\boldsymbol{\xi}}\mathbf{v} = \sqrt{g}\,\mathrm{div}_{\mathbf{x}}(\mathbf{v} \otimes C^{-1}), \tag{7.63}$$

$$\nabla_{\mathbf{x}}\mathbf{v} = \frac{1}{\sqrt{g}}\,\mathrm{div}_{\boldsymbol{\xi}}(\mathbf{v} \otimes C). \tag{7.64}$$

As an obvious corollary, note that the Jacobian matrix can be expressed as the divergence of a third-order tensor:

$$\mathcal{J} = \sqrt{g}\,\mathrm{div}_{\mathbf{x}}(\mathbf{x} \otimes C^{-1}), \tag{7.65}$$

$$\mathcal{J}^{-1} = \frac{1}{\sqrt{g}}\,\mathrm{div}_{\boldsymbol{\xi}}(\mathbf{x} \otimes C). \ \S \tag{7.66}$$

Exercise 7.3.2 Show that the second-derivatives of a transformation have the following conservative transformation rule:

$$\nabla_{\boldsymbol{\xi}}(\nabla_{\boldsymbol{\xi}}f) = \sqrt{g}\,\mathrm{div}_{\mathbf{x}}(\sqrt{g}\,C^{-1}\,\nabla_{\mathbf{x}}f \otimes C^{-1}), \tag{7.67}$$

$$\nabla_{\mathbf{x}}(\nabla_{\mathbf{x}}f) = \frac{1}{\sqrt{g}}\,\mathrm{div}_{\boldsymbol{\xi}}\left(\frac{C}{\sqrt{g}}\,\nabla_{\boldsymbol{\xi}}f \otimes C\right). \tag{7.68}$$

Compare (7.68) to (2.53)–(2.56) with α, β, and $\gamma = 1$. §

7.3.2 Divergence

Divergence of a Vector Function

In *conservative* form:

$$\mathrm{div}_{\mathbf{x}} \cdot \mathbf{v} = \frac{1}{\sqrt{g}}\,\mathrm{div}_{\boldsymbol{\xi}} \cdot (C^T\mathbf{v}). \tag{7.69}$$

Inverting:

$$\mathrm{div}_{\boldsymbol{\xi}} \cdot \mathbf{v} = \sqrt{g}\,\mathrm{div}_{\mathbf{x}} \cdot \{(C^{-1})^T\mathbf{v}\}. \tag{7.70}$$

Nonconservative forms can be expressed in terms of the trace:

$$\mathrm{div}_{\mathbf{x}} \cdot \mathbf{v} = \frac{1}{\sqrt{g}}\,\mathrm{tr}\{(\nabla_{\boldsymbol{\xi}}\mathbf{v})C^T\}, \tag{7.71}$$

$$\mathrm{div}_{\boldsymbol{\xi}} \cdot \mathbf{v} = \sqrt{g}\,\mathrm{tr}\{(\nabla_{\mathbf{x}}\mathbf{v})(C^{-1})^T\}. \tag{7.72}$$

Divergence of a Tensor Function

Let $\mathcal{T}$ be a second-order tensor function. The *conservative* form is

$$\mathrm{div}_{\mathbf{x}}\mathcal{T} = \frac{1}{\sqrt{g}}\,\mathrm{div}_{\boldsymbol{\xi}}(\mathcal{T}C). \tag{7.73}$$

Inverting:

$$\mathrm{div}_{\boldsymbol{\xi}}\mathcal{T} = \sqrt{g}\,\mathrm{div}_{\mathbf{x}}(\mathcal{T}C^{-1}). \tag{7.74}$$

Nonconservative forms are

$$\mathrm{div}_{\mathbf{x}}\mathcal{T} = \frac{1}{\sqrt{g}}[\nabla_{\boldsymbol{\xi}}\mathcal{T}]C, \tag{7.75}$$

$$\mathrm{div}_{\boldsymbol{\xi}}\mathcal{T} = \sqrt{g}\,[\nabla_{\mathbf{x}}\mathcal{T}]C^{-1}. \tag{7.76}$$

Exercise 7.3.3 Derive the transformation rule for a contraction using (7.22), (7.75) or (7.76), and (7.23):

$$[\nabla_{\boldsymbol{\xi}}S]\mathcal{T} = \sqrt{g}\,[\nabla_{\mathbf{x}}S](\mathcal{T}C^{-1}), \tag{7.77}$$

$$[\nabla_{\mathbf{x}}S]\mathcal{T} = \frac{1}{\sqrt{g}}[\nabla_{\boldsymbol{\xi}}S](\mathcal{T}C).\ \S \tag{7.78}$$

7.3.3 Curl

For $\mathbf{v} \in R^3$, define the tensor $\mathcal{T} = [\mathbf{e}_1 \times \mathbf{v}, \mathbf{e}_2 \times \mathbf{v}, \mathbf{e}_3 \times \mathbf{v}]$. Then $\mathrm{curl}_{\boldsymbol{\xi}}\mathbf{v} = \mathrm{div}_{\boldsymbol{\xi}}\mathcal{T}$; this permits use of the transformation rule for the divergence of a tensor and the rule for the divergence of a product of tensors to obtain the *nonconservative* transformation rules:

$$\mathrm{curl}_{\mathbf{x}}\mathbf{v} = [\nabla_{\boldsymbol{\xi}}\mathcal{T}](\mathcal{J}^{-1})^T, \tag{7.79}$$

$$\mathrm{curl}_{\boldsymbol{\xi}}\mathbf{v} = [\nabla_{\mathbf{x}}\mathcal{T}]\mathcal{J}^T, \tag{7.80}$$

and the *conservative* rules

$$\mathrm{curl}_{\mathbf{x}}\mathbf{v} = \frac{1}{\sqrt{g}}\mathrm{div}_{\boldsymbol{\xi}}(\mathcal{T}C), \tag{7.81}$$

$$\mathrm{curl}_{\boldsymbol{\xi}}\mathbf{v} = \sqrt{g}\,\mathrm{div}_{\mathbf{x}}(\mathcal{T}C^{-1}). \tag{7.82}$$

7.3.4 Laplacian

Using definition (7.31) of the Laplacian operator and the transformation relationships (7.69) and (7.50) previously given for the divergence and gradient, the *symmetric* form of the transformed Laplacian is readily obtained:

$$\nabla^2_{\mathbf{x}}f = \frac{1}{\sqrt{g}}\mathrm{div}_{\boldsymbol{\xi}} \cdot \left(\frac{C^TC}{\sqrt{g}}\nabla_{\boldsymbol{\xi}}f\right). \tag{7.83}$$

Since $C^TC/\sqrt{g} = \sqrt{g}\,\mathcal{G}^{-1}$, the right-hand side of (7.83) is recognized as the Beltrami operator given in (4.130), i.e.,

$$\nabla^2_{\mathbf{x}}f = \Delta_{\mathcal{B}}f. \tag{7.84}$$

Inverting (7.83) gives

$$\nabla^2_{\boldsymbol{\xi}}f = \sqrt{g}\,\mathrm{div}_{\mathbf{x}} \cdot \left\{\sqrt{g}\,(C^{-1})^TC^{-1}\nabla_{\mathbf{x}}f\right\}. \tag{7.85}$$

Exercise 7.3.4 Use (7.83) to compute the Laplacian in polar coordinates. Compare your answer to that given in a standard text. §

Nonsymmetric forms can be obtained from the identity (7.8):

$$\sqrt{g}\,\nabla_{\mathbf{x}}^2 f = \nabla_{\boldsymbol{\xi}} f \cdot \operatorname{div}_{\boldsymbol{\xi}} \frac{C^T C}{\sqrt{g}} + \operatorname{tr}\left\{ \frac{C^T C}{\sqrt{g}} \nabla_{\boldsymbol{\xi}}(\nabla_{\boldsymbol{\xi}} f) \right\} \tag{7.86}$$

and

$$\frac{1}{\sqrt{g}} \nabla_{\boldsymbol{\xi}}^2 f = \nabla_{\mathbf{x}} f \cdot \operatorname{div}_{\mathbf{x}} \left\{ \sqrt{g}\,(C^{-1})^T C^{-1} \right\} + \operatorname{tr}\{ \sqrt{g}\,(C^{-1})^T C^{-1} \nabla_{\mathbf{x}}(\nabla_{\mathbf{x}} f) \}\,. \tag{7.87}$$

The *vector* Laplacian was defined in (7.32) and transforms as

$$\nabla_{\mathbf{x}}^2 \mathbf{v} = \frac{1}{\sqrt{g}} \operatorname{div}_{\boldsymbol{\xi}} \left\{ \left(\nabla_{\boldsymbol{\xi}} \mathbf{v} \right) \frac{C^T C}{\sqrt{g}} \right\}. \tag{7.88}$$

Note the change in order of the matrices in the vector transformation rule compared to the order in the scalar rule. The inverse transformation rule is

$$\nabla_{\boldsymbol{\xi}}^2 \mathbf{v} = \sqrt{g}\,\operatorname{div}_{\mathbf{x}}\{ \sqrt{g}\,(\nabla_{\mathbf{x}} \mathbf{v})(C^{-1})^T C^{-1} \}\,. \tag{7.89}$$

7.3.5 *Derivation of the Winslow Equations*

The power of the definitions and relationships given in this chapter is illustrated in this subsection by inverting the physical-space Laplacian operator to derive the *Winslow* planar grid-generation operator. For comparison, the reader might try inverting ξ_{xx}, ξ_{yy}, etc., directly. The grid equations (5.47) given in Section 5.4.2 can be expressed as the vector Laplacian $\nabla_{\mathbf{x}}^2 \boldsymbol{\xi} = \mathbf{0}$. The inverted form (5.48) can be obtained by replacing $\mathbf{v}$ in (7.88) by $\boldsymbol{\xi}$, which yields

$$\nabla_{\mathbf{x}}^2 \boldsymbol{\xi} = \frac{1}{\sqrt{g}} \operatorname{div}_{\boldsymbol{\xi}} \left(\frac{C^T C}{\sqrt{g}} \right). \tag{7.90}$$

The Winslow grid operator $\mathcal{Q}_w \mathbf{x}$ is derived in the following sequence of steps, using (7.22), (7.14), and (7.20):

$$\frac{1}{\sqrt{g}} \operatorname{div}_{\boldsymbol{\xi}} \left(\frac{C^T C}{\sqrt{g}} \right) = \mathbf{0}, \tag{7.91}$$

$$\frac{1}{\sqrt{g}} \operatorname{div}_{\boldsymbol{\xi}} (\sqrt{g}\, G^{-1}) = \mathbf{0}, \tag{7.92}$$

$$-\sqrt{g}\, \mathcal{J} \operatorname{div}_{\boldsymbol{\xi}} (\sqrt{g}\, G^{-1}) = \mathbf{0}, \tag{7.93}$$

$$-\sqrt{g}\, \{ \operatorname{div}_{\boldsymbol{\xi}} (\sqrt{g}\, \mathcal{J} G^{-1}) - \sqrt{g}\, [\nabla_{\boldsymbol{\xi}} \mathcal{J}]\, G^{-1} \} = \mathbf{0}, \tag{7.94}$$

$$g\, [\nabla_{\boldsymbol{\xi}} \mathcal{J}]\, G^{-1} = \mathbf{0}, \tag{7.95}$$

$$\mathcal{Q}_w \mathbf{x} = \mathbf{0}\,. \tag{7.96}$$

Exercise 7.3.5 Show that $\nabla^2_{\mathbf{x}}\xi = \mathbf{0}$ is equivalent to the standard form of the grid operator given in equation (5.47). §

7.3.6 Type Invariance of the Hosted Equation

The hosted equation (2.43) in Chapter 2 has the form $\text{div}_{\mathbf{x}} \cdot (\mathcal{T}\nabla_{\mathbf{x}}f) = \tilde{g}$ (here $\tilde{g}$ is not the metric det $\mathcal{G}$), with coefficients α, β, and γ forming the elements of the symmetric matrix $\mathcal{T}$. The transformed equation (2.57) is readily re-derived using the results of this chapter. In *conservative* form,

$$\text{div}_{\mathbf{x}} \cdot (\mathcal{T}\nabla_{\mathbf{x}}f) = \frac{1}{\sqrt{g}}\,\text{div}_{\xi} \cdot \left(\frac{C^T\mathcal{T}C}{\sqrt{g}}\nabla_{\xi}f\right), \tag{7.97}$$

$$\text{div}_{\xi} \cdot (\mathcal{T}\nabla_{\xi}f) = \sqrt{g}\,\text{div}_{\mathbf{x}} \cdot (\sqrt{g}\,(C^{-1})^T\mathcal{T}C^{-1}\nabla_{\mathbf{x}}f)\,. \tag{7.98}$$

Nonconservative forms are easily derived from (7.8).

Exercise 7.3.6 Compare the result (7.97) with the transformation of the two-dimensional hosted equation in Chapter 2. §

As noted in the introduction to Chapter 2, the *type* of a partial differential equation depends on the *discriminant.* Recall that the discriminant D of the hosted equation was $D = \alpha\gamma - \beta^2$; this is nothing more than the determinant of $\mathcal{T}$ in (7.97). The discriminant D' of the transformed equation is the determinant of $\mathcal{T}' = C^T\mathcal{T}C/\sqrt{g}$. Since the determinant of the product of matrices is just the product of their determinants

$$D' = \begin{cases} D & \text{if } n = 2 \\ \frac{D}{\sqrt{g}} & \text{if } n = 3. \end{cases} \tag{7.99}$$

If $\sqrt{g} > 0$, then both discriminants have the same sign, i.e., the *type* of the transformed equation is the same as the type of the original equation, provided the transformation is nonsingular. The matrices $\mathcal{T}$ and $\mathcal{T}'$ are *congruent* since they are related by an expression of the form $\mathcal{T}' = S\mathcal{T}S^T$ (Horn and Johnson [96]). In general, congruent matrices have exactly the same number of positive eigenvalues, negative eigenvalues, and zero eigenvalues, once again explaining the invariance of type in the hosted equation.

7.3.7 Transformation of the Time Derivative

The goal here is to describe how *time-dependent* hosted PDEs such as the parabolic equation $f_t = \nabla^2_{\mathbf{x}}f$ are transformed. For two-dimensional physical space, the logical space now consists of the triple (ξ, η, τ), $0 \le \tau \le 1$; the coordinate mapping is $(x(\xi,\eta,\tau), y(\xi,\eta,\tau))$; and the dependent variable is $f = f(x(\xi,\eta,\tau), y(\xi,\eta,\tau), t(\tau))$. Note that the original physical time variable depends only on the transformed time variable τ and not on the other logical coordinates. Application of the chain rule shows

$$\begin{aligned} f_\tau &= f_x\,x_\tau + f_y\,y_\tau + f_t\,t_\tau\,, \\ &= \mathbf{x}_\tau \cdot \nabla_{\mathbf{x}}f + f_t\,t_\tau\,. \end{aligned} \tag{7.100}$$

The vector $\mathbf{x}_\tau$ is known as the grid speed, as it measures the rate-of-change of position of the grid with respect to the logical time variable. The time-dependent quantity that appears in the hosted equation is f_t, which is

$$f_t = \frac{1}{t_\tau} \{ f_\tau - \mathbf{x}_\tau \cdot \nabla_\mathbf{x} f \} . \tag{7.101}$$

This form is not particularly useful since it contains derivatives with respect to both the logical and physical variables. The fully transformed expression, in non-conservative form, is easily obtained from (7.50):

$$f_t = \frac{1}{t_\tau} \left\{ f_\tau - \frac{\mathbf{x}_\tau}{\sqrt{g}} \cdot C \nabla_{\boldsymbol{\xi}} f \right\} . \tag{7.102}$$

To obtain the conservative form, the moving-grid identity (5.122) must be proved. The proof uses (7.5), (7.71), and the chain rule:

$$\operatorname{div}_\mathbf{x} \cdot \frac{\mathbf{x}_\tau}{\sqrt{g}} = \frac{1}{\sqrt{g}} \operatorname{div}_\mathbf{x} \cdot \mathbf{x}_\tau + \mathbf{x}_\tau \cdot \nabla_\mathbf{x} \frac{1}{\sqrt{g}} , \tag{7.103}$$

$$= \frac{1}{\sqrt{g}} \operatorname{div}_\mathbf{x} \cdot \mathbf{x}_\tau - \frac{\mathbf{x}_\tau}{g} \cdot \nabla_\mathbf{x} \sqrt{g} , \tag{7.104}$$

$$= \frac{1}{g} [\operatorname{tr}\{ (\nabla_{\boldsymbol{\xi}} \mathbf{x}_\tau) C^T \} - (\sqrt{g})_\tau], \tag{7.105}$$

$$= \mathbf{0} . \tag{7.106}$$

Combining the product rule (7.5) for the divergence of a scalar times a vector with the time-dependent metric identity gives

$$\mathbf{x}_\tau \cdot \nabla_\mathbf{x} f = \sqrt{g} \operatorname{div}_\mathbf{x} \cdot \left(\frac{f}{\sqrt{g}} \mathbf{x}_\tau \right) . \tag{7.107}$$

The divergence in the metric identity may be transformed to the logical domain using the relationship (7.69) to obtain the following conservative transformation rule for the time derivative:

$$f_t = \frac{1}{t_\tau} \left\{ f_\tau - \operatorname{div}_{\boldsymbol{\xi}} \cdot \left(\frac{f}{\sqrt{g}} C^T \mathbf{x}_\tau \right) \right\} . \tag{7.108}$$

Exercise 7.3.7 Show that the parabolic equation

$$f_t = \operatorname{div}_\mathbf{x} \cdot (\mathcal{T} \nabla_\mathbf{x} f) \tag{7.109}$$

transforms to

$$f_\tau = \frac{t_\tau}{\sqrt{g}} \operatorname{div}_{\boldsymbol{\xi}} \cdot \left(\frac{C^T \mathcal{T} C}{\sqrt{g}} \nabla_{\boldsymbol{\xi}} f \right) + \operatorname{div}_{\boldsymbol{\xi}} \cdot \left(\frac{f}{\sqrt{g}} C^T \mathbf{x}_\tau \right) . \ \S \tag{7.110}$$

If one has a vector of dependent variables $\mathbf{f} = (f_1, f_2, f_3)$,

$$\frac{\partial \mathbf{f}}{\partial t} = \frac{1}{t_\tau} \left\{ \frac{\partial \mathbf{f}}{\partial \tau} - \operatorname{div}_{\boldsymbol{\xi}} \left[\mathbf{f} \otimes \left(\frac{C^T \mathbf{x}_\tau}{\sqrt{g}} \right) \right] \right\} . \tag{7.111}$$

Exercise 7.3.8 Use (7.111) to compute the quantity $\partial \mathbf{x}/\partial t$. §

The material (or substantial) derivative

$$\frac{Df}{Dt} = f_t + \mathbf{v} \cdot \nabla_{\mathbf{x}} f \tag{7.112}$$

is often encountered in fluid mechanics. Here, $\mathbf{v}$ is not an arbitrary vector, but rather the fluid velocity. In particular, note that $\mathbf{v} \neq \mathbf{x}_\tau$. Assuming a moving grid, the material derivative can be transformed to the nonconservative form

$$\frac{Df}{Dt} = \frac{1}{t_\tau}\left\{ f_\tau - \frac{1}{\sqrt{g}}(\mathbf{x}_\tau - \mathbf{v}\, t_\tau) \cdot C\, \nabla_\xi f \right\}. \tag{7.113}$$

Note that the second term drops out if $\mathbf{x}_\tau = \mathbf{v}\, t_\tau$. There is no fully conservative form of the material derivative because the moving-grid identity cannot be applied to $\mathbf{v}$.

The substantial derivative of a vector quantity $\mathbf{f}$ is defined as

$$\frac{D\mathbf{f}}{Dt} = \mathbf{f}_t + (\mathbf{v} \cdot \nabla_{\mathbf{x}})\, \mathbf{f}, \tag{7.114}$$

$$= \mathbf{f}_t + (\nabla_{\mathbf{x}} \mathbf{f})\, \mathbf{v}. \tag{7.115}$$

Applying the transformation rules (7.111) and (7.61) gives the nonconservative form

$$\frac{D\mathbf{f}}{Dt} = \frac{1}{t_\tau}\left\{ \frac{\partial \mathbf{f}}{\partial \tau} - \mathrm{div}_\xi \left[\mathbf{f} \otimes \left(\frac{C^T \mathbf{x}_\tau}{\sqrt{g}} \right) \right] \right\} + \frac{1}{\sqrt{g}} (\nabla_\xi \mathbf{f}) C^T \mathbf{v}. \tag{7.116}$$

All the results of this section readily generalize to the case R^3.

As a fitting conclusion to this chapter, the three-dimensional *Navier-Stokes* equation

$$\rho \frac{D\mathbf{v}}{Dt} = \rho \nabla_{\mathbf{x}} \phi - \nabla_{\mathbf{x}} p + \mathrm{div}_{\mathbf{x}} \mu [(\nabla_{\mathbf{x}} \mathbf{v}) + (\nabla_{\mathbf{x}} \mathbf{v})^T + \lambda (\mathrm{div}_{\mathbf{x}} \cdot \mathbf{v}) \mathcal{I}] \tag{7.117}$$

is transformed. The symbol ϕ is used for the gravity term to avoid confusion with the metric symbol g (ρ is the fluid density, p the fluid pressure, and μ the dynamic viscosity). Applying (7.116), the left-hand side of the Navier-Stokes equation transforms to

$$\rho \frac{D\mathbf{v}}{Dt} = \frac{\rho}{t_\tau}\left\{ \frac{\partial \mathbf{v}}{\partial \tau} - \mathrm{div}_\xi \left[\mathbf{v} \otimes \left(\frac{C^T \mathbf{x}_\tau}{\sqrt{g}} \right) \right] \right\} + \frac{\rho}{\sqrt{g}} (\nabla_\xi \mathbf{v}) C^T \mathbf{v} \tag{7.118}$$

while the right-hand side is transformed to

$$\frac{1}{\sqrt{g}} \Big\{ \rho\, \mathrm{div}_\xi (C\phi) - \mathrm{div}_\xi (Cp)$$

$$+ \mathrm{div}_\xi \frac{\mu}{\sqrt{g}} [(\nabla_\xi \mathbf{v}) C^T + C (\nabla_\xi \mathbf{v})^T + \lambda (\mathrm{div}_\xi \cdot C^T \mathbf{v}) \mathcal{I}] C \Big\} \tag{7.119}$$

using (7.54), (7.73), (7.61), and (7.69).

Chapter 8

Advanced Planar Variational Grid Generation

8.1 INTRODUCTION

The tensor and transformation relationships developed in the previous chapter have many applications to variational grid generation. In light of these techniques, the planar variational case, presented in Chapter 6, is reexamined in this chapter, beginning with a review of several variational principles. General planar variational principles for both physical and logical weight functions are given in Section 8.2. All the functionals of grid generation can be expressed in terms of the elements of the metric tensor and its determinant, so particular attention is paid to this case in Section 8.2.1 and several new variational principles are introduced. The concept of a homogeneous function is introduced to show that functionals based on such functions are invariant to rigid body transformations (see Section 8.2.2). The Euler-Lagrange equations of grid generation are the main focus of this chapter; they are expressed in divergence form in Section 8.3. Several forms are given before the discussion is restricted to functionals that depend on elements of the metric tensor. A compact (near-conservative) divergence form of the Euler-Lagrange equation is found and related to a generalization of the two-dimensional hosted equation. Section 8.3.3 discusses the numerical solution of this equation; the approach enables one to write a computer code that will automatically generate a grid based on a wide class of functionals of the metric tensor. Another form of the Euler-Lagrange equations, called the *Covariant Projection*, is derived in Section 8.3.4; the projected form gives a relationship between the rates-of-change of the grid metrics and is fully conservative. Excessive labor is required to explicitly calculate the Euler-Lagrange equations for arbitrary functionals, so a nonconservative expression known as "tensor form" is derived. The coefficients in the tensor form are expressed in terms of tensor products of the tangent vectors and first and second partial derivatives of the variational principle with respect to the elements of the metric tensor. The form is easy to apply and facilitates writing a computer code to solve the tensor form of the Euler-Lagrange equations for arbitrary functionals. Both physical and logical weighted cases are considered. In Section 8.4, some results are given for the second variation of the grid-generation functional. Finally, the variational grid-generation approach due to Brackbill and Saltzman is given and related to the Steinberg-Roache method favored in this book.

8.2 VARIATIONAL PRINCIPLES

The most general *planar variational* problem that is considered in this book was given in equation (6.3); it is restated here for convenience. Minimize

$$I_1[\mathbf{x}] = \int_0^1 \int_0^1 G(\boldsymbol{\xi}, \mathbf{x}, \mathbf{x}_\xi, \mathbf{x}_\eta)\, d\xi\, d\eta \tag{8.1}$$

subject to the given boundary data. If *logical* weight functions are used, it is assumed that the functional takes the form

$$I_2[\mathbf{x}] = \int_0^1 \int_0^1 \frac{\hat{G}(\mathbf{x}_\xi, \mathbf{x}_\eta)}{\phi(\boldsymbol{\xi})}\, d\xi\, d\eta, \tag{8.2}$$

where ϕ is a positive function of the logical variables. If *physical* weight functions are used, the functional takes the form

$$I_3[\mathbf{x}] = \int_0^1 \int_0^1 \frac{\hat{G}(\mathbf{x}_\xi, \mathbf{x}_\eta)}{w^2(\mathbf{x})}\, d\xi\, d\eta, \tag{8.3}$$

where w is a positive function of the physical variables. Most grid generation functionals have a yet more specific form, discussed in the next section.

8.2.1 Functionals of the Metric Tensor

Of particular interest to grid generation are variational principles that can be expressed directly in terms of the elements of the metric tensor, its determinant, and a positive physical-space weight function. In this case, the principle takes the form

$$I_4[\mathbf{x}] = \int_0^1 \int_0^1 \frac{H(g_{11}, g_{12}, g_{22}, \sqrt{g})}{w^2(\mathbf{x})}\, d\xi\, d\eta \tag{8.4}$$

where H is some smooth positive function from R^4 to R. Table 8.1 lists ten proposed grid-generation functionals having this form and also a weighted combination functional. Some of the principles have already been presented in Chapter 6; others lead to seldom-used generators that are mainly of theoretical interest. The performance of all ten is compared on a standard set of domains in the Rogue's Gallery in Appendix C. Principles in Table 8.1 that were not discussed in Chapter 6 are now briefly considered.

Orthogonality-II

The second orthogonality functional was also proposed in Steinberg and Roache [191]:

$$I_{O,II}[\mathbf{x}] = \int_0^1 \int_0^1 \frac{g_{12}^2}{g_{11}\, g_{22}}\, d\xi\, d\eta. \tag{8.5}$$

Principle	Symbol	Form of H
Length	I_L	$g_{11} + g_{22}$
Area	I_A	$(\sqrt{g})^2$
Orthogonality-I	$I_{O,I}$	g_{12}^2
Orthogonality-II	$I_{O,II}$	$\frac{g_{12}^2}{g_{11}\, g_{22}}$
Orthogonality-III	$I_{O,III}$	$\sqrt{g_{11}\, g_{22}}$
AO	I_{AO}	$g_{11}\, g_{22}$
AO-Squared	I_{AO2}	$(g_{11}\, g_{22})^2$
Winslow	I_{Wi}	$\frac{g_{11}+g_{22}}{\sqrt{g}}$
Liao	I_{Li}	$g_{11}^2 + g_{22}^2 + 2g_{12}^2$
Modified Liao	I_{ML}	$\left(\frac{g_{11}+g_{22}}{\sqrt{g}}\right)^2$
Combined	I_C	$\sum_k \alpha_k \frac{H_k}{w_k^2}$

Table 8.1 *Unweighted variational principles*

This variational principle makes geometric sense because the integrand is the square of the inner product of the two unit tangent vectors $\mathbf{x}_\xi / \sqrt{g_{11}}$ and $\mathbf{x}_\eta / \sqrt{g_{22}}$; the functional controls only the relative direction of the two tangents and not their lengths.

Orthogonality-III

Table 8.1 illustrates that different functionals can all result in orthogonal grids. The third orthogonality functional is related to the "Scaled-Laplacian" (5.30), Chapter 5, which can be obtained from the variational principle

$$I_{SL}[\mathbf{x}] = \frac{1}{2} \int_0^1 \int_0^1 \left(f\, g_{11} + \frac{g_{22}}{f} \right) d\xi\, d\eta \tag{8.6}$$

with the logical-space weight $f = f(\xi, \eta) > 0$. The Scaled-Laplacian is a special case of the weighted Length functional (6.49), with $\phi = 1/f$ and $\psi = f$. Physical-space weighting, $f = f(\mathbf{x})$, would seem more useful; however, it would not lead to the Scaled-Laplacian equation. Another interesting functional (related to the *weak constraint* approach) can be derived by inserting $f = \sqrt{g_{22}/g_{11}}$ into the variational principle (8.6); this leads to

$$I_{O,III}[\mathbf{x}] = \int_0^1 \int_0^1 \sqrt{g_{11}\, g_{22}}\, d\xi\, d\eta\,. \tag{8.7}$$

The performance of this latter functional is displayed in the Rogue's Gallery in Appendix C.

AO-Squared

As the name suggests,

$$I_{AO2}[\mathbf{x}] = \int_0^1 \int_0^1 [g_{11}\, g_{22}]^2 \, d\xi \, d\eta \,, \tag{8.8}$$

minimized subject to the given boundary conditions. This functional is included because the corresponding grids shown in the Rogue's Gallery in Appendix C are generally very good.

The Winslow Functional

The Length functional I_L leads to smooth, but possibly folded grids. A "penalty" can be introduced to move the minimum away from regions of zero Jacobian. This is accomplished by dividing the Length integrand by $\sqrt{g}$, leading to the functional

$$I_{Wi}[\mathbf{x}] = \int_0^1 \int_0^1 \frac{g_{11} + g_{22}}{\sqrt{g}} \, d\xi \, d\eta \,. \tag{8.9}$$

Since $d\xi \, d\eta = dx \, dy / \sqrt{g}$, this integral over the logical domain transforms to an integral over the physical domain having the form

$$I_{Wi}[\xi] = \int_\Omega \frac{g_{11} + g_{22}}{g} \, dx \, dy = \int_\Omega g^{11} + g^{22} \, dx \, dy \,. \tag{8.10}$$

The latter functional is the well-known variational principle for the Winslow or homogeneous TTM generator.

The Liao Functional

Liao [124] proposed the following grid-generation functional based on a proof of convexity in Dacorogna [42]. The functional has the form $H = \mid \mathcal{J} \mid^4 - 2(\det \mathcal{J})^2$, so one minimizes

$$I_{Li}[\mathbf{x}] = \int_0^1 \int_0^1 [(g_{11} + g_{22})^2 - 2g] \, d\xi \, d\eta \,. \tag{8.11}$$

The integrand is also equivalent to $H = g_{11}^2 + g_{22}^2 + 2g_{12}^2 = \mid \mathcal{G} \mid^2$. Although the functional is positive and convex (and therefore has unique minimum), the resulting grids are often folded, as is seen in the Rogue's Gallery in Appendix C.

Modified Liao

The Liao functional can be improved by introducing a penalty function to mitigate the tendency to fold. Dividing the Liao integrand by g, the resulting "Modified Liao" functional is

$$I_{ML}[\mathbf{x}] = \int_0^1 \int_0^1 \left(\frac{g_{11} + g_{22}}{\sqrt{g}} \right)^2 \, d\xi \, d\eta \,. \tag{8.12}$$

The I_{ML} integrand is just the square of the Winslow integrand. As the Rogue's Gallery shows, the penalty approach helps keep the "Modified Liao" grids from folding.

Weighted Combination

Let $\{\alpha_k\}$, $k = 1, \ldots, K$, be a set of constants whose sum is unity. Let $H_k = H_k(g_{11}, g_{12}, g_{22}, \sqrt{g})$ be K functions of the elements of the metric tensor and $w_k = w_k(\mathbf{x})$ a corresponding set of weight functions. Then the weighted combination to be minimized is

$$I_C[\mathbf{x}] = \int_0^1 \int_0^1 \sum_{k=1}^{K} \alpha_k \frac{H_k}{w_k^2} \, d\xi \, d\eta \,. \tag{8.13}$$

This general functional form can provide sufficient flexibility to generate useful grids on complicated regions.

8.2.2 Rigid Body Transformations of the Domain

Rigid body transformations in the plane consist of translations, rotations, and stretches of scale of physical space. Similar transformations on logical space will also be considered. Any useful grid generator will have the property that the grid generated after a rigid body transformation of the boundary of the *physical* domain is the same grid that would be obtained if the grid were generated using the original boundary, then the rigid body transformation applied. Criteria to guarantee such behavior are discussed in this section.

First note that physical-space weights are typically computed from the solution of a hosted PDE; because the solutions of such PDEs are typically invariant to rigid-body transformations, the weight can be invariantly defined. Thus, if $\hat{w}$ is the transformed weight, it will be the case that

$$\hat{w}(\hat{\mathbf{x}}) = w(\mathbf{x}) \,. \tag{8.14}$$

If one performs a *translation* $\hat{\mathbf{x}} = \mathbf{x} + \mathbf{x}_0$ of the physical boundary data, then the tangents $\mathbf{x}_\xi$ and $\mathbf{x}_\eta$ are independent of such a shift. Since the variational principle I_2 only depends on the tangents, the resulting grids will be invariant to translation. Variational principles such as I_3 or I_4 must have physical-space weights with the property $w(\hat{\mathbf{x}}) = w(\mathbf{x})$ to be invariant under translation. Similarly, rigid-body rotations can be represented by an orthogonal transformation $S^T = S^{-1}$, so that $\hat{\mathbf{x}} = S^T\mathbf{x}$. Rotations preserve the lengths of the tangents, so the variational principles are unaffected by rotation, provided $w(\hat{\mathbf{x}}) = w(\mathbf{x})$.

Exercise 8.2.1 Show that the Length generator (5.42) is invariant to translations and rotations of the physical domain. §

Most grid generators are not invariant to nonuniform *stretches* of the physical coordinate axes. Suppose μ and ν are positive real constants with $\hat{x} = \mu\, x$ and $\hat{y} = \nu\, y$. Then, for example, the metric $\hat{g}_{11}$ becomes $\hat{g}_{11} = \hat{x}_\xi^2 + \hat{y}_\xi^2 = \mu^2\, x_\xi^2 + \nu^2\, y_\xi^2 \neq g_{11}$. Since the metrics are not invariant, most grid generators are not invariant to such stretches. However, if the stretch is uniform, that is, if $\mu = \nu$, then the metrics *scale* with the stretch. For example $\hat{g}_{11} = \hat{x}_\xi^2 + \hat{y}_\xi^2 = \mu^2\, x_\xi^2 + \mu^2\, y_\xi^2 = \mu^2\, g_{11}$.

Exercise 8.2.2 Show that the area metric $\sqrt{g}$ *scales* with respect to a nonuniform stretch and, consequently, grids generated by the Area functional are invariant with respect to nonuniform stretches. §

In fact, most generators will be invariant to a uniform stretch $\nu = \mu$ of the physical domain because then the tangents and metrics scale with the stretch.

Exercise 8.2.3 Show that grids generated by the Length functional are invariant with respect to uniform stretches. §

Consider the effect of a nonuniform stretch of the *logical* domain on the grid generator. Let $\hat{\xi} = \mu\,\xi$ and $\hat{\eta} = \nu\,\eta$. Then $\mathbf{x}_\xi = \mu\,\mathbf{x}_{\hat{\xi}}$, $\mathbf{x}_\eta = \nu\,\mathbf{x}_{\hat{\eta}}$. Let $\hat{g}_{ij} = \mathbf{x}_{\hat{\xi}_i} \cdot \mathbf{x}_{\hat{\xi}_j}$ be the transformed metrics. Then $g_{11} = \mu^2\,\hat{g}_{11}$, $g_{12} = \mu\,\nu\,\hat{g}_{12}$, $g_{22} = \nu^2\,\hat{g}_{22}$, and $\sqrt{g} = \mu\,\nu\,\sqrt{\hat{g}}$.

Exercise 8.2.4 Show that the *Winslow* grid generator (5.48) is invariant to a nonuniform stretch of the *logical* domain, whereas the Length generator (5.42) is not. §

Since the logical domain is generally fixed, invariance of a grid generator to nonuniform stretches of the logical domain is not often required. It is useful to be aware of the property, however, since some grid-generation systems may be based on nonsquare logical domains while others are based on square logical domains.

DEFINITION 8.1 Let $\mu > 0$. A function $F : R^k \to R$ is *homogeneous* of degree n if

$$F(\mu\,x_1, \mu\,x_2, \ldots, \mu\,x_k) = \mu^n\,F(x_1, x_2, \ldots, x_k)\,. \tag{8.15}$$

Exercise 8.2.5 Verify that the functionals in Table 8.1 are homogeneous and find the degree n for each. What can be said about the weighted combination functional, I_C? Is the Area-Length combination I_W (6.69) homogeneous? §

THEOREM 8.2 If the integrand H in (8.4) is homogeneous of degree n, then the resulting unweighted grid generator is invariant to uniform stretches of both the physical and logical spaces.

Proof. The elements of the metric tensor scale as $\hat{g}_{ij} = \mu^2 g_{ij}$ under a *uniform* stretch of either the physical or logical domain. Therefore,

$$H(\hat{g}_{11}, \hat{g}_{12}, \hat{g}_{22}, \sqrt{\hat{g}}) = H(\mu^2 g_{11}, \mu^2 g_{12}, \mu^2 g_{22}, \mu^2\sqrt{g}) = \mu^{2n} H, \tag{8.16}$$

so the new functional is merely a constant times the original functional. §

8.3 THE EULER-LAGRANGE EQUATIONS

The Euler-Lagrange equations are the focus of most of this chapter since they are in fact the grid-generation equations. Euler-Lagrange equations for planar functionals were derived in Section 6.2. Using the tools of tensor analysis introduced in the previous chapter, the Euler-Lagrange equations are recast in terms of the divergence operator. Several forms are considered to cover both physical- and logical-space weight functions.

8.3.1 The General Planar Euler-Lagrange Equations

The most general form of the planar variational principle was given in Section 8.1, for which the Euler-Lagrange equations (6.11)–(6.12) apply. To write the Euler-Lagrange equations in a more compact form, define the matrix

$$\mathcal{T} = \begin{pmatrix} \frac{\partial G}{\partial x_\xi} & \frac{\partial G}{\partial x_\eta} \\ \frac{\partial G}{\partial y_\xi} & \frac{\partial G}{\partial y_\eta} \end{pmatrix} . \tag{8.17}$$

The tools developed in Chapter 7 can then be applied to write the Euler-Lagrange equations in the compact form

$$\nabla_{\mathbf{x}} G - \mathrm{div}_{\boldsymbol{\xi}} \mathcal{T} = \mathbf{0} . \tag{8.18}$$

Note the mixture of partial derivatives with respect to both physical and logical coordinates. The transformation rule (7.54) cannot be directly applied to the gradient operator because G is not strictly a function of just the physical variables, but also of the logical variables and the tangent vectors.

If the logically weighted variational principle I_2 in (8.2) is assumed, then the Euler-Lagrange equation (8.18) reduces to the conservative form

$$\mathrm{div}_{\boldsymbol{\xi}} \frac{\hat{\mathcal{T}}}{\phi} = \mathbf{0} \tag{8.19}$$

with $\hat{\mathcal{T}}$ defined by replacing G in (8.17) by $\hat{G}$. Applying the product rule (7.7), the nonconservative form of the Euler-Lagrange equations for logical weighting is just

$$\mathrm{div}_{\boldsymbol{\xi}} \hat{\mathcal{T}} = \hat{\mathcal{T}} \frac{\nabla_{\boldsymbol{\xi}} \phi}{\phi} . \tag{8.20}$$

If the physically weighted variational principle I_3 in (8.3) is assumed, then the Euler-Lagrange equation (8.18) becomes

$$\mathrm{div}_{\boldsymbol{\xi}} \frac{\hat{\mathcal{T}}}{w^2} = \hat{G} \, \nabla_{\mathbf{x}} \frac{1}{w^2} . \tag{8.21}$$

Since the weight in the inhomogeneous term is just a function of $\mathbf{x}$, the gradient can now be inverted using (7.54) to get

$$\mathrm{div}_{\boldsymbol{\xi}} \frac{\hat{\mathcal{T}}}{w^2} = \frac{\hat{G}}{\sqrt{g}} C \, \nabla_{\boldsymbol{\xi}} \frac{1}{w^2} , \tag{8.22}$$

$$= \frac{\hat{G}}{\sqrt{g}} \mathrm{div}_{\boldsymbol{\xi}} \frac{C}{w^2} . \tag{8.23}$$

Fully expanded, the nonconservative form of the physical-space weighted Euler-Lagrange equations is

$$\mathrm{div}_{\boldsymbol{\xi}} \hat{\mathcal{T}} = 2 \left\{ \hat{\mathcal{T}} - \frac{\hat{G}}{\sqrt{g}} C \right\} \frac{\nabla_{\boldsymbol{\xi}} w}{w} . \tag{8.24}$$

Notice that a fully conservative form is not obtained if physical-space weights are used. A fully conservative form of the Euler-Lagrange equations for physical-space weights is derived in Section 8.3.4.

The chain rule can be applied to the left-hand side of either (8.20) or (8.24) to obtain Euler-Lagrange equations in the *tensor* form of Section 6.4:

$$S_{11}\,\mathbf{x}_{\xi\xi} + (S_{12} + S_{12}^T)\,\mathbf{x}_{\xi\eta} + S_{22}\,\mathbf{x}_{\eta\eta} = \begin{cases} \hat{\mathcal{T}}\,\dfrac{\nabla_{\xi}\phi}{\phi}, & \text{if Logical Weight,} \\ 2\{\hat{\mathcal{T}} - \frac{\hat{G}}{\sqrt{g}}\,C\}\dfrac{\nabla_{\xi} w}{w}, & \text{if Physical Weight,} \end{cases} \tag{8.25}$$

where the matrices $S_{\mu\nu}$ are defined by

$$S_{\mu\nu} = \begin{pmatrix} \dfrac{\partial^2 \hat{G}}{\partial x_{\xi_\mu}\partial x_{\xi_\nu}} & \dfrac{\partial^2 \hat{G}}{\partial x_{\xi_\mu}\partial y_{\xi_\nu}} \\ \dfrac{\partial^2 \hat{G}}{\partial y_{\xi_\mu}\partial x_{\xi_\nu}} & \dfrac{\partial^2 \hat{G}}{\partial y_{\xi_\mu}\partial y_{\xi_\nu}} \end{pmatrix}. \tag{8.26}$$

Although this last is a powerful result, still more can be said for the functional I_4, as demonstrated in the next section.

8.3.2 *The Euler-Lagrange Equations for the Principle* I_4

The Euler-Lagrange equation (8.23) holds for a general variational principle with physical-space weights. If, in addition, the form in (8.4) is assumed, the equations become

$$\frac{H}{\sqrt{g}}\,\mathrm{div}_{\xi}\frac{C}{w^2} - \mathrm{div}_{\xi}\frac{\mathcal{T}}{w^2} = \mathbf{0}. \tag{8.27}$$

with

$$\mathcal{T} = \begin{pmatrix} \dfrac{\partial H}{\partial x_\xi} & \dfrac{\partial H}{\partial x_\eta} \\ \dfrac{\partial H}{\partial y_\xi} & \dfrac{\partial H}{\partial y_\eta} \end{pmatrix}. \tag{8.28}$$

Equation (8.27) is not significantly different from that obtained from the functional I_3, but since H has the special form in (8.4), the Euler-Lagrange equation can be developed further using the chain rule. If $A = A(\mathbf{x}_\xi, \mathbf{x}_\eta)$ is a scalar function of the tangent vectors, define the vectors

$$\frac{\partial A}{\partial \mathbf{x}_\xi} = \begin{pmatrix} \dfrac{\partial A}{\partial x_\xi} \\ \dfrac{\partial A}{\partial y_\xi} \end{pmatrix}, \quad \frac{\partial A}{\partial \mathbf{x}_\eta} = \begin{pmatrix} \dfrac{\partial A}{\partial x_\eta} \\ \dfrac{\partial A}{\partial y_\eta} \end{pmatrix}. \tag{8.29}$$

Then, for example, (8.28) becomes $\mathcal{T} = [\partial H/\partial \mathbf{x}_\xi, \partial H/\partial \mathbf{x}_\eta]$.

Exercise 8.3.1 Show that

$$\frac{\partial g_{11}}{\partial \mathbf{x}_\xi} = 2\mathbf{x}_\xi\,, \quad \frac{\partial g_{12}}{\partial \mathbf{x}_\xi} = \mathbf{x}_\eta\,, \quad \frac{\partial g_{22}}{\partial \mathbf{x}_\xi} = \mathbf{0}\,, \quad \frac{\partial \sqrt{g}}{\partial \mathbf{x}_\xi} = -\mathbf{x}_\eta^\perp, \tag{8.30}$$

$$\frac{\partial g_{11}}{\partial \mathbf{x}_\eta} = \mathbf{0}\,, \quad \frac{\partial g_{12}}{\partial \mathbf{x}_\eta} = \mathbf{x}_\xi\,, \quad \frac{\partial g_{22}}{\partial \mathbf{x}_\eta} = 2\mathbf{x}_\eta\,, \quad \frac{\partial \sqrt{g}}{\partial \mathbf{x}_\eta} = \mathbf{x}_\xi^\perp\,. \; \S \tag{8.31}$$

Principle	$\frac{\partial H}{\partial g_{11}}$	$\frac{\partial H}{\partial g_{12}}$	$\frac{\partial H}{\partial g_{22}}$	$\frac{\partial H}{\partial \sqrt{g}}$
Length	1	0	1	0
Area	0	0	0	$2\sqrt{g}$
Orthogonality-I	0	$2g_{12}$	0	0
Orthogonality-II	$-\frac{g_{12}^2}{g_{11}^2 g_{22}}$	$\frac{2g_{12}}{g_{11}g_{22}}$	$-\frac{g_{12}^2}{g_{22}^2 g_{11}}$	0
Orthogonality-III	$\frac{1}{2}\sqrt{g_{22}/g_{11}}$	0	$\frac{1}{2}\sqrt{g_{11}/g_{22}}$	0
AO	g_{22}	0	g_{11}	0
AO-Squared	$2g_{11}g_{22}^2$	0	$2g_{22}g_{11}^2$	0
Winslow	$\frac{1}{\sqrt{g}}$	0	$\frac{1}{\sqrt{g}}$	$-\frac{g_{11}+g_{22}}{g}$
Liao	$2g_{11}$	$4g_{12}$	$2g_{22}$	0
Modified Liao	$2\frac{g_{11}+g_{22}}{g}$	0	$2\frac{g_{11}+g_{22}}{g}$	$-2\frac{(g_{11}+g_{22})^2}{g^{3/2}}$

Table 8.2 *First partial derivatives of the variational principles*

The results of the previous exercise are combined with the chain rule to show

$$\frac{\partial H}{\partial \mathbf{x}_\xi} = 2\mathbf{x}_\xi \frac{\partial H}{\partial g_{11}} + \mathbf{x}_\eta \frac{\partial H}{\partial g_{12}} - \mathbf{x}_\eta^\perp \frac{\partial H}{\partial \sqrt{g}}, \tag{8.32}$$

$$\frac{\partial H}{\partial \mathbf{x}_\eta} = 2\mathbf{x}_\eta \frac{\partial H}{\partial g_{22}} + \mathbf{x}_\xi \frac{\partial H}{\partial g_{12}} + \mathbf{x}_\xi^\perp \frac{\partial H}{\partial \sqrt{g}}. \tag{8.33}$$

It is now easy to observe that the matrix $\mathcal{T}$ has the factorization $\mathcal{T} = \mathcal{J}\mathcal{B}$ where the tensor $\mathcal{B}$ is defined by

$$\mathcal{B} = \mathcal{M} + \sqrt{g}\frac{\partial H}{\partial \sqrt{g}} \mathcal{G}^{-1} \tag{8.34}$$

and $[\mathcal{M}]_{ij} = (1 + \delta_{ij})\, \partial H / \partial g_{ij}$ with δ_{ij} the Kronecker Delta. The tensor $\mathcal{B}$ thus contains partial derivatives of H with respect to $\sqrt{g}$ and the elements of the metric tensor; $\mathcal{B}$ is symmetric. Since the Jacobian matrix is the gradient of the position vector, one has the resulting nonlinear Euler-Lagrange equation

$$\mathrm{div}_\xi \frac{(\nabla_\xi \mathbf{x})\mathcal{B}}{w^2} = \frac{H}{\sqrt{g}} \mathrm{div}_\xi \frac{C}{w^2}. \tag{8.35}$$

A table of partial derivatives for the various variational principles in Table 8.1 is provided in Table 8.2.

Exercise 8.3.2 Consider the variational principle $H = \sqrt{g}$ with $w = 1$. Compute the matrix $\mathcal{B}$ and evaluate (8.35); show that the metric identity (7.14) results. Find a geometric interpretation of this result and show that for any principal H and any constant scalar, β, the principal $\tilde{H} = H + \beta\sqrt{g}$ has the same minimizing transformation as H. §

Exercise 8.3.3 Compute $\mathcal{B}$ for the two forms of the Area functional:

$$H = (\sqrt{g})^2\,, \tag{8.36}$$

$$H = g_{11}g_{22} - g_{12}^2\,, \tag{8.37}$$

and compare the results. Show that, for this $\mathcal{B}$, $\sqrt{g} = \beta w$ with β a constant is a generic property of solutions to (8.35). §

The fact that the homogeneous Euler-Lagrange equations resulting from the variational principle (8.4), with $w = 1$, can be expressed as the divergence of $\mathcal{J}\mathcal{B}$ does not need explaining, for, in general, the Euler-Lagrange equations always have the form $\text{div}_\xi \mathcal{T} = \mathbf{0}$ for some tensor $\mathcal{T}$. This can always be written as $\text{div}_\xi \mathcal{J}\tilde{\mathcal{B}} = \mathbf{0}$ with $\tilde{\mathcal{B}} = \mathcal{J}^{-1}\mathcal{T}$. However, if one expresses this same Euler-Lagrange equation as $\text{div}_\xi (\nabla_\xi \mathbf{x})\mathcal{B} = \mathbf{0}$, the analogy with the hosted equation (Section 7.3.6) is compelling. The question arises: Is there a generalization of the hosted variational principle (6.22) that can be connected to the principle I_4? The following theorem and corollary delineates such a connection.

THEOREM 8.3 If H is homogeneous of degree n, then

$$2nH = \text{tr}\{\mathcal{J}\mathcal{B}\mathcal{J}^T\} = \text{tr}\{\mathcal{G}\mathcal{B}\}\,. \tag{8.38}$$

Proof. The proof is a direct computation, based on the well-known Euler formula for homogeneous functions:

$$nH = g_{11}\frac{\partial H}{\partial g_{11}} + g_{12}\frac{\partial H}{\partial g_{12}} + g_{22}\frac{\partial H}{\partial g_{22}} + \sqrt{g}\frac{\partial H}{\partial \sqrt{g}}\,. \;\S \tag{8.39}$$

COROLLARY 8.4 The variational principle (8.4) may be rewritten

$$I_4[\mathbf{x}] = \frac{1}{2n}\int_0^1\int_0^1 \frac{\text{tr}\{(\nabla_\xi \mathbf{x})\,\mathcal{B}\,(\nabla_\xi \mathbf{x})^T\}}{w^2(\mathbf{x})}\,d\xi\,d\eta\,, \tag{8.40}$$

provided H is homogeneous of degree n, $n \neq 0$.

To make a connection between the variational principle for the hosted equation and the variational principle I_4 thus requires the additional assumption of homogeneity of the function H (equation (8.35) is derived without this assumption). The extra assumption is not a burden since homogeneity is needed anyway to achieve invariance of the grid generator under rigid-body motions, as discussed in Section 8.2.2. In general, if the principle $\text{tr}\{\mathcal{G}\tilde{\mathcal{B}}\}$ is minimized, where $\tilde{\mathcal{B}} = \tilde{\mathcal{B}}(\mathbf{x}_\xi, \mathbf{x}_\eta)$ is an arbitrary symmetric matrix, the resulting Euler-Lagrange equation has the form $\text{div}_\xi \mathcal{J}\hat{\mathcal{B}} = \mathbf{0}$, where the two matrices $\tilde{\mathcal{B}}$ and $\hat{\mathcal{B}}$ are not the same.

8.3.3 Numerical Implementation

This section shows how to write a computer code to *numerically* solve the nonlinear Euler-Lagrange equation (8.35), maintaining conservative form. The

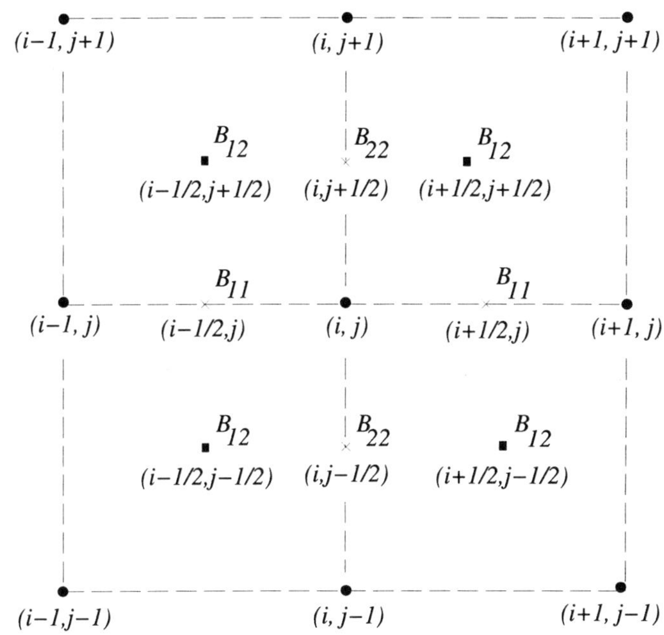

Figure 8.1 *Location of the entries of $\mathcal{B}$*

equation can be written explicitly as the pair

$$\left(\frac{B_{11}\,x_\xi + B_{12}\,x_\eta}{w^2}\right)_\xi + \left(\frac{B_{12}\,x_\xi + B_{22}\,x_\eta}{w^2}\right)_\eta = \frac{H}{\sqrt{g}}\left\{\left(\frac{y_\eta}{w^2}\right)_\xi - \left(\frac{y_\xi}{w^2}\right)_\eta\right\}, \quad (8.41)$$

$$\left(\frac{B_{11}\,y_\xi + B_{12}\,y_\eta}{w^2}\right)_\xi + \left(\frac{B_{12}\,y_\xi + B_{22}\,y_\eta}{w^2}\right)_\eta = \frac{H}{\sqrt{g}}\left\{\left(-\frac{x_\eta}{w^2}\right)_\xi + \left(\frac{x_\xi}{w^2}\right)_\eta\right\}. \quad (8.42)$$

The left-hand sides of both the first and second equations are of the same form as the two-dimensional hosted equation, so the discretization applied in Section 2.4 will result in *symmetric* stencils:

$$N_{i,j} = \frac{1}{\Delta\eta^2}\left(\frac{B_{22}}{w^2}\right)_{i,j+\frac{1}{2}}, \quad (8.43)$$

$$S_{i,j} = \frac{1}{\Delta\eta^2}\left(\frac{B_{22}}{w^2}\right)_{i,j-\frac{1}{2}}, \quad (8.44)$$

$$W_{i,j} = \frac{1}{\Delta\xi^2}\left(\frac{B_{11}}{w^2}\right)_{i-\frac{1}{2},j}, \quad (8.45)$$

$$E_{i,j} = \frac{1}{\Delta\xi^2}\left(\frac{B_{11}}{w^2}\right)_{i+\frac{1}{2},j}, \quad (8.46)$$

$$\mathrm{NE}_{i,j} = \frac{1}{4\,\Delta\xi\,\Delta\eta}\left(\frac{B_{12}}{w^2}\right)_{i+\frac{1}{2},j+\frac{1}{2}}, \quad (8.47)$$

$$\mathrm{NW}_{i,j} = -\frac{1}{4\,\Delta\xi\,\Delta\eta}\left(\frac{B_{12}}{w^2}\right)_{i-\frac{1}{2},j+\frac{1}{2}}, \tag{8.48}$$

$$\mathrm{SE}_{i,j} = -\frac{1}{4\,\Delta\xi\,\Delta\eta}\left(\frac{B_{12}}{w^2}\right)_{i+\frac{1}{2},j-\frac{1}{2}}, \tag{8.49}$$

$$\mathrm{SW}_{i,j} = \frac{1}{4\,\Delta\xi\,\Delta\eta}\left(\frac{B_{12}}{w^2}\right)_{i-\frac{1}{2},j-\frac{1}{2}}, \tag{8.50}$$

$$C_{i,j} = -\left(E_{i,j} + W_{i,j} + N_{i,j} + S_{i,j} + \mathrm{NE}_{i,j} + \mathrm{NW}_{i,j} + \mathrm{SE}_{i,j} + \mathrm{SW}_{i,j}\right). \tag{8.51}$$

The entries of the matrix $\mathcal{B}$ are located at eight positions (see Figure 8.1) $(B_{11})_{i\pm1/2,j}$, $(B_{22})_{i,j\pm1/2}$, and $(B_{12})_{i\pm1/2,j\pm1/2}$, so the tangents and metrics must be computed at these eight locations. For example,

$$(x_\xi)_{i+\frac{1}{2},j+\frac{1}{2}} \approx \frac{1}{2\,\Delta\xi}\{x_{i+1,j+1} - x_{i,j+1} + x_{i+1,j} - x_{i,j}\}. \tag{8.52}$$

The nine-point stencil equation applies to each of equations (8.41)–(8.42), but with differing right-hand sides; the right-hand sides are

$$RX_{i,j} = \frac{1}{4\,\Delta\xi\,\Delta\eta}\left(\frac{H}{\sqrt{g}}\right)_{i,j}\left\{\frac{y_{i+1,j+1} - y_{i+1,j-1}}{w^2_{i+1,j}} - \frac{y_{i-1,j+1} - y_{i-1,j-1}}{w^2_{i-1,j}}\right.$$
$$\left. - \frac{y_{i+1,j+1} - y_{i-1,j+1}}{w^2_{i,j+1}} + \frac{y_{i+1,j-1} - y_{i-1,j-1}}{w^2_{i,j-1}}\right\}, \tag{8.53}$$

$$RY_{i,j} = -\frac{1}{4\,\Delta\xi\,\Delta\eta}\left(\frac{H}{\sqrt{g}}\right)_{i,j}\left\{\frac{x_{i+1,j+1} - x_{i+1,j-1}}{w^2_{i+1,j}} - \frac{x_{i-1,j+1} - x_{i-1,j-1}}{w^2_{i-1,j}}\right.$$
$$\left. - \frac{x_{i+1,j+1} - x_{i-1,j+1}}{w^2_{i,j+1}} + \frac{x_{i+1,j-1} - x_{i-1,j-1}}{w^2_{i,j-1}}\right\}, \tag{8.54}$$

Because the central differencing used in this section follows the approach in Chapter 2, it is expected that this algorithm is a second order accurate approximation of (8.41)–(8.42). Since it can be shown that (8.35) is not elliptic in many cases, it is an important *open problem* to develop an existence and smoothness theory for the equations (8.41)–(8.42). Curiously, solution methods that apply to elliptic problems work well in some of the nonelliptic cases.

In the numerical codes described in Section B.7.3 of Appendix B, the coefficients generated by the tensor $\mathcal{B}$ are lagged in a Picard iteration like the one described in Section 5.4.2. A switch is built into the code to compute the entries of the matrix $\mathcal{B}$, depending on which variational principle has been chosen. The iterative approach outlined in this section is effective for the Length, Area, AO, and Scaled Laplace functionals, but it fails to converge on the Winslow, Orthogonality-I, Liao, Modified Liao, and AO-Squared functionals. Since several of the latter are known to be convex functionals, the lack of convergence is due to the numerical scheme and not lack of existence of a solution. If the stencil matrix is not diagonally dominant, or if the tensor $\mathcal{B}$ is not positive definite, the approach may break down. For example, if $H = g_{12}^2$, then $B_{11} = B_{22} = 0$ and the cross-derivatives in the equations dominate. Another important open problem is to develop numerical algorithms that will solve a wider class of these equations.

8.3.4 The Covariant Projections

The Euler-Lagrange equation (8.35) is a vector relationship, therefore it may be projected onto other vectors in the plane. If the projection with the *covariant* tangents is formed, the resulting equations can be integrated to obtain a conservative form that gives relationships between the rates-of-change of the elements of the metric tensor. In contrast, the *contravariant* projections are not integrable. Define the *covariant projections* to be the expressions one obtains by forming the scalar product of the vector Euler-Lagrange equation with the two tangent vectors $\mathbf{x}_\xi$ and $\mathbf{x}_\eta$. This procedure is equivalent to premultiplying (8.35) by $\mathcal{J}^T$.

The following identity is left as an exercise to the reader:

$$[\nabla_\xi \mathcal{J}^T](\mathcal{J}\mathcal{B}) = \nabla_\xi H. \tag{8.55}$$

THEOREM 8.5 The covariant projections of the Euler-Lagrange equations (8.35) may be integrated to obtain the divergence form

$$\mathrm{div}_\xi \left\{ \frac{H\mathcal{I} - \mathcal{G}\mathcal{B}}{w^2} \right\} = \mathbf{0}. \tag{8.56}$$

Proof. Begin by writing the Euler-Lagrange equation (8.35) as

$$\mathrm{div}_\xi \frac{\mathcal{J}\mathcal{B}}{w^2} = H\,(\mathcal{J}^{-1})^T\,\nabla_\xi \frac{1}{w^2}. \tag{8.57}$$

Premultiply this by $\mathcal{J}^T$ and perform the following sequence of steps:

$$\mathcal{J}^T\,\mathrm{div}_\xi \frac{\mathcal{J}\mathcal{B}}{w^2} = H\,\nabla_\xi \frac{1}{w^2}, \tag{8.58}$$

$$\mathrm{div}_\xi \frac{\mathcal{J}^T\mathcal{J}\mathcal{B}}{w^2} - [\nabla_\xi \mathcal{J}^T]\frac{\mathcal{J}\mathcal{B}}{w^2} = H\,\nabla_\xi \frac{1}{w^2}, \tag{8.59}$$

$$\mathrm{div}_\xi \frac{\mathcal{G}\mathcal{B}}{w^2} - \frac{1}{w^2}\nabla_\xi H = H\,\nabla_\xi \frac{1}{w^2}, \tag{8.60}$$

$$\mathrm{div}_\xi \frac{\mathcal{G}\mathcal{B}}{w^2} - \nabla_\xi \frac{H}{w^2} = \mathbf{0}.\ \S \tag{8.61}$$

Table 8.3 gives covariant projections for some of the functionals considered in this chapter (with $w = 1$). No simple relationship emerges between the various metrics, with the exception of the area functional. This demonstrates that, except for area, the minimizing grid does not equidistribute any property of the grid directly related to H. In particular, the weighted functional does not make $\sqrt{H}$ proportional to the weight w.

Exercise 8.3.4 Verify the result in Table 8.3 for the area functional using both $H = (\sqrt{g})^2$ and $H = g_{11}\,g_{22} - g_{12}^2$. §

Functional	$(\mathrm{div}\mathcal{J}\mathcal{B})\cdot\mathbf{x}_\xi$	$(\mathrm{div}\mathcal{J}\mathcal{B})\cdot\mathbf{x}_\eta$
Length	$(g_{11}-g_{22})_\xi + 2\,(g_{12})_\eta$	$(g_{22}-g_{11})_\eta + 2\,(g_{12})_\xi$
Area	g_ξ	g_η
Orthogonality-I	$(g_{12}^2)_\xi + 2\,(g_{11}\,g_{12})_\eta$	$(g_{12}^2)_\eta + 2\,(g_{22}\,g_{12})_\xi$
AO	$(g_{11}\,g_{22})_\xi + 2\,(g_{11}\,g_{12})_\eta$	$(g_{11}\,g_{22})_\eta + 2\,(g_{22}\,g_{12})_\xi$
Winslow	$\left(\frac{g_{22}}{\sqrt{g}}\right)_\xi - \left(\frac{g_{12}}{\sqrt{g}}\right)_\eta$	$\left(\frac{g_{11}}{\sqrt{g}}\right)_\eta - \left(\frac{g_{12}}{\sqrt{g}}\right)_\xi$

Table 8.3 *Covariant projections of the planar Euler-Lagrange equations*

Exercise 8.3.5 Show that for the Winslow functional in Table 8.1,

$$H\mathcal{I} = \frac{\mathcal{G}}{\sqrt{g}} + \sqrt{g}\,\mathcal{G}^{-1}, \tag{8.62}$$

and therefore

$$H\mathcal{I} - \mathcal{G}\mathcal{B} = 2\,\frac{C^T C}{\sqrt{g}}, \tag{8.63}$$

in agreement with (7.90). §

8.3.5 Tensor Form of the Euler-Lagrange Equations

The Euler-Lagrange equations of grid generation are usually presented and solved in the nonconservative form (6.78). Since the fully conservative approach outlined in Section 8.3.3 does not converge for some important variational principles (e.g., Winslow), the nonconservative approach remains viable. The nonconservative form can be obtained by expanding (8.35) via the product rule (7.22) in Chapter 7:

$$\mathcal{J}\,\mathrm{div}_{\boldsymbol{\xi}}\mathcal{B} + [\nabla_{\boldsymbol{\xi}}\mathcal{J}]\mathcal{B} + \frac{2}{w}\,\{H\,(\mathcal{J}^{-1})^T - \mathcal{J}\,\mathcal{B}\}\,\nabla_{\boldsymbol{\xi}} w = \mathbf{0}\,. \tag{8.64}$$

Exercise 8.3.6 Show that the last term above can be expressed as $\mathcal{J}\mathbf{b} = b_1\,\mathbf{x}_\xi + b_2\,\mathbf{x}_\eta$ where

$$\mathbf{b} = \frac{2}{w}\,\{H\mathcal{G}^{-1} - \mathcal{B}\}\,\nabla_{\boldsymbol{\xi}} w\,.\ \S \tag{8.65}$$

Using the exercise and also (7.21), equation (8.64) can be expressed as

$$\mathcal{J}\,\mathrm{div}_{\boldsymbol{\xi}}\mathcal{B} + B_{11}\,\mathbf{x}_{\xi\xi} + 2\,B_{12}\,\mathbf{x}_{\xi\eta} + B_{22}\,\mathbf{x}_{\eta\eta} + b_1\,\mathbf{x}_\xi + b_2\,\mathbf{x}_\eta = \mathbf{0}\,. \tag{8.66}$$

The first term in this expression is

$$\mathcal{J}\,\mathrm{div}_{\boldsymbol{\xi}}\mathcal{B} = \{(B_{11})_\xi + (B_{12})_\eta\}\,\mathbf{x}_\xi + \{(B_{12})_\xi + (B_{22})_\eta\}\,\mathbf{x}_\eta\,. \tag{8.67}$$

The terms on the right-hand side are expanded using the chain rule. Since $B_{11} = 2\,\partial H/\partial g_{11}$, the first term becomes

$$\left(\frac{\partial H}{\partial g_{11}}\right)_\xi \mathbf{x}_\xi = \left[(g_{11})_\xi \frac{\partial^2 H}{\partial g_{11}^2} + (g_{12})_\xi \frac{\partial^2 H}{\partial g_{11}\partial g_{12}} \right.$$

$$\left. + \; (g_{22})_\xi \frac{\partial^2 H}{\partial g_{11}\partial g_{22}} + (\sqrt{g})_\xi \frac{\partial^2 H}{\partial g_{11}\partial\sqrt{g}} \right] \mathbf{x}_\xi \,. \tag{8.68}$$

Using (4.131)-(4.136), (4.146)-(4.147), and the tensor product property (7.37), each term of the previous expression can be treated in a manner similar to the following example:

$$(g_{11})_\xi \frac{\partial^2 H}{\partial g_{11}^2}\mathbf{x}_\xi = 2\,\frac{\partial^2 H}{\partial g_{11}^2}\,(\mathbf{x}_\xi \cdot \mathbf{x}_{\xi\xi})\,\mathbf{x}_\xi, \tag{8.69}$$

$$= 2\,\frac{\partial^2 H}{\partial g_{11}^2}\,(\mathbf{x}_\xi \otimes \mathbf{x}_\xi)\,\mathbf{x}_{\xi\xi} \,. \tag{8.70}$$

Grouping all the terms, the final result can be put into nonconservative (or tensor) form $\mathcal{Q}_H\mathbf{x} = \mathbf{0}$, where

$$\mathcal{Q}_H\mathbf{x} = \mathcal{T}_{11}\,\mathbf{x}_{\xi\xi} + (\mathcal{T}_{12} + \mathcal{T}_{12}^T)\,\mathbf{x}_{\xi\eta} + \mathcal{T}_{22}\,\mathbf{x}_{\eta\eta} + b_1\,\mathbf{x}_\xi + b_2\,\mathbf{x}_\eta \,. \tag{8.71}$$

The b_i are scalars defined in the previous exercise and the $\mathcal{T}_{ij}$ are second-order tensors that are composed of the three symmetric tensor product matrices $\mathbf{x}_\xi \otimes \mathbf{x}_\xi$, $[(\mathbf{x}_\xi \otimes \mathbf{x}_\eta) + (\mathbf{x}_\eta \otimes \mathbf{x}_\xi)]$, $\mathbf{x}_\eta \otimes \mathbf{x}_\eta$, and the 2×2 identity $\mathcal{I}$. Abbreviated formulas for the case in which H does not depend on $\sqrt{g}$ are given here; complete formulas are given in Appendix A.

$$\begin{aligned} \mathcal{T}_{11} = {} & \frac{\partial H}{\partial g_{11}}\mathcal{I} \\ & + 2\,\frac{\partial^2 H}{\partial g_{11}^2}\,(\mathbf{x}_\xi \otimes \mathbf{x}_\xi) \\ & + \frac{\partial^2 H}{\partial g_{11}\partial g_{12}}\,[(\mathbf{x}_\xi \otimes \mathbf{x}_\eta) + (\mathbf{x}_\eta \otimes \mathbf{x}_\xi)] \\ & + \frac{1}{2}\,\frac{\partial^2 H}{\partial g_{12}^2}\,(\mathbf{x}_\eta \otimes \mathbf{x}_\eta) \,, \end{aligned} \tag{8.72}$$

$$\begin{aligned} \mathcal{T}_{12} + \mathcal{T}_{12}^T = {} & \frac{\partial H}{\partial g_{12}}\mathcal{I} \\ & + 2\,\frac{\partial^2 H}{\partial g_{11}\partial g_{12}}\,(\mathbf{x}_\xi \otimes \mathbf{x}_\xi) \\ & + 2\,\frac{\partial^2 H}{\partial g_{11}\partial g_{22}}\,[(\mathbf{x}_\xi \otimes \mathbf{x}_\eta) + (\mathbf{x}_\eta \otimes \mathbf{x}_\xi)] \\ & + 2\,\frac{\partial^2 H}{\partial g_{12}\partial g_{22}}\,(\mathbf{x}_\eta \otimes \mathbf{x}_\eta) \,, \end{aligned} \tag{8.73}$$

and

$$\begin{aligned}\mathcal{T}_{22} = & \frac{\partial H}{\partial g_{22}} \mathcal{I} \\ & + \frac{1}{2} \frac{\partial^2 H}{\partial g_{12}^2} (\mathbf{x}_\xi \otimes \mathbf{x}_\xi) \\ & + \frac{\partial^2 H}{\partial g_{22} \partial g_{12}} [(\mathbf{x}_\xi \otimes \mathbf{x}_\eta) + (\mathbf{x}_\eta \otimes \mathbf{x}_\xi)] \\ & + 2 \frac{\partial^2 H}{\partial g_{22}^2} (\mathbf{x}_\eta \otimes \mathbf{x}_\eta) .\end{aligned} \tag{8.74}$$

Exercise 8.3.7 Compute the tensors $\mathcal{T}_{ij}$ for $H = g_{12}^2$, $H = g_{11} g_{22}$, and compare the results to (6.67) and (6.72). §

Tensor form is lengthy, but quite useful for computing the Euler-Lagrange equation of a complicated functional. For example, try computing the nonconservative form of $H = \sqrt{g_{11} g_{22}}$ directly, without using (8.72)-(8.74). In addition, the formulas for $\mathcal{T}_{ij}$ can be used in a planar variational grid-generation code to numerically solve (8.71) using the formulas developed in Section 6.4.

For some functionals (such as Area and Winslow), it is convenient to use an alternate form of the expressions (8.72)-(8.74), namely that obtained by expressing the tensor products in terms of contravariant tangent vectors. This is accomplished using the identities (7.45)-(7.48) in Chapter 7. The result is given in Appendix A.

Exercise 8.3.8 Use the contravariant form of the tensor Euler-Lagrange equations in Appendix A to show that the $\mathcal{T}_{ij}$ for $H = (\sqrt{g})^2$ are

$$\mathcal{T}_{11} = (\mathbf{x}_\eta^\perp \otimes \mathbf{x}_\eta^\perp) , \tag{8.75}$$

$$\mathcal{T}_{12} + \mathcal{T}_{12}^T = -[(\mathbf{x}_\xi^\perp \otimes \mathbf{x}_\eta^\perp) + (\mathbf{x}_\eta^\perp \otimes \mathbf{x}_\xi^\perp)] , \tag{8.76}$$

$$\mathcal{T}_{22} = (\mathbf{x}_\xi^\perp \otimes \mathbf{x}_\xi^\perp) . \tag{8.77}$$

Compare these to what is obtained for the area functional using (8.72)-(8.74) and compare them to (6.61) in Chapter 6. §

Another reason for introducing the contravariant tensor form is to show that the Winslow grid generator is indeed obtained from the functional $H = (g_{11} + g_{22})/\sqrt{g}$. This is readily done using (A.4)-(A.6) in conjunction with (7.48).

Exercise 8.3.9 Show that the tensors for the Winslow functional are

$$g^{\frac{3}{2}} \mathcal{T}_{11} = g_{22} [(\mathbf{x}_\xi^\perp \otimes \mathbf{x}_\xi^\perp) + (\mathbf{x}_\eta^\perp \otimes \mathbf{x}_\eta^\perp)] , \tag{8.78}$$

$$g^{\frac{3}{2}} (\mathcal{T}_{12} + \mathcal{T}_{12}^T) = -2 g_{12} [(\mathbf{x}_\xi^\perp \otimes \mathbf{x}_\xi^\perp) + (\mathbf{x}_\eta^\perp \otimes \mathbf{x}_\eta^\perp)] , \tag{8.79}$$

$$g^{\frac{3}{2}} \mathcal{T}_{22} = g_{11} [(\mathbf{x}_\xi^\perp \otimes \mathbf{x}_\xi^\perp) + (\mathbf{x}_\eta^\perp \otimes \mathbf{x}_\eta^\perp)] . \text{ §} \tag{8.80}$$

All three tensors in the previous exercise are proportional to the tensor $g\,(\mathcal{J}^{-1})^T\mathcal{J}^{-1}$. If $\sqrt{g} \neq 0$, $g\,(\mathcal{J}^{-1})^T\mathcal{J}^{-1}$ is invertible [see (7.43)-(7.44)]; premultiplying the tensor form by $\frac{\mathcal{J}\mathcal{J}^T}{g}$ results in the Winslow operator $\mathcal{Q}_w\mathbf{x}$:

$$g_{22}\,\mathbf{x}_{\xi\xi} - 2\,g_{12}\,\mathbf{x}_{\xi\eta} + g_{11}\,\mathbf{x}_{\eta\eta} = \mathbf{0}. \tag{8.81}$$

Exercise 8.3.10 Form the projections $\mathbf{x}_\xi \cdot \mathcal{Q}_w\mathbf{x}$ and $\mathbf{x}_\eta \cdot \mathcal{Q}_w\mathbf{x}$ for the TTM operator and show [using (4.131)-(4.136)] that

$$\frac{1}{2}\left(\frac{g_{22}}{g_{11}}\right)_\xi = \left(\frac{g_{12}}{g_{11}}\right)_\eta, \tag{8.82}$$

$$\frac{1}{2}\left(\frac{g_{11}}{g_{22}}\right)_\eta = \left(\frac{g_{12}}{g_{22}}\right)_\xi. \S \tag{8.83}$$

8.3.6 *Logical-Space Weighting*

If a *logical-space weight* function $\phi(\xi,\eta)$ is used in the variational principle

$$I_5[\mathbf{x}] = \int_0^1\int_0^1 \frac{H}{\phi}\,d\xi\,d\eta, \tag{8.84}$$

then the first term in the Euler-Lagrange equation (gradient with respect to the physical variables) drops out, leaving just

$$\mathrm{div}_\xi \frac{\mathcal{J}\mathcal{B}}{\phi} = \mathbf{0}. \tag{8.85}$$

The inhomogeneous form is

$$\mathrm{div}_\xi \mathcal{J}\mathcal{B} = \mathcal{J}\mathcal{B}\frac{\nabla_\xi\phi}{\phi}. \tag{8.86}$$

In contrast to the physical-weight case, the covariant projection of the logical-space equation is not fully integrable:

$$\mathrm{div}_\xi \frac{\mathcal{G}\mathcal{B}}{\phi} = \frac{1}{\phi}\nabla_\xi H. \tag{8.87}$$

The tensor form (8.71) still holds, but the vector for the first-order derivatives is

$$\mathbf{b} = \mathcal{B}\frac{\nabla_\xi\phi}{\phi}. \tag{8.88}$$

As discussed in Chapter 3, logical-space weighting is not recommended since the effect of the weight occurs at an unpredictable location in physical space.

8.4 *THE SECOND VARIATION*

As noted in Section 6.2, if it is possible to prove that the *second variation* of a functional is positive then it is possible to prove that a unique minimum exists.

It was shown that for the two-dimensional hosted equation the second variation was positive if the matrix $\mathcal{T}$ had positive eigenvalues; the analogous situation for the variational principle involving H would be for $\mathcal{B}$ to have positive eigenvalues. However, since $\mathcal{B}$ in general is a function of the elements of the metric tensor, the problem is nonlinear and the criterion that applies to the hosted equation is insufficient for the grid-generation equations (but it is a good start).

The second variation of the homogeneous grid-generation functional (8.4) may be computed using (6.5). Terms of the following type occur in the integrand:

$$(\mathbf{c}_\xi \cdot \mathbf{x}_\xi)\,(\mathbf{x}_\eta \cdot \mathbf{c}_\eta)\,\frac{\partial^2 H}{\partial g_{11}\partial g_{22}}\,. \tag{8.89}$$

The result (7.40) can be applied to obtain the tensor product form

$$\mathbf{c}_\xi \cdot [(\mathbf{x}_\xi \otimes \mathbf{x}_\eta)\,\mathbf{c}_\eta]\frac{\partial^2 H}{\partial g_{11}\partial g_{22}}\,, \tag{8.90}$$

which motivates the following result.

THEOREM 8.6 The second variation of the functional involving H has the form

$$\frac{1}{2}D^2_{\mathbf{c}}I_4 = \int_0^1\int_0^1 [\mathbf{c}_\xi, \mathbf{c}_\eta]^T \begin{pmatrix} \mathcal{T}_{11} & \mathcal{T}_{12} \\ \mathcal{T}_{12}^T & \mathcal{T}_{22} \end{pmatrix} \begin{bmatrix} \mathbf{c}_\xi \\ \mathbf{c}_\eta \end{bmatrix} d\xi\, d\eta\,. \tag{8.91}$$

The proof is a lengthy and tedious vector and tensor analysis calculation. §

Let the 4×4 block matrix $\mathcal{H}$ be denoted by

$$\mathcal{H} = \begin{pmatrix} \mathcal{T}_{11} & \mathcal{T}_{12} \\ \mathcal{T}_{12}^T & \mathcal{T}_{22} \end{pmatrix}. \tag{8.92}$$

The 2×2 blocks are just the matrices $\mathcal{T}_{ij}$ that appear in the tensor form (8.71) of the Euler-Lagrange equations; $\mathcal{T}_{12}$ is obtained from formula (8.73) for $\mathcal{T}_{12} + \mathcal{T}_{12}^T$ by dividing the tensor by two and eliminating terms involving $\mathbf{x}_\eta \otimes \mathbf{x}_\xi$. Theorem 8.6 gives a necessary condition for the existence of a minimum of I_4, namely that the second variation be positive. A sufficient condition for the second variation to be positive is that the block matrix above have positive eigenvalues. For example, if $H = g_{11} + g_{22}$, then $\mathcal{H}$ is just the identity matrix. Therefore, the second variation of the Length functional is positive.

Unfortunately, this approach is not helpful in analyzing any of the other functionals in Table 8.1. For example, the eigenvalues of the matrix $\mathcal{H}$ can be computed for the AO functional:

$$2\,\lambda_1 = (g_{11} + g_{22}) + [(g_{11} + g_{22})^2 + 12\,g_{11}\,g_{22}]^{\frac{1}{2}}\,, \tag{8.93}$$

$$2\,\lambda_2 = (g_{11} + g_{22}) - [(g_{11} + g_{22})^2 + 12\,g_{11}\,g_{22}]^{\frac{1}{2}}\,, \tag{8.94}$$

$$\lambda_3 = g_{22}\,, \tag{8.95}$$

$$\lambda_4 = g_{11}\,. \tag{8.96}$$

The second eigenvalue is seen to be negative, while the others are positive. Therefore, one cannot conclude anything certain about the positivity of the second variation of the AO functional from this approach.

More sophisticated approaches to convexity can be found in Dacorogna [42]. Many functionals are not convex in the classic sense, but have the property of "polyconvexity." A function $G = G(x_\xi, y_\xi, x_\eta, y_\eta, \sqrt{g})$ is *polyconvex* if it is formally convex in each of its arguments. Polyconvexity, in turn, can be linked to notions of ellipticity. Liao [124] has studied the polyconvexity of the weighted functionals (6.69) and (8.11). Generalizations of the Winslow functional are treated in Dacorogna [42, p. 134].

A connection between convexity and ellipticity is made in reference [42], namely that under the proper assumptions, convexity is equivalent to the Legendre-Hadamard ellipticity condition. The theory given there is not directly applicable to the grid generation functionals; it is an open question as to whether it can be shown that the ellipticity test in Chapter 6 is a consequence of the convexity condition given in the present section.

8.5 CONTRAVARIANT APPROACH

Variational grid generation may also be formulated in terms of functionals whose domain is the set of *inverse mappings* $(\xi(x,y), \eta(x,y))$; this is the approach taken in the well-known variational method of Brackbill and Saltzman [21]. The *Brackbill-Saltzman* variational principles have the general form (6.19), for which the Euler-Lagrange equations are (6.20)–(6.21). To compare the Brackbill-Saltzman method to the variational method described in this chapter, assume the specific functional form

$$I_{BS}[\xi(x,y), \eta(x,y)] = \int_\Omega \frac{1}{w^2} \hat{H}\left(g^{11}, g^{12}, g^{22}, \frac{1}{\sqrt{g}}\right) dx\, dy\,, \tag{8.97}$$

which is to be minimized to find the functions $\xi(x,y)$ and $\eta(x,y)$. The integrand $\hat{H}$ is a function of the elements of the inverse metric tensor; a physical-space weight function $w = w(x,y)$ is assumed. The goal is to derive the Euler-Lagrange equations in the manner used in this chapter and to invert them using the transformation rules given in the previous chapter.

The Euler-Lagrange equation for (8.97) is

$$\mathrm{div}_{\mathbf{x}} \frac{\hat{S}}{w^2} = \mathbf{0}\,, \tag{8.98}$$

where

$$\hat{S} = \begin{pmatrix} \frac{\partial \hat{H}}{\partial \xi_x} & \frac{\partial \hat{H}}{\partial \xi_y} \\ \frac{\partial \hat{H}}{\partial \eta_x} & \frac{\partial \hat{H}}{\partial \eta_y} \end{pmatrix}. \tag{8.99}$$

The right-hand side of the Euler-Lagrange equation is zero due to use of a physical-space weight; if a logical-space weight had been used, a nonzero right-hand side would appear. This is just the opposite of the situation that occurred in the direct mapping approach; see equations (8.19) and (8.22).

Following the approach used in Section 8.3.2, the tensor $\hat{S}$ may be factored to obtain $\hat{S} = \hat{\mathcal{B}}(\nabla_{\mathbf{x}}\boldsymbol{\xi})$, where

$$\hat{\mathcal{B}} = \begin{pmatrix} 2\,\frac{\partial \hat{H}}{\partial g^{11}} & \frac{\partial \hat{H}}{\partial g^{12}} \\ \frac{\partial \hat{H}}{\partial g^{12}} & 2\,\frac{\partial \hat{H}}{\partial g^{22}} \end{pmatrix} + \frac{1}{\sqrt{g}}\,\frac{\partial \hat{H}}{\partial\left(\frac{1}{\sqrt{g}}\right)}\,\mathcal{G}\,. \tag{8.100}$$

$\hat{\mathcal{B}}$ is symmetric. The Brackbill-Saltzman Euler-Lagrange equation is then

$$\mathrm{div}_{\mathbf{x}} \frac{\hat{\mathcal{B}}\,(\nabla_{\mathbf{x}}\boldsymbol{\xi})}{w^2} = \mathbf{0}\,. \tag{8.101}$$

Compared to the direct mapping Euler-Lagrange equation (8.35), the order of the matrices has been reversed, $\mathcal{B} \to \hat{\mathcal{B}}$, and the Jacobian has become its inverse.

The Euler-Lagrange equations resulting from such a minimization must be inverted to obtain a computationally useful scheme. The transformation relationship (7.73) derived in the previous chapter is applied to (8.101) to yield

$$\frac{1}{\sqrt{g}}\,\mathrm{div}_{\boldsymbol{\xi}}\left(\frac{\sqrt{g}\,\hat{\mathcal{B}}\,\mathcal{G}^{-1}}{w^2}\right) = \mathbf{0}\,. \tag{8.102}$$

Example 8.7 Let $\hat{H} = g^{11} + g^{22}$ be the homogeneous TTM functional. The Euler-Lagrange equation is $\mathrm{div}_{\mathbf{x}}\mathcal{J}^{-1} = \mathbf{0}$, i.e., the vector Laplacian of $\boldsymbol{\xi}$. Transformed, this reads $\mathrm{div}_{\boldsymbol{\xi}}(\sqrt{g}\,\mathcal{G}^{-1}) = \mathbf{0}$. §

Example 8.8 Let $\hat{H} = (1/\sqrt{g})^{-1}$ be the area functional. The Euler-Lagrange equation is $\mathrm{div}_{\mathbf{x}}\sqrt{g}\mathcal{J}^T = \mathbf{0}$, which, when transformed, reads $\mathrm{div}_{\boldsymbol{\xi}} g\,\mathcal{I} = \mathbf{0}$. §

The *contravariant* projection of the Brackbill-Saltzman equation (8.101) is derived next.

Exercise 8.5.1 Verify that

$$[\nabla_{\mathbf{x}}(\mathcal{J}^{-1})^T]\,(\hat{\mathcal{B}}\,\mathcal{J}^{-1}) = \nabla_{\mathbf{x}}\hat{H}\,.\ \S \tag{8.103}$$

The contravariant projection of (8.101) is found by premultiplying (8.101) by $(\mathcal{J}^{-1})^T$. Using (8.103), the following projection is obtained:

$$\mathrm{div}_{\mathbf{x}}\left\{\frac{(\mathcal{J}^{-1})^T\,\hat{\mathcal{B}}\,\mathcal{J}^{-1}}{w^2}\right\} = \frac{1}{w^2}\,\nabla_{\mathbf{x}}\hat{H}\,. \tag{8.104}$$

The contravariant projection is nonintegrable due to the use of a physical-space weighting function; if a logical-space weighting is used, the contravariant projection can be shown to be integrable. This situation is the opposite as in the direct mapping approach, for which the covariant projection is integrable if a physical-space weighting is used but not if a logical space weighting is used. The contravariant projection may be rewritten to obtain an expression similar in form to (8.56):

$$\mathrm{div}_{\mathbf{x}}\left\{\frac{\hat{H}\,\mathcal{I} - (\mathcal{J}^{-1})^T\,\hat{\mathcal{B}}\,\mathcal{J}^{-1}}{w^2}\right\} = \hat{H}\,\nabla_{\mathbf{x}}\frac{1}{w^2}\,. \tag{8.105}$$

The Brackbill-Saltzman approach can be related to the Steinberg-Roache approach. If the variational principle (8.97) is transformed to the logical domain, one obtains a variational principle involving H, with $H = \sqrt{g}\hat{H}$. The chain rule can be applied to show that

$$\frac{\partial \hat{H}}{\partial g^{11}} = \sqrt{g}\frac{\partial H}{\partial g_{22}}, \tag{8.106}$$

$$\frac{\partial \hat{H}}{\partial g^{12}} = -\sqrt{g}\frac{\partial H}{\partial g_{12}}, \tag{8.107}$$

$$\frac{\partial \hat{H}}{\partial g^{22}} = \sqrt{g}\frac{\partial H}{\partial g_{11}}, \tag{8.108}$$

$$\frac{\partial \hat{H}}{\partial\left(\frac{1}{\sqrt{g}}\right)} = H - 2\left(g_{11}\frac{\partial H}{\partial g_{11}} + g_{12}\frac{\partial H}{\partial g_{12}} + g_{22}\frac{\partial H}{\partial g_{22}} + \frac{\sqrt{g}}{2}\frac{\partial H}{\partial \sqrt{g}}\right) \tag{8.109}$$

This result can be used to obtain the following relationship between the matrices $\hat{\mathcal{B}}$ and $\mathcal{B}$:

$$\hat{\mathcal{B}} = \frac{H\mathcal{G} - \mathcal{G}\mathcal{B}\mathcal{G}}{\sqrt{g}}. \tag{8.110}$$

THEOREM 8.9 The covariant projection (8.56) of the Steinberg-Roache Euler-Lagrange equations, transformed to physical space, equals $\sqrt{g}$ times the Brackbill-Saltzman Euler-Lagrange equation (8.101). Conversely, the contravariant projection (8.105) of the Brackbill-Saltzman Euler-Lagrange equations, transformed to logical space, equals $1/\sqrt{g}$ times the Steinberg-Roache Euler-Lagrange equation (8.35).

Proof. The first statement is evident from the following sequence, based on (8.110) and the transformation rules for the divergence of a tensor:

$$\operatorname{div}_{\boldsymbol{\xi}}\left\{\frac{H\mathcal{I} - \mathcal{G}\mathcal{B}}{w^2}\right\} = \operatorname{div}_{\boldsymbol{\xi}}\left\{\frac{\sqrt{g}\,\hat{\mathcal{B}}\,\mathcal{G}^{-1}}{w^2}\right\}, \tag{8.111}$$

$$= \sqrt{g}\operatorname{div}_{\mathbf{x}}\left\{\frac{\hat{\mathcal{B}}\,\mathcal{J}^{-1}}{w^2}\right\}. \tag{8.112}$$

The second statement is based on

$$\hat{H}\operatorname{div}_{\mathbf{x}}\frac{\mathcal{I}}{w^2} - \operatorname{div}_{\mathbf{x}}\left\{\frac{\hat{H}\mathcal{I} - (\mathcal{J}^{-1})^T\hat{\mathcal{B}}\mathcal{J}^{-1}}{w^2}\right\}$$

$$= \frac{H}{\sqrt{g}}\operatorname{div}_{\mathbf{x}}\frac{\mathcal{I}}{w^2} - \operatorname{div}_{\mathbf{x}}\left\{\frac{\mathcal{J}\mathcal{B}\mathcal{J}^T}{\sqrt{g}w^2}\right\}, \tag{8.113}$$

$$= \frac{1}{\sqrt{g}}\left\{\frac{H}{\sqrt{g}}\operatorname{div}_{\boldsymbol{\xi}}\frac{C}{w^2} - \operatorname{div}_{\boldsymbol{\xi}}\frac{\mathcal{J}\mathcal{B}}{w^2}\right\}. \ \S \tag{8.114}$$

Another way to view the first statement of the theorem is to say that one can obtain the Brackbill-Saltzman Euler-Lagrange equation by premultiplying the Steinberg-Roache Euler-Lagrange equation by $\mathcal{J}^T$ and then "post-multiplying" the result, again by $\mathcal{J}^T$. From (7.74) in Chapter 7, post-multiplication is accomplished by transforming from logical to physical coordinates.

It is clear from this theorem that the two approaches to variational grid generation lead to basically the same grid-generation equations, so the two methods are essentially equivalent. Although the covariant approach has a natural geometric interpretation, the contravariant method (discussed further in Chapter 11) is preferred because of the utility of physical-space weight functions. Such weights directly give a symmetric form for the Euler-Lagrange equations in the contravariant case, whereas in the covariant case, one must first form the projection of the Euler-Lagrange equations to obtain the symmetric form.

Chapter 9

Grid Generation in Three Dimensions

Considerably less is known about robust methods of grid generation in three dimensions than in the plane. A large part of the reason for this is that the topology of three-dimensional (3D) objects is significantly more complex than in two dimensions, and also the computing cost of generating a grid in 3D is much greater than in 2D. While many objects in the plane can be viewed as moderate distortions of the unit square, few objects in three-dimensional space are readily envisioned as moderate distortions of a cube. Most three-dimensional objects are better thought of as the union of a number of separate cube-like parts. If a set of maps, $X^3_{\ell,3} : U_3 \rightarrow \Omega^3_{\ell,3}$ from the unit cube to $\ell = 1, 2, \ldots, L$ three-dimensional volumes in E^3 is constructed, there remains the difficulty of putting the separate maps together in such a way as to ensure continuity of slopes across the interfaces between the blocks. This patching problem is currently most effectively addressed by the cubic spline approach (see Shih [169]) of algebraic grid generation, by use of Neumann boundary conditions in an elliptic method, or by changing the shape of the logical domain to avoid patching of multiple maps.

The intention in this chapter is to provide a brief overview of the basic problem of three-dimensional grid generation, namely finding a single map X^3_3 from the unit cube to a block-like domain Ω^3_3. Relevant relationships from Chapter 4 are thus reviewed and extended in Section 9.1. To solve the basic problem, an initial boundary map ∂X^3_3 must be given; generally the boundary is defined in terms of six surface parameterizations described in Section 9.2.1. Frequently, the boundary of the object is given as a set of data points from which a surface parameterization must be constructed. Often the given parameterization of the boundary results in poor grids when the former is extended into the interior. If the given boundary parameterization is inadequate, the curve and surface algorithms described in the next chapter may be applied to perform a re-parameterization. Once the not inconsequential task of determining the boundary parameterization is completed, the three-dimensional transfinite interpolation formula given in Section 9.2.2 can be used to extend the grid into the interior. If the transfinite interpolation grid is not adequate, then inhomogeneous grid generators based on partial differential equations may be used to improve the grid; this approach is quite costly and is currently taken only as a last resort.

Although many of the methods discussed in the planar chapter extend to three dimensions, most are not widely used in practice. Noticeably absent from the list of widely used three-dimensional methods are conformal maps, hyperbolic grid generators, the biharmonic approach, and variational methods (including the direct optimization method). Algebraic methods currently reign as the most successful approach to three-dimensional structured grid generation due to the speed with which they may be computed. Only when algebraic methods fail does one consider differential equation-based grid generators, due to the relative slowness at which the latter are computed. The most widely used of the elliptic generators is the 3D inhomogeneous Thompson-Thames-Mastin generator, as modified in the Thomas-Middlecoff or Steger-Sorenson approach (see Section 9.2.3). Some success has also been had with 3D orthogonal grid generators. Papers that contain discussions of 3D grid generation include Arina and Casella [12], Eiseman [63], Mastin and Thompson [127], Shubin, Stephens, and Bell [171], Sorenson and Steger [182], Steger [189], Theodoropoulos and Bergeles [202], Warsi and Ziebarth [225], and Warsi [224]. Numerous applications of three-dimensional grid generation may be found in the conference proceedings edited by Arcilla, Häuser, Eiseman, and Thompson [9], Häuser and Taylor [89], Sengupta, Häuser, Eiseman, and Thompson [167], and Thompson [210].

The development in Chapter 8 is followed to extend variational methods to three-dimensional volumes in Section 9.3. Variational methods have scarcely been used in three-dimensional grid generation, primarily due to excessive storage and computational requirements, but the efficient variational algorithms presented in this chapter may overcome some of the limitations.

9.1 VOLUME DIFFERENTIAL GEOMETRY

Relationships from Chapter 4 are reviewed and extended in this section. The reader may wish to review Sections 4.2.3 and 4.3 for related material. Let

$$\mathbf{x} = (x(\xi,\eta,\zeta),\, y(\xi,\eta,\zeta),\, z(\xi,\eta,\zeta)) \in \Omega_3^3, \tag{9.1}$$

where the coordinates are smooth functions of $(\xi,\eta,\zeta) \in U_3$. The three *covariant tangents* are

$$\mathbf{x}_\xi = (x_\xi,\, y_\xi,\, z_\xi), \tag{9.2}$$

$$\mathbf{x}_\eta = (x_\eta,\, y_\eta,\, z_\eta), \tag{9.3}$$

$$\mathbf{x}_\zeta = (x_\zeta,\, y_\zeta,\, z_\zeta). \tag{9.4}$$

The Jacobian matrix is $\mathcal{J} = \nabla_{\boldsymbol{\xi}}\mathbf{x}$, i.e.,

$$\mathcal{J} = \begin{pmatrix} x_\xi & x_\eta & x_\zeta \\ y_\xi & y_\eta & y_\zeta \\ z_\xi & z_\eta & z_\zeta \end{pmatrix}. \tag{9.5}$$

The determinant is the root of the metric $\sqrt{g} = \det(\mathcal{J}) = \mathbf{x}_\xi \cdot (\mathbf{x}_\eta \times \mathbf{x}_\zeta)$.

Exercise 9.1.1 Following the notation introduced in (8.29), show that

$$\frac{\partial\sqrt{g}}{\partial \mathbf{x}_\xi} = \mathbf{x}_\eta \times \mathbf{x}_\zeta\,, \quad \frac{\partial\sqrt{g}}{\partial \mathbf{x}_\eta} = \mathbf{x}_\zeta \times \mathbf{x}_\xi\,, \quad \frac{\partial\sqrt{g}}{\partial \mathbf{x}_\zeta} = \mathbf{x}_\xi \times \mathbf{x}_\eta\,. \ \S \tag{9.6}$$

The inverse Jacobian matrix is $\mathcal{J}^{-1} = \nabla_\mathbf{x}\boldsymbol{\xi}$, i.e.,

$$\mathcal{J}^{-1} = \begin{pmatrix} \xi_x & \xi_y & \xi_z \\ \eta_x & \eta_y & \eta_z \\ \zeta_x & \zeta_y & \zeta_z \end{pmatrix}. \tag{9.7}$$

For reference, recall the relations (4.83)

$$\nabla_\mathbf{x}\xi = \frac{\mathbf{x}_\eta \times \mathbf{x}_\zeta}{\sqrt{g}}\,, \quad \nabla_\mathbf{x}\eta = \frac{\mathbf{x}_\zeta \times \mathbf{x}_\xi}{\sqrt{g}}\,, \quad \nabla_\mathbf{x}\zeta = \frac{\mathbf{x}_\xi \times \mathbf{x}_\eta}{\sqrt{g}}. \tag{9.8}$$

These relations are analogous to equations (4.122)–(4.123); the cross products of the covariant tangents play the role of the "perp" vectors in that section.

Exercise 9.1.2 Apply the identity (4.11) in Chapter 4 to (9.8) to obtain

$$\mathbf{x}_\xi = \sqrt{g}\,(\nabla_\mathbf{x}\eta \times \nabla_\mathbf{x}\zeta), \tag{9.9}$$

$$\mathbf{x}_\eta = \sqrt{g}\,(\nabla_\mathbf{x}\zeta \times \nabla_\mathbf{x}\xi), \tag{9.10}$$

$$\mathbf{x}_\zeta = \sqrt{g}\,(\nabla_\mathbf{x}\xi \times \nabla_\mathbf{x}\eta)\,. \ \S \tag{9.11}$$

The covariant *metric tensor* is defined by $\mathcal{G} = \mathcal{J}^T\mathcal{J}$, i.e.,

$$\mathcal{G} = \begin{pmatrix} g_{11} & g_{12} & g_{13} \\ g_{12} & g_{22} & g_{23} \\ g_{13} & g_{23} & g_{33} \end{pmatrix}. \tag{9.12}$$

The entries of $\mathcal{G}$ consist of scalar products of the form

$$g_{ij} = \mathbf{x}_{\xi_i} \cdot \mathbf{x}_{\xi_j}\,, \tag{9.13}$$

with $i, j = 1, 2, 3$, $\xi_1 = \xi$, $\xi_2 = \eta$, and $\xi_3 = \zeta$ (e.g., $g_{13} = \mathbf{x}_\xi \cdot \mathbf{x}_\zeta$). The determinant of $\mathcal{G}$ is denoted by g.

Exercise 9.1.3 Compute $g = \det\mathcal{G}$ from (9.12). §

The *inverse* metric tensor has the form

$$\mathcal{G}^{-1} = \begin{pmatrix} g^{11} & g^{12} & g^{13} \\ g^{12} & g^{22} & g^{23} \\ g^{13} & g^{23} & g^{33} \end{pmatrix}. \tag{9.14}$$

The relationship between the elements of the inverse metric tensor and the metric tensor were given in (4.110) and in Exercise 4.3.1.

The metric identities (7.12) and (7.13) hold for the 3×3 auxiliary matrices $C = \sqrt{g}\,(\mathcal{J}^{-1})^T$ and $C^{-1} = \mathcal{J}^T/\sqrt{g}$.

Exercise 9.1.4 Show that in three dimensions, $\text{div}_{\boldsymbol{\xi}} C = \mathbf{0}$ is equivalent to the identity

$$\frac{\partial}{\partial \xi}(\mathbf{x}_\eta \times \mathbf{x}_\zeta) + \frac{\partial}{\partial \eta}(\mathbf{x}_\zeta \times \mathbf{x}_\xi) + \frac{\partial}{\partial \zeta}(\mathbf{x}_\xi \times \mathbf{x}_\eta) = \mathbf{0}\,. \ \S \tag{9.15}$$

Exercise 9.1.5 Show that

$$\mathbf{x}_\eta \times \mathbf{x}_\zeta = \sqrt{g}\,\{g^{11}\mathbf{x}_\xi + g^{12}\mathbf{x}_\eta + g^{13}\mathbf{x}_\zeta\}, \tag{9.16}$$

$$\mathbf{x}_\zeta \times \mathbf{x}_\xi = \sqrt{g}\,\{g^{12}\mathbf{x}_\xi + g^{22}\mathbf{x}_\eta + g^{23}\mathbf{x}_\zeta\}, \tag{9.17}$$

$$\mathbf{x}_\xi \times \mathbf{x}_\eta = \sqrt{g}\,\{g^{13}\mathbf{x}_\xi + g^{23}\mathbf{x}_\eta + g^{33}\mathbf{x}_\zeta\}\,. \tag{9.18}$$

These relationships are analogous to (4.126)–(4.127). §

Exercise 9.1.6 Explicitly write out the three-dimensional metric identity $\text{div}_{\boldsymbol{\xi}}(\sqrt{g}\,\mathcal{J}\,\mathcal{G}^{-1}) = \mathbf{0}$. §

The three-dimensional version of the planar Gauss relations (4.137)–(4.139) is

$$\mathbf{x}_{\xi_i \xi_j} = \sum_{\ell=1}^{3} \Gamma_{ij}^{\ell}\,\mathbf{x}_{\xi_\ell} \tag{9.19}$$

where Γ_{ij}^{ℓ} is the space Christoffel symbol of the second kind. The latter is defined in terms of the space Christoffel symbol of the first kind, $[ij,k]$, through the relation

$$\Gamma_{ij}^{\ell} = \sum_{k=1}^{3} g^{\ell k}\,[ij,k]\,, \tag{9.20}$$

with the definition

$$[ij,k] = \frac{1}{2}\left\{\frac{\partial g_{ik}}{\partial \xi_j} + \frac{\partial g_{jk}}{\partial \xi_i} - \frac{\partial g_{ij}}{\partial \xi_k}\right\}. \tag{9.21}$$

Exercise 9.1.7 Show that the space Christoffel symbol of the first kind is merely the following inner product:

$$[ij,k] = \mathbf{x}_{\xi_k} \cdot \mathbf{x}_{\xi_i \xi_j}\,. \tag{9.22}$$

Thus, relation (9.21) is analogous to the planar relations (4.131)–(4.136). §

Exercise 9.1.8 Verify that the *gradient* in 3D transforms as in (7.50). The other transformation relations in Section 7.3 also generalize directly to 3D volumes. §

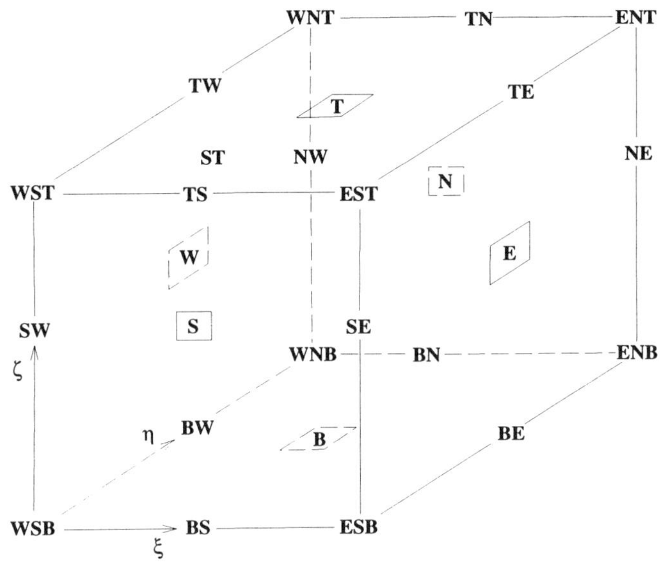

Figure 9.1 *The volume grid-generation problem*

9.2 APPROACHES TO THREE-DIMENSIONAL GRID GENERATION

9.2.1 The Volume Grid-Generation Problem

The *basic problem* can be stated: given a simply connected region $\Omega \subset R^3$ in physical space, find a mapping $\mathbf{x} = \mathbf{x}(\xi, \eta, \zeta)$ from the unit cube U_3 in logical space E^3 to the region Ω. The physical region is specified by giving its boundary; this task is considerably more complicated in three dimensions than in two. Six bounding surfaces

$$\mathbf{x}_W(s,t), \quad \mathbf{x}_E(s,t), \quad 0 \le s,t \le 1, \tag{9.23}$$

$$\mathbf{x}_N(r,t), \quad \mathbf{x}_S(r,t), \quad 0 \le r,t \le 1, \tag{9.24}$$

$$\mathbf{x}_T(r,s), \quad \mathbf{x}_B(r,s), \quad 0 \le r,s \le 1, \tag{9.25}$$

are required (see Figure 9.1). The subscripts on $\mathbf{x}$ stand for the *west, east, north, south, top,* and *bottom* surfaces of the logical domain. There are twelve edges

$$\mathbf{x}_{SW}(t), \quad \mathbf{x}_{SE}(t), \quad \mathbf{x}_{NW}(t), \quad \mathbf{x}_{NE}(t), \tag{9.26}$$

$$\mathbf{x}_{BW}(s), \quad \mathbf{x}_{TW}(s), \quad \mathbf{x}_{BE}(s), \quad \mathbf{x}_{TE}(s), \tag{9.27}$$

$$\mathbf{x}_{BS}(r), \quad \mathbf{x}_{TS}(r), \quad \mathbf{x}_{BN}(r), \quad \mathbf{x}_{TN}(r), \tag{9.28}$$

and eight corner points

$$\mathbf{x}_{WSB}, \quad \mathbf{x}_{WST}, \quad \mathbf{x}_{WNB}, \quad \mathbf{x}_{WNT}, \tag{9.29}$$

$$\mathbf{x}_{ESB}, \quad \mathbf{x}_{EST}, \quad \mathbf{x}_{ENB}, \quad \mathbf{x}_{ENT}. \tag{9.30}$$

The surface boundary conditions are then

$$\begin{aligned}
\mathbf{x}(0,s,t) &= \mathbf{x}_W(s,t)\,,\\
\mathbf{x}(1,s,t) &= \mathbf{x}_E(s,t)\,,\\
\mathbf{x}(r,0,t) &= \mathbf{x}_S(r,t)\,,\\
\mathbf{x}(r,1,t) &= \mathbf{x}_N(r,t)\,,\\
\mathbf{x}(r,s,0) &= \mathbf{x}_B(r,s)\,,\\
\mathbf{x}(r,s,1) &= \mathbf{x}_T(r,s)\,.
\end{aligned} \tag{9.31}$$

The edges also need boundary conditions:

$$\begin{aligned}
\mathbf{x}(0,0,t) &= \mathbf{x}_{SW}(t)\,,\\
\mathbf{x}(0,1,t) &= \mathbf{x}_{NW}(t)\,,\\
\mathbf{x}(1,0,t) &= \mathbf{x}_{SE}(t)\,,\\
\mathbf{x}(1,1,t) &= \mathbf{x}_{NE}(t)\,,\\
\mathbf{x}(0,s,0) &= \mathbf{x}_{BW}(s)\,,\\
\mathbf{x}(0,s,1) &= \mathbf{x}_{TW}(s)\,,\\
\mathbf{x}(1,s,0) &= \mathbf{x}_{BE}(s)\,,\\
\mathbf{x}(1,s,1) &= \mathbf{x}_{TE}(s)\,,\\
\mathbf{x}(r,0,0) &= \mathbf{x}_{BS}(r)\,,\\
\mathbf{x}(r,0,1) &= \mathbf{x}_{TS}(r)\,,\\
\mathbf{x}(r,1,0) &= \mathbf{x}_{BN}(r)\,,\\
\mathbf{x}(r,1,1) &= \mathbf{x}_{TN}(r)\,,
\end{aligned} \tag{9.32}$$

as do the corners:

$$\begin{aligned}
\mathbf{x}(0,0,0) &= \mathbf{x}_{WSB}\,,\\
\mathbf{x}(1,0,0) &= \mathbf{x}_{ESB}\,,\\
\mathbf{x}(0,1,0) &= \mathbf{x}_{WNB}\,,\\
\mathbf{x}(0,0,1) &= \mathbf{x}_{WST}\,,\\
\mathbf{x}(1,1,1) &= \mathbf{x}_{ENT}\,,\\
\mathbf{x}(0,1,1) &= \mathbf{x}_{WNT}\,,\\
\mathbf{x}(1,0,1) &= \mathbf{x}_{EST}\,,\\
\mathbf{x}(1,1,0) &= \mathbf{x}_{ENB}\,.
\end{aligned} \tag{9.33}$$

Exercise 9.2.1 Work out a few of the consistency conditions for the edges and corners of the physical domain similar to the planar conditions given in Table 1.3. §

Before the grid can be extended into the interior, adequate grids must be generated on the twelve bounding edges, and subsequently, the six bounding surfaces. This may involve re-parameterizing the original data (9.23)-(9.28); this is the subject of Chapter 10. The technique of introducing cuts to reduce the connectivity of the physical domain is useful, as is modifying the topology of the logical domain (see Thompson, Warsi, and Mastin [215]).

Exercise 9.2.2 Work out the 3D extension of the discretization (5.6) at the end of Section 5.2. §

9.2.2 *3D Transfinite Interpolation*

Algebraic grid generation is perhaps the most successful approach to three-dimensional grid generation due to the relative speed with which such grids may be calculated. The methods described in Section 5.3.1 have natural extensions to three dimensions; the references cited in that section can be consulted for details. Only the simplest of the transfinite interpolation formulas is given in this brief survey. The transfinite interpolation formula (1.45) in Section 1.5 has the following extension to three dimensions:

$$\mathbf{x}(\xi,\eta,\zeta) = \mathbf{x}_1 + \mathbf{x}_2 + \mathbf{x}_3 - \mathbf{x}_{12} - \mathbf{x}_{13} - \mathbf{x}_{23} + \mathbf{x}_{123}\,, \tag{9.34}$$

where

$$\mathbf{x}_1 = (1-\xi)\,\mathbf{x}_W(\eta,\zeta) + \xi\,\mathbf{x}_E(\eta,\zeta)\,, \tag{9.35}$$

$$\mathbf{x}_2 = (1-\eta)\,\mathbf{x}_S(\xi,\zeta) + \eta\,\mathbf{x}_N(\xi,\zeta)\,, \tag{9.36}$$

$$\mathbf{x}_3 = (1-\zeta)\,\mathbf{x}_B(\xi,\eta) + \zeta\,\mathbf{x}_T(\xi,\eta)\,, \tag{9.37}$$

$$\begin{aligned}\mathbf{x}_{12} = {} & (1-\xi)(1-\eta)\,\mathbf{x}_{SW}(\zeta) \\ & + (1-\xi)\,\eta\,\mathbf{x}_{NW}(\zeta) \\ & + \xi\,(1-\eta)\,\mathbf{x}_{SE}(\zeta) \\ & + \xi\,\eta\,\mathbf{x}_{NE}(\zeta)\,,\end{aligned} \tag{9.38}$$

$$\begin{aligned}\mathbf{x}_{13} = {} & (1-\xi)(1-\zeta)\,\mathbf{x}_{BW}(\eta) \\ & + (1-\xi)\,\zeta\,\mathbf{x}_{TW}(\eta) \\ & + \xi\,(1-\zeta)\,\mathbf{x}_{BE}(\eta) \\ & + \xi\,\zeta\,\mathbf{x}_{TE}(\eta)\,,\end{aligned} \tag{9.39}$$

$$\begin{aligned}\mathbf{x}_{23} = {} & (1-\eta)(1-\zeta)\,\mathbf{x}_{BS}(\xi) \\ & + (1-\eta)\,\zeta\,\mathbf{x}_{TS}(\xi) \\ & + \eta\,(1-\zeta)\,\mathbf{x}_{BN}(\xi) \\ & + \eta\,\zeta\,\mathbf{x}_{TN}(\xi)\,,\end{aligned} \tag{9.40}$$

$$\begin{aligned}\mathbf{x}_{123} = &(1-\xi)(1-\eta)(1-\zeta)\,\mathbf{x}_{WSB}\\ &+(1-\xi)(1-\eta)\,\zeta\,\mathbf{x}_{WST}\\ &+(1-\xi)\,\eta\,(1-\zeta)\,\mathbf{x}_{WNB}\\ &+(1-\xi)\,\eta\,\zeta\,\mathbf{x}_{WNT}\\ &+\xi\,(1-\eta)(1-\zeta)\,\mathbf{x}_{ESB}\\ &+\xi\,(1-\eta)\,\zeta\,\mathbf{x}_{EST}\\ &+\xi\,\eta\,(1-\zeta)\,\mathbf{x}_{ENB}\\ &+\xi\,\eta\,\zeta\,\mathbf{x}_{ENT}\,. \qquad (9.41)\end{aligned}$$

The transfinite formula is often adequate if the physical domain has been divided into several cube-like subdomains. If not, this approach may serve to generate the initial guess in iterative solution procedures for solving the partial differential equations of 3D grid generation described in the following sections.

9.2.3 3D Thompson-Thames-Mastin

The 3D inhomogeneous Thompson-Thames-Mastin (TTM) equation is based on the map

$$\boldsymbol{\xi} = (\xi(x,y,z),\eta(x,y,z),\zeta(x,y,z)) \qquad (9.42)$$

from physical to logical space. The planar relationship (5.75) generalizes to:

$$\nabla^2_{\mathbf{x}}\boldsymbol{\xi} = \mathbf{f}\,, \qquad (9.43)$$

where the components of $\mathbf{f}$ are $f_i = g^{ii} P_i$ (no sum on the index i). The P_i are logical-space control functions. The system (9.43) is easily inverted using the approach in Section 7.3.5:

$$[\nabla_{\boldsymbol{\xi}} J]\, G^{-1} = -J\mathbf{f}\,, \qquad (9.44)$$

or, explicitly,

$$\begin{aligned}&g^{11}\,\mathbf{x}_{\xi\xi} + g^{22}\,\mathbf{x}_{\eta\eta} + g^{33}\,\mathbf{x}_{\zeta\zeta} +\\ &2\,g^{12}\,\mathbf{x}_{\xi\eta} + 2\,g^{13}\,\mathbf{x}_{\xi\zeta} + 2\,g^{23}\,\mathbf{x}_{\eta\zeta} +\\ &g^{11}\,P_1\,\mathbf{x}_{\xi} + g^{22}\,P_2\,\mathbf{x}_{\eta} + g^{33}\,P_3\,\mathbf{x}_{\zeta} = \mathbf{0}. \qquad (9.45)\end{aligned}$$

The operator notation for the 3D TTM method is $\mathcal{Q}_{TTM}\mathbf{x} = -J\mathbf{f}$ where

$$\begin{aligned}\mathcal{Q}_{TTM}\mathbf{x} = &\,g^{11}\,\mathbf{x}_{\xi\xi} + g^{22}\,\mathbf{x}_{\eta\eta} + g^{33}\,\mathbf{x}_{\zeta\zeta}\\ &+2\,g^{12}\,\mathbf{x}_{\xi\eta} + 2\,g^{13}\,\mathbf{x}_{\xi\zeta} + 2\,g^{23}\,\mathbf{x}_{\eta\zeta}\,. \qquad (9.46)\end{aligned}$$

It is not appropriate to refer to the three-dimensional TTM generator as the Winslow generator since the generalization originated later (see, for example, Warsi [224]). As noted in previous chapters, there is no guarantee against folding with the three-dimensional homogeneous TTM generator since the maximum principle does not hold in this setting. The source term **f** is generally modified in applications to achieve control over the grid at the boundary following Thomas [206] or Sorenson and Steger [182]. The latter essentially enforces Neumann boundary conditions to permit grid lines to intersect bounding surfaces at right angles so that coordinate lines will be smooth across blocks. Section 5.6.2 gives the planar version of the latter algorithm.

9.3 THE VARIATIONAL METHOD

9.3.1 3D Variational Principles

Variational methods have seen little application to three-dimensional grid generation, primarily due to the complexity of the equations that arise and the resulting demand on storage and computational effort. In spite of these difficulties, variational methods may prove useful given the relatively large amount of control over the grid that is attained. The theory developed in Chapter 8 carries over readily to the three-dimensional case. Based on the results of the previous variational chapters, it is proposed to minimize the functional

$$I[\mathbf{x}] = \int_0^1 \int_0^1 \int_0^1 \frac{H}{w^2(\mathbf{x})}\, d\xi\, d\eta\, d\zeta \tag{9.47}$$

with $H = H(g_{11}, g_{12}, g_{13}, g_{22}, g_{23}, g_{33}, \sqrt{g})$ being a smooth positive function from R^7 to R and w a weight function of the physical variables. The function H is assumed to be homogeneous of some degree to achieve invariance of the generator to rigid-body motions. Considerable uncertainty exists at present concerning the choice of variational principle.

Some of the planar grid generators in Table 8.1 have natural extensions to 3D, while generalizations of others are more difficult to construct (there being numerous possibilities). Note that any difficulties with planar generators will extend to 3D because any planar region can be naturally extended to a 3D volume. Table 9.1 lists some proposed 3D variational principles. Length is easily generalized; its tendency to generate folded grids on nonconvex planar domains will carry over to three dimensions. The planar equal-area functional extends readily to the three-dimensional equal-volume functional; as was true in the plane, the equal-volume functional (or powers thereof) is the only generator for which $\sqrt{H}$ is directly proportional to the weight function when H is evaluated at a solution grid. The volume functional will generate nonsmooth grids.

Orthogonality has several generalizations to 3D; strictly orthogonal grids exist on only a limited set of box-like domains having "triply-orthogonal" coordinate systems (Struik [199]) so the variational approach, with its least-squares solutions, makes considerable sense. Three orthogonality functionals are proposed

Principle	Symbol	Form of H
"Length"	I_L	$\mathrm{tr}G$
Area	I_A	$(\sqrt{g})^2$
Orthogonality-I	$I_{O,I}$	$g_{33}\,g_{12}^2 + g_{11}\,g_{23}^2 + g_{22}\,g_{13}^2$
Orthogonality-II	$I_{O,II}$	$\frac{g_{12}^2}{g_{11}\,g_{22}} + \frac{g_{23}^2}{g_{22}\,g_{33}} + \frac{g_{13}^2}{g_{11}\,g_{33}}$
Orthogonality-III	$I_{O,III}$	$\sqrt{g_{11}\,g_{22}\,g_{33}}$
AO	I_{AO}	$g_{11}\,g_{22}\,g_{33} - g_{12}\,g_{23}\,g_{13}$
Smoothness	I_{Wi}	$\sqrt{g}\,\mathrm{tr}(G^{-1})$
Liao	I_{Li}	$\mid G \mid^2$

Table 9.1 *Three-dimensional variational principles*

in Table 9.1. The orthogonality functional g_{12}^2 in two dimensions may be written $g_{12}^2 = -g\,g^{12}\,g_{12}$, suggesting the generalization

$$H = -g\,(g_{12}\,g^{12} + g_{23}\,g^{23} + g_{13}\,g^{13}) = g_{33}\,g_{12}^2 + g_{11}\,g_{23}^2 + g_{22}\,g_{13}^2 \tag{9.48}$$

for Orthogonality-I in the table (this one is proposed primarily for its use in the extension of AO to three dimensions). Orthogonality II is a straightforward generalization of the planar functional. The Orthogonality III functional results in an Euler-Lagrange equation similar to several nonvariational 3D orthogonal grid generation systems that have been proposed (see Warsi [224], Arina and Casselas [12], and Theodoropoulos and Bergeles [202]).

Multiple generalizations of the planar AO functional are possible due to its dependence on the orthogonality concept. The most obvious generalization would be $g_{11}\,g_{22}\,g_{33}$, but numerical experiments suggest that this functional lacks sufficient convexity. If AO is viewed as the sum of the area and Orthogonality-I functionals, one obtains $H = g_{11}\,g_{22}\,g_{33} - g_{12}\,g_{23}\,g_{13}$, which appears to perform better in numerical tests. A variational principle for the three-dimensional homogeneous Thompson-Thames-Mastin generator is readily constructed, as is a 3D Liao functional. Undoubtedly, there are many other 3D functionals that might be constructed; there is little experience with any of the functionals in this chapter to suggest which are best. Preliminary numerical results suggest that more uniform volumes and angles can be obtained using the AO functional as opposed to transfinite interpolation or Length.

Exercise 9.3.1 Show that the functions H in Table 9.1 are homogeneous and determine the degree homogeneity. Compare the degree of homogeneity of each functional to their planar counterparts. §

9.3.2 The 3D Euler-Lagrange Equations

With the definitions in Chapter 7, the approach of Chapter 8 carries over directly to three dimensions. Assuming a functional of the form (9.47), the three-dimensional Euler-Lagrange equation has exactly the same form as it does in the plane:

$$\frac{H}{\sqrt{g}}\,\mathrm{div}_{\boldsymbol{\xi}}\frac{\mathcal{C}}{w^2} - \mathrm{div}_{\boldsymbol{\xi}}\frac{\mathcal{J}\mathcal{B}}{w^2} = \mathbf{0}, \tag{9.49}$$

where

$$\mathcal{J}\mathcal{B} = \left[\frac{\partial H}{\partial \mathbf{x}_\xi}, \frac{\partial H}{\partial \mathbf{x}_\eta}, \frac{\partial H}{\partial \mathbf{x}_\zeta}\right], \tag{9.50}$$

and

$$\mathcal{B} = \mathcal{M} + \sqrt{g}\,\frac{\partial H}{\partial \sqrt{g}}\,\mathcal{G}^{-1}, \tag{9.51}$$

and

$$[\mathcal{M}]_{ij} = (1 + \delta_{ij})\,\frac{\partial H}{\partial g_{ij}}. \tag{9.52}$$

Exercise 9.3.2 Show that

$$\frac{\partial H}{\partial \mathbf{x}_\xi} = 2\,\mathbf{x}_\xi\frac{\partial H}{\partial g_{11}} + \mathbf{x}_\eta\,\frac{\partial H}{\partial g_{12}} + \mathbf{x}_\zeta\,\frac{\partial H}{\partial g_{13}} + (\mathbf{x}_\eta \times \mathbf{x}_\zeta)\,\frac{\partial H}{\partial \sqrt{g}}, \tag{9.53}$$

$$\frac{\partial H}{\partial \mathbf{x}_\eta} = \mathbf{x}_\xi\frac{\partial H}{\partial g_{12}} + 2\,\mathbf{x}_\eta\,\frac{\partial H}{\partial g_{22}} + \mathbf{x}_\zeta\,\frac{\partial H}{\partial g_{23}} + (\mathbf{x}_\zeta \times \mathbf{x}_\xi)\,\frac{\partial H}{\partial \sqrt{g}}, \tag{9.54}$$

$$\frac{\partial H}{\partial \mathbf{x}_\zeta} = \mathbf{x}_\xi\frac{\partial H}{\partial g_{13}} + \mathbf{x}_\eta\,\frac{\partial H}{\partial g_{23}} + 2\,\mathbf{x}_\zeta\,\frac{\partial H}{\partial g_{33}} + (\mathbf{x}_\xi \times \mathbf{x}_\eta)\,\frac{\partial H}{\partial \sqrt{g}}\,.\ \S \tag{9.55}$$

The covariant projection of (9.49) is again,

$$\mathrm{div}_{\boldsymbol{\xi}}\left\{\frac{H\mathcal{I} - \mathcal{G}\mathcal{B}}{w^2}\right\} = \mathbf{0}. \tag{9.56}$$

Exercise 9.3.3 Show that for the 3D TTM functional,

$$H = \frac{1}{\sqrt{g}}\,[g_{11}\,g_{22} + g_{22}\,g_{33} + g_{33}\,g_{11} - g_{12}^2 - g_{13}^2 - g_{23}^2]\,. \tag{9.57}$$

Show that the functional (9.57) reduces to the H in the two-dimensional Winslow functional, plus the term $\sqrt{g}$ in the limiting case $g_{13} \to 0$, $g_{23} \to 0$, and $g_{33} \to 1$. Show that for H as in (9.57), $\mathcal{B} = \mathcal{M} - H\mathcal{G}^{-1}$ where

$$\mathcal{M} = \frac{2}{\sqrt{g}}\begin{pmatrix} g_{22} + g_{33} & -g_{12} & -g_{13} \\ -g_{12} & g_{11} + g_{33} & -g_{23} \\ -g_{13} & -g_{23} & g_{11} + g_{22} \end{pmatrix}.\ \S \tag{9.58}$$

The nonconservative, tensor form $\mathcal{Q}_H \mathbf{x} = \mathbf{0}$ is

$$\begin{aligned} \mathcal{T}_{11}\,\mathbf{x}_{\xi\xi} + \mathcal{T}_{22}\,\mathbf{x}_{\eta\eta} + \mathcal{T}_{33}\,\mathbf{x}_{\zeta\zeta} + & \\ (\mathcal{T}_{12} + \mathcal{T}_{12}^T)\,\mathbf{x}_{\xi\eta} + & \\ (\mathcal{T}_{23} + \mathcal{T}_{23}^T)\,\mathbf{x}_{\eta\zeta} + & \\ (\mathcal{T}_{13} + \mathcal{T}_{13}^T)\,\mathbf{x}_{\xi\zeta} + & \\ b_1\,\mathbf{x}_\xi + b_2\,\mathbf{x}_\eta + b_3\,\mathbf{x}_\zeta = \mathbf{0}. & \end{aligned} \tag{9.59}$$

The following expressions give $\mathcal{T}_{11}$ and $(\mathcal{T}_{12} + \mathcal{T}_{12}^T)$ for the case in which H does not depend on $\sqrt{g}$. Expressions for the other tensors in (9.59) can be obtained from the following, by cyclic permutation of the indices:

$$\begin{aligned} \mathcal{T}_{11} = & \left(\frac{\partial H}{\partial g_{11}}\right) \mathcal{I} \\ & + 2\,\frac{\partial^2 H}{\partial g_{11}^2}(\mathbf{x}_\xi \otimes \mathbf{x}_\xi) \\ & + \frac{1}{2}\,\frac{\partial^2 H}{\partial g_{12}^2}(\mathbf{x}_\eta \otimes \mathbf{x}_\eta) \\ & + \frac{\partial^2 H}{\partial g_{11}\partial g_{12}}[(\mathbf{x}_\xi \otimes \mathbf{x}_\eta) + (\mathbf{x}_\eta \otimes \mathbf{x}_\xi)] \\ & + \frac{\partial^2 H}{\partial g_{11}\partial g_{13}}[(\mathbf{x}_\xi \otimes \mathbf{x}_\zeta) + (\mathbf{x}_\zeta \otimes \mathbf{x}_\xi)] \\ & + \frac{1}{2}\,\frac{\partial^2 H}{\partial g_{13}^2}(\mathbf{x}_\zeta \otimes \mathbf{x}_\zeta) \\ & + \frac{1}{2}\,\frac{\partial^2 H}{\partial g_{12}\partial g_{13}}[(\mathbf{x}_\eta \otimes \mathbf{x}_\zeta) + (\mathbf{x}_\zeta \otimes \mathbf{x}_\eta)], \end{aligned} \tag{9.60}$$

$$\begin{aligned} \mathcal{T}_{12} + \mathcal{T}_{12}^T = & \frac{\partial H}{\partial g_{12}} \mathcal{I} \\ & + \frac{\partial^2 H}{\partial g_{11}\partial g_{23}}[(\mathbf{x}_\xi \otimes \mathbf{x}_\zeta) + (\mathbf{x}_\zeta \otimes \mathbf{x}_\xi)] \\ & + 2\frac{\partial^2 H}{\partial g_{11}\partial g_{12}}(\mathbf{x}_\xi \otimes \mathbf{x}_\xi) \\ & + 2\frac{\partial^2 H}{\partial g_{11}\partial g_{22}}[(\mathbf{x}_\xi \otimes \mathbf{x}_\eta) + (\mathbf{x}_\eta \otimes \mathbf{x}_\xi)] \\ & + \frac{1}{2}\frac{\partial^2 H}{\partial g_{12}^2}[(\mathbf{x}_\xi \otimes \mathbf{x}_\eta) + (\mathbf{x}_\eta \otimes \mathbf{x}_\xi)] \\ & + \frac{1}{2}\frac{\partial^2 H}{\partial g_{12}\partial g_{13}}[(\mathbf{x}_\xi \otimes \mathbf{x}_\zeta) + (\mathbf{x}_\zeta \otimes \mathbf{x}_\xi)] \end{aligned}$$

$$+ 2\frac{\partial^2 H}{\partial g_{12}\partial g_{22}}(\mathbf{x}_\eta \otimes \mathbf{x}_\eta)$$

$$+ \frac{1}{2}\frac{\partial^2 H}{\partial g_{12}\partial g_{23}}[(\mathbf{x}_\eta \otimes \mathbf{x}_\zeta) + (\mathbf{x}_\zeta \otimes \mathbf{x}_\eta)]$$

$$+ \frac{\partial^2 H}{\partial g_{13}\partial g_{22}}[(\mathbf{x}_\eta \otimes \mathbf{x}_\zeta) + (\mathbf{x}_\zeta \otimes \mathbf{x}_\eta)]$$

$$+ \frac{\partial^2 H}{\partial g_{13}\partial g_{23}}(\mathbf{x}_\zeta \otimes \mathbf{x}_\zeta) . \tag{9.61}$$

If the tensor form of the TTM equations is calculated using the contravariant form, results similar to (8.78)–(8.80) apply.

The second variation in three dimensions is the obvious generalization of the result (8.91) with

$$\mathcal{H} = \begin{pmatrix} \mathcal{T}_{11} & \mathcal{T}_{12} & \mathcal{T}_{13} \\ \mathcal{T}_{12}^T & \mathcal{T}_{22} & \mathcal{T}_{23} \\ \mathcal{T}_{13}^T & \mathcal{T}_{23}^T & \mathcal{T}_{33} \end{pmatrix} . \tag{9.62}$$

The numerical algorithms presented in Sections 6.4 and 8.3.3 extend directly to three dimensions. If the former approach is used, there are a prohibitive 162 three-dimensional stencil arrays to be stored for the most general functionals. This can be made more manageable by storing instead the elements of the tensors T_{ij}, which reduces the number of three-dimensional arrays to thirty-six (plus three right-hand side arrays). The stencils must then be computed from the thirty-six arrays as needed. Considerably fewer arrays may be needed for specific variational principles, as many of the off-diagonal terms may then be zero. The conservative scheme outlined in Section 8.3.3 fares considerably better; only fourteen three-dimensional arrays (plus three more for the right-hand sides) need be stored for the most general functionals. Thus, conservative forms of the three-dimensional Euler-Lagrange equations appear to offer the best hope of performing 3D variational grid generation in practical applications.

9.3.3 A Variational Approach to the Steger-Sorenson Algorithm

A planar version of the Steger-Sorenson algorithm for the nonconservative operator $\mathcal{Q}\mathbf{x}$ was described in Section 5.6.2. In this section, it is shown that the algorithm may be combined with the variational approach described in the present 3D chapter (it can also be combined with the planar variational algorithm). Assume the following boundary condition on the $\zeta = 0$ boundary:

$$\mathbf{x}_\zeta = \sigma_1(\xi,\eta)\frac{\mathbf{x}_\xi}{\sqrt{g_{11}}} + \sigma_2(\xi,\eta)\frac{\mathbf{x}_\eta}{\sqrt{g_{22}}} + \sigma_3(\xi,\eta)\frac{(\mathbf{x}_\xi \times \mathbf{x}_\eta)}{\sqrt{gg^{33}}} , \tag{9.63}$$

where the σ_i are user-specified functions, with $\sigma_3 > 0$. For example, the choice $\sigma_1 = \sigma_2 = 0$ and $\sigma_3 = \sqrt{g^{33}s^2}$ gives $g_{23} = g_{13} = 0$ and $\sqrt{g_{33}} = s$ on the $\zeta = 0$ boundary. Equation (9.49) can be written

$$\mathrm{div}_{\boldsymbol{\xi}} \mathcal{J}\mathcal{B} = \mathcal{J}\mathbf{b} \tag{9.64}$$

with $\mathbf{b} = -(2/w)\{H\mathcal{G}^{-1} - \mathcal{B}\}\nabla_{\xi} w$. Rather than use this form, replace $\mathbf{b}$ with $\mathbf{b} = \exp(-\lambda\zeta)\,\mathbf{b}_0(\xi,\eta)$, so that

$$\mathrm{div}_{\xi}\mathcal{J}\mathcal{B} = e^{-\lambda\zeta}\,\mathcal{J}\,\mathbf{b}_0\,. \tag{9.65}$$

If the last equation is evaluated at $\zeta = 0$, and projected onto $\mathcal{J}_0^T$, one finds $\mathbf{b}_0$ in terms of the boundary metrics and their rates-of-change. The grid generation equation in conservative form is

$$\mathrm{div}_{\xi}\mathcal{J}\,\mathcal{B} = e^{-\lambda\zeta}\,\mathcal{J}\,\{\mathcal{G}^{-1}\,\mathrm{div}_{\xi}(\mathcal{G}\mathcal{B} - H\,\mathcal{I})\}_0\,. \tag{9.66}$$

The subscript "0" denotes that the portion within brackets must be evaluated at $\zeta = 0$, with the boundary condition (9.63) imposed (formulas for the 3D boundary metrics and their rates-of-change may be derived in the manner of Section 5.6.1). The 3D extension of the numerical algorithm described in Section 8.3.3 is used to solve (9.66); the steps in the algorithm may be summarized as follows:

1. Compute an initial guess for the grid,
2. Use the initial guess and the boundary condition (9.63) to compute the right-hand side of (9.66),
3. Solve (9.66) for the updated grid,
4. Apply a convergence test; return to second step if not satisfied.

Chapter 10

Variational Grid Generation on Curves and Surfaces

The goal of the present chapter is to present a variational approach to curve and surface grid generation. A significant difference between curve and line, or surface and planar, grid generation is that the Jacobian matrix is not square. As a result, constraints must be imposed on the minimization to ensure that the resulting grid points lie upon the given physical object. One way to impose these constraints is to convert the minimization to the form of the previous chapters by the use of a *parameter space.* The parameter space has no curvature and thus the techniques of the previous chapters apply. Although the parametric approach is successful when a conservative form of the equations is used, an unexpected difficulty arises with the nonconservative form of the parametric Euler-Lagrange equations; namely, solutions to the discrete equations may bifurcate on objects having large curvature. The other approach to imposing the constraint is through the use of Lagrange Multipliers. The Lagrange Multiplier approach is also not without difficulty as it is subject to off-object truncation error and problems handling the multiplier. Other approaches to curve and surface grid generation include those studied by Eyler and White [71], Garon and Camarero [76], Saltzman [162], Warsi [226], and Whitney and Thomas [230].

10.1 CURVES

The study of curves provides an excellent place to begin this chapter as the problem possesses most of the difficult features encountered in the problem of surface grid generation. Curves are of interest to surface and volume grid generation because the boundaries of these more complicated objects must be parameterized by a curve grid-generation algorithm before the surface or volume algorithm can be applied. There is little to be found in the literature on the subject of curve grid generation; stretching and other re-parameterizations of curves are widely used.

10.1.1 Differential Geometry of Curves

A review and extension of the relevant relations from Chapter 4 pertaining to the differential geometry of curves is useful in the development of variational curve grid generation. Key concepts are introduced in this section, including arc-length, curvature, and the Frenet formulas.

Let $\mathbf{x}(\xi) = (x(\xi), y(\xi), z(\xi))$ be a parameterization of a $C^2[0,1]$ curve in R^3. The tangent vector is $\mathbf{x}_\xi = (x_\xi, y_\xi, z_\xi)$. Define $g_{11} = \mathbf{x}_\xi \cdot \mathbf{x}_\xi$ so that $\sqrt{g_{11}}$ is the length of the tangent. Let s be the arc-length parameter, defined by

$$s(\xi) = \int_0^\xi \sqrt{g_{11}}\, d\mu \,. \tag{10.1}$$

Applying the chain rule to the tangent gives $\mathbf{x}_\xi = s_\xi \mathbf{x}_s$, so $\mathbf{x}_\xi = \sqrt{g_{11}}\,\mathbf{x}_s$. This shows that $\mathbf{x}_s$ must be a unit tangent vector. Following Struik [199], let $\hat{\mathbf{t}} = \mathbf{x}_s$ be the unit tangent. Since $\hat{\mathbf{t}} \cdot \hat{\mathbf{t}} = 1$, differentiation of this relationship with respect to s shows $\mathbf{t}_s$ is normal to $\hat{\mathbf{t}}$. Let $\hat{\mathbf{n}}$ be a unit vector in the direction of $\mathbf{t}_s$. Then there exists a scalar, denoted by κ, such that

$$\frac{d\hat{\mathbf{t}}}{ds} = \kappa\, \hat{\mathbf{n}} \,. \tag{10.2}$$

The scalar κ is known as the *curvature* and (10.2) is the first *Frenet* formula. It follows that the magnitude of the curvature is

$$\kappa(s) = (\mathbf{x}_{ss} \cdot \mathbf{x}_{ss})^{\frac{1}{2}} \,. \tag{10.3}$$

The sign of κ is a matter of convention since it is not uniquely defined by this relationship.

One can also define the *torsion* unit vector $\hat{\mathbf{b}} = \hat{\mathbf{t}} \times \hat{\mathbf{n}}$ to form a local orthogonal reference frame $(\hat{\mathbf{t}}, \hat{\mathbf{n}}, \hat{\mathbf{b}})$ known as the *moving trihedron*. Three Frenet formulas relating these quantities are well known:

$$\frac{d\hat{\mathbf{t}}}{ds} = \kappa\, \hat{\mathbf{n}} \,, \tag{10.4}$$

$$\frac{d\hat{\mathbf{n}}}{ds} = -\kappa\, \hat{\mathbf{t}} + \tau\, \hat{\mathbf{b}} \,, \tag{10.5}$$

$$\frac{d\hat{\mathbf{b}}}{ds} = -\tau\, \hat{\mathbf{n}} \,, \tag{10.6}$$

where τ is the *torsion*.

If one differentiates the relation $\mathbf{x}_\xi = s_\xi \hat{\mathbf{t}}$ with respect to ξ, the *curve identity* is obtained:

$$\mathbf{x}_{\xi\xi} = s_{\xi\xi}\, \hat{\mathbf{t}} + s_\xi^2\, \kappa\, \hat{\mathbf{n}} \,, \tag{10.7}$$

which may also be written in the more useful form

$$\mathbf{x}_{\xi\xi} = \frac{1}{2} \frac{(g_{11})_\xi}{g_{11}} \mathbf{x}_\xi + g_{11}\, \kappa\, \hat{\mathbf{n}} \,. \tag{10.8}$$

The curve identity bears the same relationship to curves as the *Gauss* equations do to surfaces, with the coefficient of the first term in (10.8) playing the role of the surface Christoffel symbol. While (10.4) shows that for the equal arc-length parameterization, $\mathbf{x}_{ss}$ lies in the direction of $\hat{\mathbf{n}}$, (10.8) shows that for arbitrary parameterizations, $\mathbf{x}_{\xi\xi}$ lies in the $\hat{\mathbf{n}}$-$\hat{\mathbf{t}}$ plane. Note that $\mathbf{x}_{\xi\xi}$ is independent of $\hat{\mathbf{b}}$. It is emphasized that this relationship holds for any parameterization of the curve and therefore does not imply any particular grid (i.e., it is not a grid-generation equation). Note that (10.8) shows $\hat{\mathbf{n}} \cdot \mathbf{x}_{\xi\xi} = g_{11}\kappa$; this fact is useful later. The covariant projection of (10.8) is the identity

$$\mathbf{x}_\xi \cdot \mathbf{x}_{\xi\xi} = \frac{1}{2}(g_{11})_\xi . \tag{10.9}$$

Exercise 10.1.1 Use (10.8) to obtain the following expression for the curvature:

$$\kappa^2 = \frac{(\mathbf{x}_\xi \times \mathbf{x}_{\xi\xi}) \cdot (\mathbf{x}_\xi \times \mathbf{x}_{\xi\xi})}{(\mathbf{x}_\xi \cdot \mathbf{x}_\xi)^3} . \tag{10.10}$$

Apply the chain rule (i.e., $\mathbf{x}_\xi = r_\xi \mathbf{x}_r$, etc.) to show that the curvature (10.10) is an invariant of the parameterization, i.e., the curvature at a fixed point $\mathbf{x}$ on the curve in physical space is independent of the parameterization. §

An expression similar to (10.10) holds for the torsion. The torsion is also an invariant of the parameterization, but it does not play a significant role in second-order grid generation methods since it depends on third derivatives of the parameterization. In the special case of planar curves, the curvature and torsion relations simplify to

$$\kappa = \frac{x_\xi y_{\xi\xi} - y_\xi x_{\xi\xi}}{(x_\xi^2 + y_\xi^2)^{\frac{3}{2}}} , \tag{10.11}$$

$$\tau = 0 . \tag{10.12}$$

Exercise 10.1.2 Use (10.5)–(10.6) to show

$$\hat{\mathbf{n}}_\xi = -\kappa\, \mathbf{x}_\xi + \tau\, (\mathbf{x}_\xi \times \hat{\mathbf{n}}) , \tag{10.13}$$

$$\hat{\mathbf{b}}_\xi = -\sqrt{g_{11}}\, \tau\, \hat{\mathbf{n}} . \text{ §} \tag{10.14}$$

Project 10.1.3 Given a parameterization $\mathbf{x}(\xi)$ of a curve, (10.1) can be used to compute the equal arc-length grid. First, obtain or write your own numerical integration subroutine. Let

$$L = \int_0^1 \sqrt{g_{11}}\, d\mu \tag{10.15}$$

be the total length of the curve. Write a computer code to compute L and then the set of points $\{r_i\}$, $i = 1, \dots M$, such that

$$\int_{r_{i-1}}^{r_i} \sqrt{g_{11}}\, d\mu = \frac{L}{M} . \tag{10.16}$$

Compute $\mathbf{x}_i = \mathbf{x}(r_i)$. Use the parabolic curve $x = \xi$, $y = \alpha\xi(1 - \xi)$, with α a positive parameter to test the code. §

10.1.2 Transformation Relations on a Curve

The transformation relationships from Chapter 7 concerning gradient, divergence, and Laplacian are extended to curves. A difficulty arises from the fact that the Jacobian matrix

$$\mathcal{J} = \begin{pmatrix} x_\xi \\ y_\xi \\ z_\xi \end{pmatrix} \tag{10.17}$$

is not square and therefore its inverse does not exist. Matrices are denoted by brackets, for example, $\mathcal{J} = [\mathbf{x}_\xi]$, while $\mathbf{x}_\xi$ is reserved for the column vector.

The metric tensor is $\mathcal{G} = \mathcal{J}^T\mathcal{J}$, which has just the single element $g_{11} = x_\xi^2 + y_\xi^2 + z_\xi^2$. Also, if $g = \det\mathcal{G}$, then $g = g_{11}$ in this highly degenerate case. The inverse metric tensor is $\mathcal{G}^{-1}$, which has the single element $1/g_{11}$. Recall that the *null space* of a matrix is the set of all elements of the vector space that map to the zero element under the linear transformation given by the matrix. The null space of $\mathcal{J}^T$ is the set of all linear combinations of the vectors $\hat{\mathbf{n}}$ and $\hat{\mathbf{b}}$. The *tangent space*, **T**, is defined to be the complement of the null space, i.e., the set of all vectors that lie in the direction of $\mathbf{x}_\xi$.

Since $\mathcal{G}$ is invertible, a generalized inverse can be defined by

$$(\mathcal{J}^{-1})^T = \mathcal{J}\mathcal{G}^{-1}, \tag{10.18}$$

$$= \frac{[\mathbf{x}_\xi]}{g_{11}}. \tag{10.19}$$

If the definitions $C = \sqrt{g}\,(\mathcal{J}^{-1})^T$ and $C^{-1} = \mathcal{J}^T/\sqrt{g}$ from the previous chapters are applied to curves,

$$C = \frac{[\mathbf{x}_\xi]}{\sqrt{g_{11}}}, \tag{10.20}$$

$$= [\mathbf{x}_s], \tag{10.21}$$

i.e., C has the same entries as the unit tangent vector. On the other hand,

$$C^{-1} = \frac{[\mathbf{x}_\xi]^T}{\sqrt{g_{11}}}, \tag{10.22}$$

$$= [\mathbf{x}_s]^T, \tag{10.23}$$

has the same entries as does the unit tangent (row) vector.

The divergence of a 3×1 matrix S is the column vector $(\text{div}_{\boldsymbol{\xi}} S)_i = \partial S_i/\partial\xi$, $i = 1, 2, 3$. For example, the divergence of C yields the curve *metric* identity:

$$\text{div}_{\boldsymbol{\xi}} C = \sqrt{g_{11}}\,\kappa\,\hat{\mathbf{n}}. \tag{10.24}$$

This is the conservative form of the curve identity (10.8). Note the appearance of an additional term proportional to the curvature compared to the planar formula (7.12).

The next goal is to obtain transformation relationships for the gradient. Suppose one has the scalar function $f = f(x(\xi), y(\xi), z(\xi))$ defined on the curve. Then the logical space gradient (or covariant derivative) of f is just $\nabla_{\boldsymbol{\xi}} f = f_\xi$. The physical-space gradient is not well defined because the function f is not smoothly defined in a neighborhood of the curve. Using (7.50) for an analog, *define* the physical-space gradient in terms of the generalized inverse to be the column vector

$$\nabla_{\mathbf{x}} f = (\mathcal{J}^{-1})^T f_\xi \,, \tag{10.25}$$

$$= \frac{\mathbf{x}_\xi}{g_{11}} f_\xi \,. \tag{10.26}$$

The transformed physical-space gradient belongs to the tangent space $\mathbf{T}$. Both the chain rule $f_\xi = f_x\, x_\xi + f_y\, y_\xi + f_z\, z_\xi$, and the transformation rule

$$f_\xi = \mathcal{J}^T \, \nabla_{\mathbf{x}} f \tag{10.27}$$

formally hold when this definition is adopted. Observe that the generalized inverse permits construction of relations that formally agree with the relations given in Chapter 7. The additional term proportional to curvature in the metric identity (10.24) gives rise to an additional term in the "conservative" form of the gradient:

$$\nabla_{\mathbf{x}} f = \frac{1}{\sqrt{g_{11}}} \operatorname{div}_{\boldsymbol{\xi}}(Cf) - \kappa\, f\, \hat{\mathbf{n}} \,. \tag{10.28}$$

Let $\mathbf{v}^T \in R^3$. To achieve consistency with the definition of the gradient, the divergence of $\mathbf{v}$ must transform as

$$\operatorname{div}_{\mathbf{x}} \cdot \mathbf{v} = \frac{1}{\sqrt{g}} \operatorname{div}_{\boldsymbol{\xi}} \cdot C^T \mathbf{v}^T \,, \tag{10.29}$$

$$= \frac{1}{\sqrt{g_{11}}} \left(\frac{\mathbf{x}_\xi \cdot \mathbf{v}}{\sqrt{g_{11}}} \right)_\xi \,. \tag{10.30}$$

This can be proved using the definition (10.26) of the gradient and the fact that the divergence of $\mathbf{v}$ is the trace of the physical-space gradient of $\mathbf{v}$. If one lets $\mathbf{v}^T = \mathbf{x}_\xi / \sqrt{g_{11}}$, then (10.30) gives

$$\operatorname{div}_{\mathbf{x}} \cdot \mathbf{x}_s^T = 0 \,, \tag{10.31}$$

which is the analog of (7.13).

Using (10.26) and (10.29), the transformed Laplacian in conservative form is

$$\nabla_{\mathbf{x}}^2 f = \frac{1}{\sqrt{g_{11}}} \left(\frac{f_\xi}{\sqrt{g_{11}}} \right)_\xi \,. \tag{10.32}$$

Exercise 10.1.4 Let $f = \xi$ in (10.32) and set $\nabla_{\mathbf{x}}^2 \xi = 0$. Perform the analog to the sequence of steps in Section 7.3.5 to obtain "Winslow" equations for a curve. §

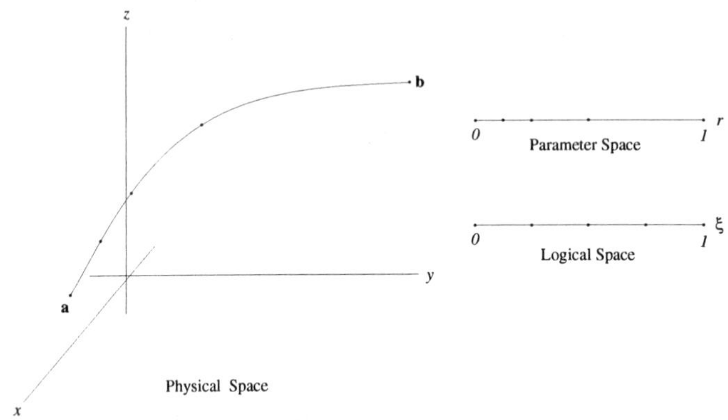

Figure 10.1 *Curve parameterization*

10.1.3 Parametric Approach to Variational Curve Grid Generation

Assume that an initial parameterization $\mathbf{x}(r)$, $0 \le r \le 1$, of a C^2 curve is given and the curve is to be re-parameterized using a variational grid generator. The line grid functional (3.67) in Chapter 3 has a natural generalization to a curve functional and preserves the property that the local tangent is proportional to the physical-weight function. To be consistent with the variational theory outlined in Chapter 8, minimize the curve functional

$$I[\mathbf{x}] = \int_0^1 \frac{H(g_{11})}{w^2(\mathbf{x})}\, d\xi, \tag{10.33}$$

where $H : R \to R$ is a arbitrary positive real homogeneous function, $g_{11} = \mathbf{x}_\xi \cdot \mathbf{x}_\xi$ is the length metric, and w is a physical-space weight function. The set of admissible functions must satisfy the end-point constraints $\mathbf{x}(0) = \mathbf{a} \in R^3$ and $\mathbf{x}(1) = \mathbf{b} \in R^3$. This functional is more general than necessary since the obvious choice is $H = g_{11}$. As will be demonstrated, if $H = g_{11}$, the unit weight $w = 1$ produces the equal arc-length parameterization. The difference between the line functional and (10.33) is that here $\mathbf{x} = (x, y, z)$, i.e., there are three functions $x(\xi)$, $y(\xi)$, and $z(\xi)$ to be found. As a result, a constraint must be imposed on the minimization to ensure that the new parameterization $\mathbf{x}(\xi)$ is a re-parameterization of the original curve $\mathbf{x}(r)$; i.e., the new points must lie upon the curve implied by the old parameterization.

There are two approaches to implementing the constraint; one is referred to as the Lagrange Multiplier approach (treated in Section 10.1.4) and the other as the parametric approach (treated in this section). The concept of *parameter space* is introduced in the parametric approach to ensure that grid points lie upon the given curve. Parameter space consists of an intermediate domain $0 \le r \le 1$ such that $r = r(\xi)$ is a mapping from logical space to parameter space and $\mathbf{x} = \mathbf{x}(r)$ is the user-specified mapping from parameter space to physical space (see

Figure 10.1). The following transformation rules apply to the composite map from logical to physical space:

$$\mathbf{x}_\xi = \mathbf{x}_r \, r_\xi \,, \tag{10.34}$$

$$\mathcal{J} = \hat{\mathcal{J}} \, r_\xi \,, \tag{10.35}$$

and

$$g_{11} = \hat{g}_{11} \, r_\xi^2 \,, \tag{10.36}$$

with $\hat{\mathcal{J}}^T = (x_r, y_r, z_r)$ and $\hat{g}_{11} = \mathbf{x}_r \cdot \mathbf{x}_r$. One may also define $\hat{\mathcal{G}} = \hat{\mathcal{J}}^T \hat{\mathcal{J}} = [\hat{g}_{11}]$; then $\mathcal{G} = r_\xi^2 \, \hat{\mathcal{G}}$. The minimization (10.33) over the set of admissible functions $\mathbf{x}$ becomes a minimization over the set of admissible functions $r(\xi)$:

$$I[r] = \int_0^1 \frac{H}{\hat{w}^2(r)} \, d\xi \tag{10.37}$$

with $r(0) = 0, r(1) = 1$, and the weight function $\hat{w}(r) = w(\mathbf{x}(r))$ defined in terms of the parameter r. The constraint is automatically satisfied in the parametric approach because the given parameterization $\mathbf{x}(r)$ ensures that grid points will lie upon the given curve for any $r(\xi)$ that minimizes (10.37).

The Euler-Lagrange equation gives the grid generation equation for the curve; it has the form

$$\frac{\partial}{\partial r}\left(\frac{H}{\hat{w}^2}\right) - \frac{d}{d\xi}\left(\frac{\mathcal{T}}{\hat{w}^2}\right) = 0 \,, \tag{10.38}$$

where $\mathcal{T} = [\partial H / \partial r_\xi]$.

Exercise 10.1.5 Define $\mathcal{B} = 2\,[\partial H / \partial g_{11}]$. Apply the chain rule to $\mathcal{T}$ to show that

$$\mathcal{T} = r_\xi \, \hat{\mathcal{G}} \, \mathcal{B} \,. \; \S \tag{10.39}$$

Exercise 10.1.6 To obtain an efficient computational scheme, the first term in (10.38) must be expressed in terms of the logical space derivative. This is somewhat delicate since H is not directly a function of r. Apply the chain rule to show that

$$\left(\frac{H}{\hat{w}^2}\right)_r = \frac{1}{r_\xi}\left(\frac{H}{\hat{w}^2}\right)_\xi - \frac{\hat{\mathcal{G}} \, r_{\xi\xi} \, \mathcal{B}}{\hat{w}^2} \,. \; \S \tag{10.40}$$

From the results of the previous exercises, the Euler-Lagrange equation now reads

$$\left(\frac{H}{\hat{w}^2}\right)_\xi - \frac{\hat{\mathcal{G}} \, r_\xi \, r_{\xi\xi} \, \mathcal{B}}{\hat{w}^2} - r_\xi \left(\frac{\hat{\mathcal{G}} \, r_\xi \, \mathcal{B}}{\hat{w}^2}\right)_\xi = 0 \,. \tag{10.41}$$

This equation may be expressed in *conservative* form:

$$\left(\frac{H - r_\xi^2 \, \hat{\mathcal{G}} \, \mathcal{B}}{\hat{w}^2}\right)_\xi = 0 \,. \tag{10.42}$$

Exercise 10.1.7 Let $H = g_{11}$ and show that the conservative form (10.42) of the Euler-Lagrange equation becomes

$$\left(\frac{\hat{g}_{11}\, r_\xi^2}{\hat{w}^2}\right)_\xi = 0 \tag{10.43}$$

and that therefore $\sqrt{g_{11}} = \beta\,\hat{w}(r)$ is a generic property of the solution with β a constant. Note that the equal arc-length solution is obtained when $\hat{w} = 1$. §

Numerical Implementation

The nonlinear equation (10.43) is solved in the same manner as the line equation (3.16) using (3.45) of Section 3.4. Define $\Gamma = \hat{g_{11}} r_\xi / \hat{w}^2$ to obtain the stencil equation

$$\frac{L_i}{(\Delta\xi)^2}\, r_{i-1} + \frac{C_i}{(\Delta\xi)^2}\, r_i + \frac{R_i}{(\Delta\xi)^2}\, r_{i+1} = 0\,, \tag{10.44}$$

where

$$L_i = \Gamma_{i-\frac{1}{2}}\,, \tag{10.45}$$

$$R_i = \Gamma_{i+\frac{1}{2}}\,, \tag{10.46}$$

$$C_i = -(L_i + R_i)\,. \tag{10.47}$$

The lagged coefficients are evaluated as follows:

$$\Gamma_{i+\frac{1}{2}} = \left(\frac{\hat{g}_{11}}{\hat{w}^2}\right)_{i+\frac{1}{2}} (r_{i+1} - r_i)\,. \tag{10.48}$$

The first factor on the right is merely a function of the parameter r, so it may be evaluated at the half-point by

$$r_{i+\frac{1}{2}} = (r_{i+1} + r_i)/2.$$

As usual, $i = 1, \ldots, M-1$ are the interior points and $r_0 = 0$, $r_M = 1$ are the boundary conditions. The stencils are symmetric since $L_{i+1} = R_i$. The approximation is second-order accurate.

Project 10.1.8 Convert the code written for Project 3.4.6 from Chapter 3 to a curve grid generator that solves (10.43). Use the parabola $\mathbf{x}(r) = (r, \alpha r\,(1-r), 0)$, $0 \le r \le 1$, with $\alpha \ge 0$ for a model problem. §

Nonconservative Form and Bifurcation

A nonconservative form of the curve grid generator is obtained by expanding (10.43):

$$r_{\xi\xi} + \left\{\frac{\mathbf{x}_r \cdot \mathbf{x}_{rr}}{\mathbf{x}_r \cdot \mathbf{x}_r} - \frac{\hat{w}_r}{\hat{w}}\right\} r_\xi^2 = 0\,. \tag{10.49}$$

This equation can be discretized and solved in a manner analogous to (3.17) in Chapter 3; however, that approach leads to a serious difficulty, as described in

Steinberg and Roache [193]. Briefly summarized, in the continuum the equal arc-length problem has a unique solution and so the Euler-Lagrange equation (10.49) with $\hat{w} = 1$ is expected to have a unique solution. But when centered differences are used to discretize this equation the solution bifurcates and uniqueness is lost.

The planar parabolic curve in Project 10.1.8 is used as a model problem. If there are just $M = 2$ segments, there is one unknown parameter value $r = r(\frac{1}{2})$ to be calculated. Central differences at this location gives the derivatives $r_\xi = 1$ and $r_{\xi\xi} = 4\,(1 - 2\,r)$. The discretized equal arc-length equation (10.49 with $\hat{w} = 1$) becomes

$$4\,(1 - 2\,r) - \frac{2\,\alpha^2\,(1 - 2\,r)}{1 + \alpha^2\,(1 - 2\,r)^2} = 0\,. \tag{10.50}$$

For $\alpha < \sqrt{2}$, there is the unique solution $r = 1/2$, while for $\alpha > \sqrt{2}$, there are three real solutions, namely,

$$r = \frac{1}{2} \quad \text{and} \quad r = \frac{1}{2} \pm \frac{1}{2}\sqrt{\frac{1}{2} - \frac{1}{\alpha^2}}\,. \tag{10.51}$$

Only the first solution actually divides the curve into equal arc-length. Unfortunately, it is an unstable fixed point in the iteration used to solve the nonlinear finite difference equation. Thus, for essentially all initial guesses, the iteration does not converge to the desired root but rather to one of the two spurious, unequal arc-length, roots. The problem persists even if M is large. Alternate differencing schemes applied to (10.49) do not alleviate the bifurcation problem. Numerical experiments show that one cannot get past the bifurcation point using continuation procedures. Similar behavior was reported for parametric formulations of surface variational principles. If the parametric approach to curve and surface grid generation is taken, the nonconservative form (10.49) should be avoided; the conservative equation (10.43) is strongly preferred since it does not exhibit bifurcation behavior when discretized.

Project 10.1.9 Discretize (10.49) using centered differences and write a numerical code to solve the resulting equations. Replicate the bifurcation behavior described above (see Steinberg and Roache [193]). §

10.1.4 *Lagrange Multiplier Approach to Variational Curve Grid Generation*

The second way to impose the constraint on the minimization (10.33) is to use Lagrange Multipliers. Again, the problem is to minimize

$$I[\mathbf{x}] = \int_0^1 G(\xi, \mathbf{x}, \mathbf{x}_\xi)\, d\xi, \tag{10.52}$$

where $G : R^7 \to R$ is a smooth function, over all parameterizations of the curve $\mathbf{x} = \mathbf{x}(\xi)$, $0 \le \xi \le 1$.

The following derivation of the Euler-Lagrange equation follows the definition (6.4) of the first variation of a functional. Now, because most curves are not linear, $\mathbf{x} + \epsilon\,\mathbf{c}$ will not lie on the curve for nonzero c. So a generalization is needed. Let

$\mathbf{x}(\xi,\epsilon)$ be a family of parameterizations of the given curve such that $\mathbf{x}(\xi,0) = \mathbf{x}(\xi)$, $\mathbf{x}(0,\epsilon) = \mathbf{x}(0)$ and $\mathbf{x}(1,\epsilon) = \mathbf{x}(1)$. Thus

$$\frac{\mathbf{x}(\xi,\epsilon) - \mathbf{x}(\xi,0)}{\epsilon} \tag{10.53}$$

is a secant vector, and its limit as ϵ goes to zero is a tangent vector, that is, a multiple of the unit tangent:

$$\frac{\partial}{\partial\epsilon}\mathbf{x}(\xi,\epsilon)|_{\epsilon=0} = C\,\hat{\mathbf{t}}\,. \tag{10.54}$$

Note that $C = C(\xi)$. Now define

$$F(\epsilon) = \int_0^1 G(\xi, \mathbf{x}(\xi,\epsilon), \mathbf{x}_\xi(\xi,\epsilon))\, d\xi\,. \tag{10.55}$$

For the curve $\mathbf{x}(\xi)$ to minimize I, $\epsilon = 0$ must be a minimum of of $F(\epsilon)$. This derivative defines the first variation of I:

$$D_{\hat{\mathbf{t}}} I = \frac{\partial F}{\partial \epsilon}(0)\,. \tag{10.56}$$

The chain rule then gives

$$D_{\hat{\mathbf{t}}} I = \int_0^1 \left(\frac{\partial G}{\partial \mathbf{x}} - \frac{d}{d\xi}\frac{\partial G}{\partial \mathbf{x}_\xi}\right)\cdot \hat{\mathbf{t}}\, C\, d\xi\,. \tag{10.57}$$

In order for the first variation to be zero, the first part of the integrand must be normal to the tangent space, i.e., there exist functions $\lambda_1(\xi)$, $\lambda_2(\xi)$ such that

$$\frac{\partial G}{\partial \mathbf{x}} - \frac{d}{d\xi}\frac{\partial G}{\partial \mathbf{x}_\xi} = \lambda_1\,\hat{\mathbf{n}} + \lambda_2\,\hat{\mathbf{b}}\,. \tag{10.58}$$

This is the Lagrange Multiplier form of the Euler-Lagrange equations for a curve.

Applying (10.58) to the curve functional (10.33), the constrained Euler-Lagrange equation is

$$\frac{\partial}{\partial \mathbf{x}}\left(\frac{H}{w^2}\right) - \frac{d}{d\xi}\frac{\mathcal{T}}{w^2} = \lambda_1\,\hat{\mathbf{n}} + \lambda_2\,\hat{\mathbf{b}} \tag{10.59}$$

where the Lagrange Multipliers are, as yet, unknown and $\mathcal{T} = \partial H/\partial \mathbf{x}_\xi = \mathcal{J}\mathcal{B}$. The components of $\mathbf{x}$ are not independent of one another since $\mathbf{x}$ lies on the curve; thus, the first term of (10.59) must be inverted using the generalized inverse relationship (10.28). Equation (10.59) then becomes

$$\frac{H}{\sqrt{g_{11}}}\left(\frac{\mathbf{x}_\xi}{\sqrt{g_{11}}\,w^2}\right)_\xi - \left(\frac{\mathbf{x}_\xi\,\mathcal{B}}{w^2}\right)_\xi = \left(\lambda_1 + \kappa\frac{H}{w^2}\right)\hat{\mathbf{n}} + \lambda_2\,\hat{\mathbf{b}}\,. \tag{10.60}$$

Exercise 10.1.10 Form the inner products of equation (10.60) with $\hat{\mathbf{n}}$ and $\hat{\mathbf{b}}$. Use the identities (10.13)–(10.14) to show

$$\lambda_1 = -\frac{\mathcal{B}}{w^2}\,g_{11}\,\kappa, \tag{10.61}$$

$$\lambda_2 = 0\,.\ \S \tag{10.62}$$

Pulling these results together, the Euler-Lagrange equation in multiplier form is

$$\frac{H}{\sqrt{g_{11}}}\left(\frac{\mathbf{x}_\xi}{\sqrt{g_{11}}\,w^2}\right)_\xi - \left(\frac{\mathbf{x}_\xi\,\mathcal{B}}{w^2}\right)_\xi = \frac{\kappa}{w^2}\,(H - g_{11}\mathcal{B})\,\hat{\mathbf{n}}. \tag{10.63}$$

If $H = g_{11}$, the previous result becomes

$$\left(\frac{\mathbf{x}_\xi}{w^2}\right)_\xi = -\frac{w_\xi}{w^3}\mathbf{x}_\xi + g_{11}\frac{\kappa}{w^2}\hat{\mathbf{n}}, \tag{10.64}$$

which further simplifies to the *conservative* form

$$\left(\frac{\mathbf{x}_\xi}{w}\right)_\xi = \frac{g_{11}}{w}\,\kappa\,\hat{\mathbf{n}}. \tag{10.65}$$

The *nonconservative* form is

$$\mathbf{x}_{\xi\xi} - \frac{w_\xi}{w}\mathbf{x}_\xi = g_{11}\,\kappa\,\hat{\mathbf{n}}. \tag{10.66}$$

Exercise 10.1.11 Use (10.63) to derive the Euler-Lagrange equations for the variational principle $H = \sqrt{g_{11}}$ with $w = 1$. Compare the Euler-Lagrange equation to the curve metric identity and interpret the results. §

Exercise 10.1.12 Project (10.66) onto $\mathbf{x}_\xi$ and use the parametric relation $\mathbf{x}_\xi = \mathbf{x}_r\, r_\xi$ to derive (10.49). Show $\sqrt{g_{11}} = \beta\, w$ is a solution for this H. §

The *covariant projection* is obtained by forming the inner product of (10.63) with $\mathbf{x}_\xi$,

$$\left(\frac{H - \mathcal{G}\mathcal{B}}{w^2}\right)_\xi = 0. \tag{10.67}$$

Since $\mathcal{G} = r_\xi^2\,\hat{\mathcal{G}}$, the covariant projection can be immediately converted into the parametric formulation (10.42). For $H = g_{11}$, the covariant projection of the weighted curve grid generation equation is

$$\left(\frac{g_{11}}{w^2}\right)_\xi = 0. \tag{10.68}$$

Numerical Algorithms for the Lagrange Multiplier Form

In planar numerical grid generation it was noted that truncation error shifts points from their continuum location in the plane to the location of the discrete node. As a result, the discrete planar grid will not exactly have the properties of the continuum grid; for example, the continuum solution may be orthogonal while the numerical solution is not exactly so, especially if coarse resolutions are used. As shown in Section 5.4.3, another effect of truncation error is that the discrete grid can be folded when the continuum grid is not.

An additional truncation error effect can occur in the case of numerical curve and surface grid generation. Suppose one numerically solves the curve grid gen-

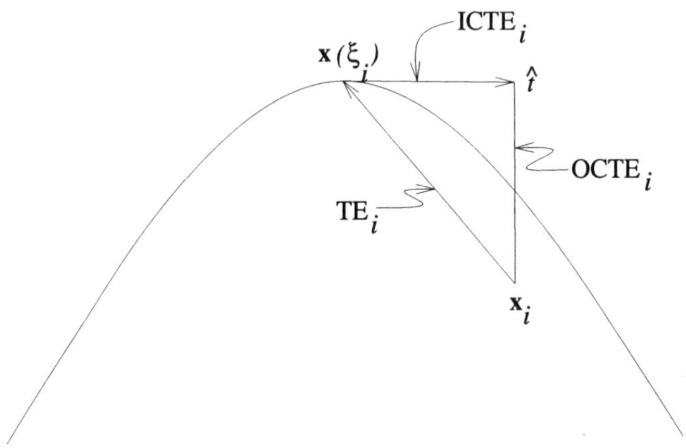

Figure 10.2 *Geometry of curve-grid truncation errors*

eration equation (10.65) or (10.66) using some finite difference approximation. Let the exact solution to the curve grid equation at $\xi = \xi_i$ be $\mathbf{x}(\xi_i)$ and the discrete solution at that point be $\mathbf{x}_i$ (see Figure 10.2). The continuum solution point $\mathbf{x}(\xi_i)$ lies exactly upon the given curve, while the discrete point $\mathbf{x}_i$ may not. As the discretization becomes finer, it is expected that the discrete point approaches the point $\mathbf{x}(\xi_i)$ on the curve. Let $\mathbf{v}_i = \mathbf{x}(\xi_i) - \mathbf{x}_i$ and define the local truncation error to be

$$TE_i = \sqrt{\mathbf{v}_i \cdot \mathbf{v}_i}\,. \tag{10.69}$$

Local *in-curve truncation error*, ICTE, is defined by

$$ICTE_i = \mathbf{v}_i \cdot \hat{\mathbf{t}}\,. \tag{10.70}$$

To measure the normal displacement of the discrete grid node from the given curve, one may define the local *off-curve truncation error*, OCTE, by

$$OCTE_i = \sqrt{TE_i^2 - ICTE_i^2}\,, \tag{10.71}$$

$$= |\, \mathbf{v}_i \times \hat{\mathbf{t}} \,|\,. \tag{10.72}$$

Note that both TE_i and $OCTE_i$ are nonnegative.

Exercise 10.1.13 Show that if $\mathbf{v}_i \in \mathbf{T}$, then $OCTE_i = 0$ and $ICTE_i = TE_i$. Show that if $\mathbf{v}_i$ is in the null space of J^T, then $OCTE_i = TE_i$ and $ICTE_i = 0$. §

The parametric formulation of the curve grid generator (Section 10.1.3) guarantees that $OCTE_i = 0$; however, ICTE may prevent the discrete curve grid from being exactly equal-arc. The Lagrange Multiplier formulation does not guarantee zero OCTE. In this case, prohibitively large resolutions may be needed for the grid points to converge toward the given curve, especially if the curvature is large (Knupp [113]). In planar grid generation, off-planar truncation error (OPTE) is zero by definition, so $TE_i = IPTE_i$, i.e., all the observed truncation error lies in the plane. In general, in-object truncation error may result in inexact metric relationships and sometimes in grid folding, while off-object truncation error results in discrete grids that do not lie on the target object.

Numerical algorithms for solving the systems of three nonlinear equations (10.65) or (10.66) are now considered. Both have vector finite difference stencils of the form

$$L_i \frac{\mathbf{x}_{i-1}}{(\Delta\xi)^2} + C_i \frac{\mathbf{x}_i}{(\Delta\xi)^2} + R_i \frac{\mathbf{x}_{i+1}}{(\Delta\xi)^2} = \mathbf{G}_i \tag{10.73}$$

with $C_i = -(L_i + R_i)$. The systems are solved by successively applying a tridiagonal solver to each of the three equations and iterating on the nonlinearity. For equation (10.65), the stencil coefficients are

$$L_i = \frac{1}{w_{i-\frac{1}{2}}}, \tag{10.74}$$

$$R_i = \frac{1}{w_{i+\frac{1}{2}}}, \tag{10.75}$$

while for equation (10.66), the coefficients are

$$L_i = 1 + \frac{\Delta\xi}{2}\left(\frac{w_\xi}{w}\right)_i, \tag{10.76}$$

$$R_i = 1 - \frac{\Delta\xi}{2}\left(\frac{w_\xi}{w}\right)_i. \tag{10.77}$$

To obtain second-order accuracy, the weights should be averaged to the half-points as, for example, $w_{i+\frac{1}{2}} = (w_i + w_{i+1})/2$. The right-hand side $\mathbf{G}_i$ of these equations are the delicate part of the algorithm, as is demonstrated in the following project.

Project 10.1.14 Write a computer code to solve (10.65) using the algorithm described above for the model parabola in Project 10.1.8. The right-hand side is

$$\mathbf{G}_i = \left(\frac{g_{11}}{w}\kappa\hat{\mathbf{n}}\right)_i. \tag{10.78}$$

Use the analytic expression for planar curves

$$\kappa\hat{\mathbf{n}} = \frac{x_r y_{rr} - y_r x_{rr}}{(x_r^2 + y_r^2)^2}\mathbf{x}_r^\perp \tag{10.79}$$

derived from (10.11). The right-hand side $\mathbf{G}_i$ must be evaluated at $\mathbf{x}_i$; in general this requires inverting the original analytic parameterization of the curve, i.e., given $\mathbf{x}_i$, find r_i to evaluate (10.79). For the parabolic curve this is easy since

$r = x$. In the general case, the given original parameterization most likely must be inverted numerically. Evaluate g_{11} on the right-hand side using the centered finite difference expression for $\mathbf{x}_\xi$. Set $\alpha = 1$, $w = 1$ and compute the solution grid, the chord lengths between adjacent nodes, and the maximum off-curve truncation error. Perform a grid convergence study and verify that $OSTE \to 0$ in second-order fashion. Try larger values of α. It is tempting to eliminate OSTE by setting $y_i = y(x_i)$, but then (10.65) is not satisfied (i.e., the grid is not equal arc-length) and, in any case, the approach only works for curves of the form $y = y(x)$ and not more general cases. §

The numerical results for the nonconservative parametric equation (10.49) suggest that (10.66) would exhibit bifurcation behavior, but it does not (Knupp [113]). Evidently, there is enough information in the Lagrange Multiplier form to avoid the ambiguities inherent in the projected form (10.49) that lead to bifurcations.

The projected equation (10.67) may be solved via a Newton iterative technique described in Knupp [113]. The approach is effective in that it avoids bifurcation and "off-curve" truncation error. The main limitation of the Newton approach is in the complexity of its implementation. If one reintroduces the parameter-space variable so that the tangent transforms as in (10.34), (10.67) reduces to (10.42) and the solution methods employed in Project 10.1.8 can be used.

The best variational approach to generating grids on curves appears to be to solve the conservative form (10.43) of the parametric equations. The parametric approach is a natural way to impose the constraint that the grid points lie on the given curve. The approach simultaneously avoids off-curve truncation error and bifurcation. Use of the nonconservative parametric equation should be avoided since this leads to bifurcations. Bifurcations for the conservative form were reported in Steinberg and Roache [195], but a different form, not fully conservative, was used in that study. Numerical experiments using the model problem on the fully conservative form (10.43) indicate that bifurcations do not occur. Bifurcations also do not occur if the projected equation (10.67) is solved, using the Newton iteration technique. The Lagrange multiplier forms are unsatisfactory because they permit nonzero off-curve truncation error, especially if the resolution is coarse or the curve has regions of high curvature. In addition, they require inversion of the original parameterization in order to determine the curvature at the required location.

10.2 SURFACES

The following discussion of variational surface grid generation is modeled after the results obtained for curves. The surface differential relations from Chapter 4 are reviewed and extended. Transformations for hosted equations on a surface are described. Parametric and Lagrange Multiplier approaches to variational surface grid generation are presented. Euler-Lagrange equations for a general functional with physical-space weighting are developed.

10.2.1 Differential Geometry of Surfaces

First-Order Relationships

Let $S \subset R^3$ be a smooth, simply connected, orientable surface. A point on S is specified by a mapping from logical space U_2 to physical space X_2^3 via three coordinate functions

$$\mathbf{x}(\xi,\eta) = (x(\xi,\eta),\, y(\xi,\eta),\, z(\xi,\eta)\,)\,. \tag{10.80}$$

The two surface tangent vectors, $\mathbf{x}_\xi = (x_\xi, y_\xi, z_\xi)$ and $\mathbf{x}_\eta = (x_\eta, y_\eta, z_\eta)$, serve as covariant basis vectors for the *tangent plane*. The tangent plane is the subspace $\mathbf{T}$ spanned by these tangent vectors. Let $\mathcal{J} = \nabla_{\boldsymbol{\xi}}\mathbf{x}$ be the 3×2 Jacobian transformation matrix,

$$\mathcal{J} = \begin{pmatrix} x_\xi & x_\eta \\ y_\xi & y_\eta \\ z_\xi & z_\eta \end{pmatrix}. \tag{10.81}$$

Assume the surface is nondegenerate, i.e., $\mathcal{J}$ has rank 2, or, $|\, \mathbf{x}_\xi \times \mathbf{x}_\eta \,| \neq 0$. The metric tensor as defined in Chapter 4 is $\mathcal{G} = \mathcal{J}^T\mathcal{J}$. The elements of $\mathcal{G}$ are

$$g_{11} = \mathbf{x}_\xi \cdot \mathbf{x}_\xi\,, \quad g_{12} = \mathbf{x}_\xi \cdot \mathbf{x}_\eta\,, \quad g_{22} = \mathbf{x}_\eta \cdot \mathbf{x}_\eta. \tag{10.82}$$

The determinant of the metric tensor is $g = g_{11}\,g_{22} - g_{12}^2$. The assumption on the rank of the Jacobian matrix ensures that $g \neq 0$ so $\mathcal{G}$ is invertible.

Exercise 10.2.1 Show that $g = |\, \mathbf{x}_\xi \times \mathbf{x}_\eta \,|^2$. §

A surface *normal* vector is $\mathbf{J} = \mathbf{x}_\xi \times \mathbf{x}_\eta$ and a *unit* normal vector is $\hat{\mathbf{n}} = \mathbf{J}/\sqrt{g}$.

Exercise 10.2.2 Verify the following relationship, which can be derived from the vector identity (4.11):

$$\mathbf{x}_\eta \times \hat{\mathbf{n}} = \frac{g_{22}\,\mathbf{x}_\xi - g_{12}\,\mathbf{x}_\eta}{\sqrt{g}}\,, \tag{10.83}$$

$$\hat{\mathbf{n}} \times \mathbf{x}_\xi = \frac{-g_{12}\,\mathbf{x}_\xi + g_{11}\,\mathbf{x}_\eta}{\sqrt{g}}\,.\ \S \tag{10.84}$$

The scalar product identities in Table 4.5 hold if the proper analogies are made.

Second-Order Relationships

Second-order quantities are introduced to develop the Gauss and metric identities. Following Stoker [197], define the elements of the *curvature tensor* $\mathcal{W}$ as

$$\mathcal{W} = \begin{pmatrix} L & M \\ M & N \end{pmatrix} \tag{10.85}$$

where

$$L = \hat{\mathbf{n}} \cdot \mathbf{x}_{\xi\xi}\,, \quad M = \hat{\mathbf{n}} \cdot \mathbf{x}_{\xi\eta}\,, \quad N = \hat{\mathbf{n}} \cdot \mathbf{x}_{\eta\eta}\,. \tag{10.86}$$

The well-known *Gauss identities* for a surface can be derived by resolving the second-order vectors into combinations of the tangent plane and surface normal vectors:

$$\mathbf{x}_{\xi\xi} = \Gamma^1_{11}\,\mathbf{x}_\xi + \Gamma^2_{11}\,\mathbf{x}_\eta + L\,\hat{\mathbf{n}}\,, \tag{10.87}$$

$$\mathbf{x}_{\xi\eta} = \Gamma^1_{12}\,\mathbf{x}_\xi + \Gamma^2_{12}\,\mathbf{x}_\eta + M\,\hat{\mathbf{n}}\,, \tag{10.88}$$

$$\mathbf{x}_{\eta\eta} = \Gamma^1_{22}\,\mathbf{x}_\xi + \Gamma^2_{22}\,\mathbf{x}_\eta + N\,\hat{\mathbf{n}}\,, \tag{10.89}$$

with surface *Schwarz-Christoffel* symbols as defined in (4.140)–(4.145). Terms proportional to the unit surface normal are added to the planar relations. The Gauss identities are satisfied by every parameterization of a surface.

Various cross-product relationships can be derived from (10.87)–(10.89) which are useful in obtaining the *Weingarten* equations:

$$\hat{\mathbf{n}}_\xi = \left(\frac{g_{12}\,M - g_{22}\,L}{g}\right)\mathbf{x}_\xi + \left(\frac{g_{12}\,L - g_{11}\,M}{g}\right)\mathbf{x}_\eta\,, \tag{10.90}$$

$$\hat{\mathbf{n}}_\eta = \left(\frac{g_{12}\,N - g_{22}\,M}{g}\right)\mathbf{x}_\xi + \left(\frac{g_{12}\,M - g_{11}\,N}{g}\right)\mathbf{x}_\eta\,. \tag{10.91}$$

These show that $\hat{\mathbf{n}}_\xi$ and $\hat{\mathbf{n}}_\eta$ lie in the tangent plane.

Exercise 10.2.3 Show that one can rewrite the elements of the curvature tensor as $L = -\mathbf{n}_\xi \cdot \mathbf{x}_\xi$, $M = -\mathbf{n}_\xi \cdot \mathbf{x}_\eta = -\mathbf{n}_\eta \cdot \mathbf{x}_\xi$, and $N = -\mathbf{n}_\eta \cdot \mathbf{x}_\eta$. §

The *mean surface curvature,*

$$\bar{\kappa} = \frac{g_{11}\,N - 2\,g_{12}\,M + g_{22}\,L}{2\,g}\,, \tag{10.92}$$

plays an important role in surface grid generation since it is an invariant of the surface parameterization.

Exercise 10.2.4 Show that $\bar{\kappa}$ is invariant to the choice of surface parameterization. §

Exercise 10.2.5 Verify that

$$\mathrm{tr}\{G^{-1}\mathcal{W}\} = 2\,\bar{\kappa}\,.\ \S \tag{10.93}$$

THEOREM 10.1 The surface *metric identity*

$$\frac{\partial}{\partial\xi}\left(\frac{g_{22}\,\mathbf{x}_\xi - g_{12}\,\mathbf{x}_\eta}{\sqrt{g}}\right) + \frac{\partial}{\partial\eta}\left(\frac{g_{11}\,\mathbf{x}_\eta - g_{12}\,\mathbf{x}_\xi}{\sqrt{g}}\right) = 2\,\bar{\kappa}\,\sqrt{g}\,\hat{\mathbf{n}} \tag{10.94}$$

holds for any parameterization of the surface.

Proof.

$$\frac{\partial}{\partial\xi}\left(\frac{g_{22}\,\mathbf{x}_\xi - g_{12}\,\mathbf{x}_\eta}{\sqrt{g}}\right) + \frac{\partial}{\partial\eta}\left(\frac{-g_{12}\,\mathbf{x}_\xi + g_{11}\,\mathbf{x}_\eta}{\sqrt{g}}\right) = \tag{10.95}$$

$$\frac{\partial}{\partial\xi}(\mathbf{x}_\eta \times \hat{\mathbf{n}}) + \frac{\partial}{\partial\eta}(\hat{\mathbf{n}} \times \mathbf{x}_\xi) = \tag{10.96}$$

$$\mathbf{x}_\eta \times \hat{\mathbf{n}}_\xi + \hat{\mathbf{n}}_\eta \times \mathbf{x}_\xi = 2\,\bar{\kappa}\,\sqrt{g}\,\hat{\mathbf{n}} \tag{10.97}$$

where the last step is obtained by forming the appropriate cross products with the Weingarten equations (10.90)–(10.91) and the definition (10.92). §

Higher order relationships of differential geometry such as the Codazzi equations, the *Theorema Egregium*, and the theory of geodesics could be presented, but these have seen little application in grid generation to date.

Exercise 10.2.6 A simple derivation of a widely used nonvariational surface grid generator (Warsi [226]) can be obtained from the surface Gauss relations. Expand the surface vector $g_{22}\,\mathbf{x}_{\xi\xi} - 2\,g_{12}\,\mathbf{x}_{\xi\eta} + g_{11}\,\mathbf{x}_{\eta\eta}$ in the linear combination $P\,\mathbf{x}_\xi + Q\,\mathbf{x}_\eta + \lambda\,\hat{\mathbf{n}}$ and show $\lambda = 2\,g\,\bar{\kappa}$. To convert this identity into a grid-generation equation, the coefficients P and Q are replaced by weight functions, while the coefficient λ ensures that points lie on the given surface (to within truncation error). §

10.2.2 Transformation Relations on a Surface

Since the surface Jacobian matrix is not square, its determinant is undefined and $\mathcal{J}$ has no inverse. Furthermore, the matrix $\mathcal{J}^T$ takes vectors in R^3 to R^2 so that if $\mathbf{v} \in \mathbf{T}$, then $\mathcal{J}^T\mathbf{v} \in R^2$, i.e., the image of a vector in the tangent plane is a vector in R^2. Observe that the *null space* of $\mathcal{J}^T$ consists of scalar multiples of the unit surface normal vector. In particular,

$$\mathcal{J}^T\hat{\mathbf{n}} = \mathbf{0}. \tag{10.98}$$

This is made clear by observing that the components of this matrix-vector product is the vector having components $\mathbf{x}_\xi \cdot \hat{\mathbf{n}}$ and $\mathbf{x}_\eta \cdot \hat{\mathbf{n}}$.

Although the inverse of the Jacobian matrix does not exist, a *generalized inverse* can be constructed using the definition

$$(\mathcal{J}^{-1})^T = \mathcal{J}\,\mathcal{G}^{-1}. \tag{10.99}$$

This is well defined since the surface metric tensor is square and is assumed to have nonzero determinant. Explicit forms of the inverse are

$$(\mathcal{J}^{-1})^T = \frac{1}{g}[g_{22}\,\mathbf{x}_\xi - g_{12}\,\mathbf{x}_\eta, -g_{12}\,\mathbf{x}_\xi + g_{11}\,\mathbf{x}_\eta], \tag{10.100}$$

$$= \frac{1}{\sqrt{g}}[\mathbf{x}_\eta \times \hat{\mathbf{n}}, \hat{\mathbf{n}} \times \mathbf{x}_\xi]. \tag{10.101}$$

The generalized inverse takes vectors in R^2 to vectors in $\mathbf{T}$.

Exercise 10.2.7 Let $\mathbf{v} \in \mathbf{T}$. Show that

$$(\mathcal{J}^{-1})^T(\mathcal{J}^T\mathbf{v}) = \mathbf{v} \tag{10.102}$$

and verify the entries in Table 10.1. §

It can be shown from the definitions that $\mathcal{J}\mathcal{J}^{-1} = \mathcal{I}_{3\times3}$ and $\mathcal{J}^{-1}\mathcal{J} = \mathcal{I}_{2\times2}$. The null space of $\mathcal{J}^{-1}$ is the same as the null space of $\mathcal{J}^T$, i.e.,

$$\mathcal{J}^{-1}\hat{\mathbf{n}} = \mathbf{0}. \tag{10.103}$$

Transformation relationships for hosted equations on surfaces can be developed in terms of the generalized inverse. The auxiliary matrices, $C = \sqrt{g}\,(\mathcal{J}^{-1})^T$

Matrix	Dimensions	Domain	Range	Null Space
$\mathcal{J}$	3×2	R^2	$\mathbf{T}$	$\mathbf{0}$
$\mathcal{J}^T$	2×3	R^3	R^2	$\hat{\mathbf{n}}$
$\mathcal{J}^{-1}$	2×3	R^3	R^2	$\hat{\mathbf{n}}$
$(\mathcal{J}^{-1})^T$	3×2	R^2	$\mathbf{T}$	$\mathbf{0}$

Table 10.1 *Properties of the Jacobian matrix and its generalized inverse*

and $C^{-1} = \mathcal{J}^T / \sqrt{g}$ are defined as in previous chapters. The metric identity (10.94) is simply stated

$$\text{div}_{\boldsymbol{\xi}} C = 2\,\bar{\kappa}\,\sqrt{g}\hat{\mathbf{n}}\,. \tag{10.104}$$

Assume for the moment that

$$\text{div}_{\mathbf{x}} C^{-1} = \mathbf{0}\,. \tag{10.105}$$

This will be proven later in this section to be a consequence of definitions to be adopted. No curvature term appears in (10.105) due to the fact that logical space has zero curvature.

Let $f = f(\mathbf{x}(\xi, \eta))$ be a scalar function defined on S. The logical space *gradient* is well defined

$$\nabla_{\boldsymbol{\xi}} f = \begin{pmatrix} f_\xi \\ f_\eta \end{pmatrix}, \tag{10.106}$$

with $\nabla_{\boldsymbol{\xi}} f \in R^2$. The physical-space derivatives of f are not well defined since f is defined only on the surface. Therefore, the following *definition* of the physical-space gradient of f is adopted:

$$\nabla_{\mathbf{x}} f = (\mathcal{J}^{-1})^T\,\nabla_{\boldsymbol{\xi}} f\,, \tag{10.107}$$

$$= \frac{1}{\sqrt{g}}\,C\,\nabla_{\boldsymbol{\xi}} f\,. \tag{10.108}$$

The definition is based on analogy with the planar transformation relationship and employs the generalized inverse of the Jacobian matrix. The gradient $\nabla_{\mathbf{x}} f = (f_x, f_y, f_z)^T$ lies in the tangent plane, i.e., $\nabla_{\mathbf{x}} f \in \mathbf{T}$. This definition can be inverted using the result (10.102) to obtain nonconservative forms

$$\nabla_{\boldsymbol{\xi}} f = \mathcal{J}^T\,\nabla_{\mathbf{x}} f\,, \tag{10.109}$$

$$= \sqrt{g}\,C^{-1}\,\nabla_{\mathbf{x}} f\,, \tag{10.110}$$

which formally agree with the planar relation.

The "conservative" form of the surface gradient contains extra terms

$$\operatorname{div}_{\boldsymbol{\xi}}(Cf) = C\,\nabla_{\boldsymbol{\xi}} f + f\,\operatorname{div}_{\boldsymbol{\xi}} C, \tag{10.111}$$

$$= C\,\nabla_{\boldsymbol{\xi}} f + 2\,f\,\sqrt{g}\,\bar{\kappa}\,\hat{\mathbf{n}}. \tag{10.112}$$

Accepting (10.105),

$$\operatorname{div}_{\mathbf{x}}(C^{-1}f) = C^{-1}\,\nabla_{\mathbf{x}} f. \tag{10.113}$$

These give the *conservative* forms of the gradient:

$$\nabla_{\boldsymbol{\xi}} f = \sqrt{g}\,\operatorname{div}_{\mathbf{x}}(C^{-1}f), \tag{10.114}$$

$$\nabla_{\mathbf{x}} f = \frac{1}{\sqrt{g}}\operatorname{div}_{\boldsymbol{\xi}}(Cf) - 2\,f\,\bar{\kappa}\,\hat{\mathbf{n}}. \tag{10.115}$$

As $\kappa \to 0$, the surface formula approaches the planar result.

The transformation relations involving the gradient of a vector are again based on an analogy with the planar relations. Let $\mathbf{v} \in R^3$ and $\hat{\mathbf{v}} \in R^2$. Then $\nabla_{\boldsymbol{\xi}}\mathbf{v}$ is a 3×2 matrix, $\nabla_{\mathbf{x}}\mathbf{v}$ is 3×3, $\nabla_{\boldsymbol{\xi}}\hat{\mathbf{v}}$ is 2×2, and $\nabla_{\mathbf{x}}\hat{\mathbf{v}}$ is a 2×3 matrix. Their transformation rules read

$$\nabla_{\mathbf{x}}\hat{\mathbf{v}} = \frac{1}{\sqrt{g}}\,(\nabla_{\boldsymbol{\xi}}\hat{\mathbf{v}})\,C^T, \tag{10.116}$$

$$\nabla_{\boldsymbol{\xi}}\mathbf{v} = \sqrt{g}\,(\nabla_{\mathbf{x}}\mathbf{v})\,(C^{-1})^T. \tag{10.117}$$

Exercise 10.2.8 Show that $\nabla_{\mathbf{x}}\boldsymbol{\xi} = \mathcal{J}^{-1}$. §

Let $\mathbf{v} \in R^3$. The conservative form of the physical-space *divergence* can be constructed from definition (10.108) of the gradient:

$$\operatorname{div}_{\mathbf{x}} \cdot \mathbf{v} = \frac{1}{\sqrt{g}}\operatorname{div}_{\boldsymbol{\xi}} \cdot (C^T\mathbf{v}). \tag{10.118}$$

This definition makes sense because $C^T\mathbf{v} \in R^2$; it is consistent with the planar case.

Exercise 10.2.9 Expand the right-hand side of (10.118) and show that this relationship for the divergence is consistent with (10.108). §

The result can be inverted by defining $\hat{\mathbf{v}} = C^T\mathbf{v} \in R^2$:

$$\operatorname{div}_{\boldsymbol{\xi}} \cdot \hat{\mathbf{v}} = \sqrt{g}\,\operatorname{div}_{\mathbf{x}} \cdot (C^{-1})^T\,\hat{\mathbf{v}}. \tag{10.119}$$

The divergence relation (10.118) can be used to obtain the transformation relationships for the divergence of a *tensor.* Let $\hat{\mathcal{T}}_{2\times 3}$ take vectors in R^3 to vectors in R^2. Then $(\hat{\mathcal{T}}\,C)_{2\times 2}$ and

$$\operatorname{div}_{\mathbf{x}}\hat{\mathcal{T}} = \frac{1}{\sqrt{g}}\operatorname{div}_{\boldsymbol{\xi}}(\hat{\mathcal{T}}\,C) \tag{10.120}$$

is a vector in R^2. If $\mathcal{T}_{3\times 2}$ takes vectors in R^2 to vectors in R^3, then $(\mathcal{T} C^{-1})_{3\times 3}$ and

$$\mathrm{div}_{\boldsymbol{\xi}} \mathcal{T} = \sqrt{g}\, \mathrm{div}_{\mathbf{x}} (\mathcal{T} C^{-1}) \tag{10.121}$$

is a vector in R^3.

The inverse metric identity (10.105) can now be proved as a consequence of the formal definition (10.120):

$$\mathrm{div}_{\mathbf{x}} C^{-1} = \frac{1}{\sqrt{g}}\, \mathrm{div}_{\boldsymbol{\xi}} (C^{-1} C) . \tag{10.122}$$

The term on the right-hand side is the zero vector in R^2.

The *Laplacian* can be defined using the gradient and divergence results and noting that the physical-space gradient of f lies in the tangent plane,

$$\nabla_{\mathbf{x}}^2 f = \frac{1}{\sqrt{g}}\, \mathrm{div}_{\boldsymbol{\xi}} \cdot \left(\frac{C^T C}{\sqrt{g}} \nabla_{\boldsymbol{\xi}} f \right) . \tag{10.123}$$

The vector Laplacian transforms as

$$\nabla_{\mathbf{x}}^2 \hat{\mathbf{v}} = \frac{1}{\sqrt{g}}\, \mathrm{div}_{\boldsymbol{\xi}} \left\{ \frac{(\nabla_{\boldsymbol{\xi}} \hat{\mathbf{v}})\, C^T C}{\sqrt{g}} \right\} , \tag{10.124}$$

with $\hat{\mathbf{v}} \in R^2$.

Exercise 10.2.10 Invert the relation $\nabla_{\mathbf{x}}^2 \boldsymbol{\xi} = \mathbf{P}$ to obtain the Warsi surface grid generator given in Exercise 10.2.6. §

If $\mathcal{T}$ takes vectors in R^3 to vectors in R^3, the elliptic form is

$$\mathrm{div}_{\mathbf{x}} \cdot \mathcal{T} \nabla_{\mathbf{x}} f = \frac{1}{\sqrt{g}}\, \mathrm{div}_{\boldsymbol{\xi}} \cdot \left(\frac{C^T \mathcal{T} C}{\sqrt{g}} \nabla_{\boldsymbol{\xi}} f \right) . \tag{10.125}$$

To summarize, once the generalized inverse has been defined, the physical operators gradient, divergence, and Laplacian can be defined by analogy to the planar relationships. The results are consistent with one another and with the planar operations.

10.2.3 Parametric Approach to Variational Surface Grid Generation

Based on the analogy to curve grid generation (Section 10.1.3), the machinery just developed is used to derive the Euler-Lagrange equations for a surface variational principle. Two approaches are described, the first employs the parameter space concept, while the second is based on Lagrange Multipliers. The general surface functional is of the form

$$I[\mathbf{x}] = \int_0^1 \int_0^1 G(\boldsymbol{\xi}, \mathbf{x}, \mathbf{x}_\xi, \mathbf{x}_\eta)\, d\xi\, d\eta \tag{10.126}$$

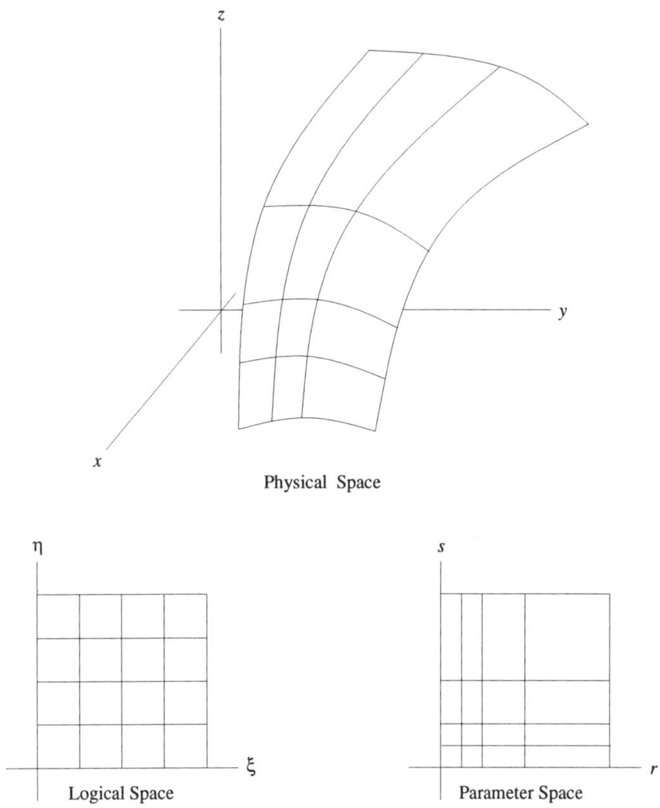

Figure 10.3 *Surface parameter space*

with $G : R^{11} \to R^1$ being a positive, real, smooth, homogeneous function and $\mathbf{x} \in S$. The minimization is, of course, constrained so that the mapping takes points to the given surface.

The parametric form of the surface Euler-Lagrange equations is derived in this section. Introduce the parameter space $\mathbf{r} = (r, s)$ so that $\mathbf{x} = \mathbf{x}(r, s)$ (see Figure 10.3). The transformed tangents are

$$\mathbf{x}_\xi = \mathbf{x}_r\, r_\xi + \mathbf{x}_s\, s_\xi\,, \tag{10.127}$$

$$\mathbf{x}_\eta = \mathbf{x}_r\, r_\eta + \mathbf{x}_s\, s_\eta\,. \tag{10.128}$$

The surface Jacobian transforms as

$$\mathcal{J} = \hat{\mathcal{J}}(\nabla_{\boldsymbol{\xi}}\mathbf{r}) \tag{10.129}$$

with $\hat{\mathcal{J}} = (\nabla_{\mathbf{r}}\mathbf{x})$ being a 3×2 matrix. Define the transformed metric tensor $\hat{\mathcal{G}}$ to be

$$\hat{\mathcal{G}} = \hat{\mathcal{J}}^T \hat{\mathcal{J}} \tag{10.130}$$

and define $\hat{g} = \det \hat{\mathcal{G}}$ so that $\sqrt{\hat{g}} = | \mathbf{x}_r \times \mathbf{x}_s |$. The original metric tensor $\mathcal{G} = \mathcal{J}^T \mathcal{J}$ transforms as

$$\mathcal{G} = (\nabla_{\boldsymbol{\xi}} \mathbf{r})^T \hat{\mathcal{G}} (\nabla_{\boldsymbol{\xi}} \mathbf{r}) . \tag{10.131}$$

Define the parameter-space Jacobian determinant

$$\sqrt{p} = \det(\nabla_{\boldsymbol{\xi}} \mathbf{r}) , \tag{10.132}$$

$$= r_\xi \, s_\eta - r_\eta \, s_\xi . \tag{10.133}$$

Using the fact that the determinant of the product of two matrices is the product of their determinants, relation (10.131) shows

$$\sqrt{g} = \sqrt{p}\sqrt{\hat{g}} . \tag{10.134}$$

The surface functional (10.126) is converted to

$$I[\mathbf{r}] = \int_0^1 \int_0^1 \frac{H(g_{11}, g_{12}, g_{22}, \sqrt{g})}{\hat{w}^2(\mathbf{r})} \, d\xi \, d\eta \tag{10.135}$$

in the parametric formulation. The constraint that the mapping lie on the given surface is automatically satisfied.

Exercise 10.2.11 Use (10.131) to verify that

$$\frac{\partial g_{11}}{\partial \mathbf{r}_\xi} = 2 \hat{\mathcal{G}} \mathbf{r}_\xi , \quad \frac{\partial g_{12}}{\partial \mathbf{r}_\xi} = \hat{\mathcal{G}} \mathbf{r}_\eta , \quad \frac{\partial g_{22}}{\partial \mathbf{r}_\xi} = \mathbf{0}, \tag{10.136}$$

$$\frac{\partial g_{11}}{\partial \mathbf{r}_\eta} = \mathbf{0} , \quad \frac{\partial g_{12}}{\partial \mathbf{r}_\eta} = \hat{\mathcal{G}} \mathbf{r}_\xi , \quad \frac{\partial g_{22}}{\partial \mathbf{r}_\eta} = 2 \hat{\mathcal{G}} \mathbf{r}_\eta, \tag{10.137}$$

and use (10.134) to show that

$$\frac{\partial \sqrt{g}}{\partial \mathbf{r}_\xi} = -\sqrt{\hat{g}} \, \mathbf{r}_\eta^\perp , \quad \frac{\partial \sqrt{g}}{\partial \mathbf{r}_\eta} = \sqrt{\hat{g}} \, \mathbf{r}_\xi^\perp . \ \S \tag{10.138}$$

The Euler-Lagrange equation for the principle (10.126) is

$$\frac{\partial}{\partial \mathbf{r}} \left(\frac{H}{\hat{w}^2} \right) - \mathrm{div}_{\boldsymbol{\xi}} \frac{\mathcal{T}}{\hat{w}^2} = \mathbf{0} \tag{10.139}$$

where $\mathcal{T}_{2\times 2} = [\partial H / \partial \mathbf{r}_\xi, \partial H / \partial \mathbf{r}_\eta]$ and $\hat{w} = \hat{w}(\mathbf{r})$. Using (10.136), (10.137), and (10.138), the chain rule can be applied in the manner of Chapter 8 to find

$$\mathcal{T} = \hat{\mathcal{G}} (\nabla_{\boldsymbol{\xi}} \mathbf{r}) \mathcal{M} + \sqrt{\hat{g}} \frac{\partial H}{\partial \sqrt{g}} C_{\mathbf{r}} \tag{10.140}$$

where

$$[\mathcal{M}]_{ij} = (1 + \delta_{ij}) \frac{\partial H}{\partial g_{ij}} \tag{10.141}$$

and

$$C_r = \sqrt{\rho}\,[(\nabla_{\xi}\mathbf{r})^{-1}]^T, \tag{10.142}$$

$$= \begin{pmatrix} s_\eta & -s_\xi \\ -r_\eta & r_\xi \end{pmatrix}. \tag{10.143}$$

Exercise 10.2.12 Begin with (10.140) and show that $\mathcal{T} = [\partial H/\partial \mathbf{r}_\xi, \partial H/\partial \mathbf{r}_\eta]$. §

Exercise 10.2.13 Let

$$\mathcal{B} = \mathcal{M} + \sqrt{g}\,\frac{\partial H}{\partial \sqrt{g}}\,G^{-1} \tag{10.144}$$

as in Chapter 8. Show that

$$\mathcal{T} = \hat{\mathcal{G}}\,(\nabla_{\xi}\mathbf{r})\,\mathcal{B}. \text{ §} \tag{10.145}$$

As was true in the case of curves, the first term in the Euler-Lagrange equation cannot be directly converted to a conservative, logical-space form. The following relationship applies:

$$\frac{\partial}{\partial \xi}\left(\frac{H}{\hat{w}^2}\right) = \left(\nabla_{\xi}\mathbf{r}\right)^T \frac{\partial}{\partial \mathbf{r}}\left(\frac{H}{\hat{w}^2}\right) + \frac{1}{\hat{w}^2}\,[\nabla_{\xi}(\nabla_{\xi}\mathbf{r})^T]\,\hat{\mathcal{G}}\,(\nabla_{\xi}\mathbf{r})\,\mathcal{B}. \tag{10.146}$$

The Euler-Lagrange equation in parametric form is then

$$\frac{C_r}{\sqrt{\rho}}\left\{\frac{\partial}{\partial \xi}\left(\frac{H}{\hat{w}^2}\right) - \frac{1}{\hat{w}^2}\,[\nabla_{\xi}(\nabla_{\xi}\mathbf{r})^T]\,\hat{\mathcal{G}}\,(\nabla_{\xi}\mathbf{r})\,\mathcal{B}\right\} - \mathrm{div}_{\xi}\left(\frac{\hat{\mathcal{G}}\,(\nabla_{\xi}\mathbf{r})\,\mathcal{B}}{\hat{w}^2}\right) = \mathbf{0}. \tag{10.147}$$

Covariant Projection

If one multiplies the previous Euler-Lagrange equation by $(\nabla_{\xi}\mathbf{r})^T = \sqrt{\rho}\,C_r^{-1}$, it is easy to show that a fully conservative form is obtained:

$$\mathrm{div}_{\xi}\left\{\frac{H\mathcal{I} - (\nabla_{\xi}\mathbf{r})^T\,\hat{\mathcal{G}}\,(\nabla_{\xi}\mathbf{r})\,\mathcal{B}}{\hat{w}^2}\right\} = \mathbf{0}. \tag{10.148}$$

Numerical solutions to this equation may be developed in a manner similar to Section 8.3.3, linearizing (10.148) by factoring out the leading matrix:

$$\mathrm{div}_{\xi}\frac{(\nabla_{\xi}\mathbf{r})^T}{\hat{w}^2}\,\{H\,[(\nabla_{\xi}\mathbf{r})^{-1}]^T - \hat{\mathcal{G}}\,(\nabla_{\xi}\mathbf{r})\,\mathcal{B}\} = \mathbf{0}. \tag{10.149}$$

10.2.4 Lagrange Multiplier Approach to Variational Surface Grid Generation

The Lagrange Multiplier form of the surface Euler-Lagrange equations for an arbitrary functional of the following form is now derived:

$$I[\mathbf{x}] = \int_0^1\int_0^1 \frac{H(g_{11}, g_{12}, g_{22}, \sqrt{g})}{w^2(\mathbf{x})}\,d\xi\,d\eta \tag{10.150}$$

with the constraints that the boundary data is satisfied and the minimizing transformation lies on the given surface. The Lagrange Multiplier λ is introduced to ensure the constraint is met. The Euler-Lagrange equations are proportional to a vector in the null space of $\mathcal{J}^T$; this can be demonstrated in a manner similar to the derivation supplied in Section 10.1.4 for curves, or one can refer to equations (6.14)-(6.15) from Chapter 6.

Let $\mathcal{T} = [\partial H/\partial \mathbf{x}_\xi, \partial H/\partial \mathbf{x}_\eta]$ so that the Euler-Lagrange equation has the form

$$\frac{\partial}{\partial \mathbf{x}}\left(\frac{H}{w^2}\right) - \mathrm{div}_{\boldsymbol{\xi}}\frac{\mathcal{T}}{w^2} = \lambda\,\hat{\mathbf{n}} \tag{10.151}$$

with λ to be determined. The first term can be converted to a logical-space gradient using the gradient relation (10.115) to get

$$\frac{\partial}{\partial \mathbf{x}}\left(\frac{H}{w^2}\right) = H\frac{\partial}{\partial \mathbf{x}}\left(\frac{1}{w^2}\right), \tag{10.152}$$

$$= H\left\{\frac{1}{\sqrt{g}}\,\mathrm{div}_{\boldsymbol{\xi}}\frac{C}{w^2} - \frac{2}{w^2}\,\bar{\kappa}\,\hat{\mathbf{n}}\right\}. \tag{10.153}$$

The second term in (10.151) is $\mathcal{T} = \mathcal{J}\mathcal{B}$, with the matrix $\mathcal{B}$ defined as in Chapter 8. The Euler-Lagrange equation (10.151) now reads

$$\frac{H}{\sqrt{g}}\mathrm{div}_{\boldsymbol{\xi}}\left(\frac{C}{w^2}\right) - \mathrm{div}_{\boldsymbol{\xi}}\left\{\frac{\mathcal{J}\mathcal{B}}{w^2}\right\} = \left(\lambda + \frac{2H\bar{\kappa}}{w^2}\right)\hat{\mathbf{n}}. \tag{10.154}$$

The multiplier λ is determined by projecting this equation onto the unit surface normal vector. To do so requires a few lemmas.

LEMMA 10.2 The gradient of the unit surface normal vector is

$$\nabla_{\boldsymbol{\xi}}\hat{\mathbf{n}} = -\frac{C}{\sqrt{g}}\,\mathcal{W}. \tag{10.155}$$

Proof.

$$\sqrt{g}\,C^{-1}\,\nabla_{\boldsymbol{\xi}}\hat{\mathbf{n}} = \mathcal{J}^T\,[\hat{\mathbf{n}}_\xi, \hat{\mathbf{n}}_\eta], \tag{10.156}$$

$$= \begin{pmatrix} \hat{\mathbf{n}}_\xi \cdot \mathbf{x}_\xi & \hat{\mathbf{n}}_\eta \cdot \mathbf{x}_\xi \\ \hat{\mathbf{n}}_\xi \cdot \mathbf{x}_\eta & \hat{\mathbf{n}}_\eta \cdot \mathbf{x}_\eta \end{pmatrix}, \tag{10.157}$$

$$= -\mathcal{W}.\ \S \tag{10.158}$$

LEMMA 10.3 The projection of the first term in (10.154) onto the unit surface normal is

$$\hat{\mathbf{n}}\cdot\mathrm{div}_{\boldsymbol{\xi}}\left(\frac{C}{w^2}\right) = \frac{2}{w^2}\sqrt{g}\,\bar{\kappa}. \tag{10.159}$$

Proof. Observe that $\mathcal{C}^T\hat{\mathbf{n}} = \mathbf{0}$ due to the result (10.103). Using (7.8) from Chapter 7,

$$\hat{\mathbf{n}} \cdot \text{div}_{\xi}\left(\frac{\mathcal{C}}{w^2}\right) = \text{div}_{\xi} \cdot \left(\frac{\mathcal{C}^T\hat{\mathbf{n}}}{w^2}\right) - \text{tr}\left\{\frac{\mathcal{C}^T}{w^2}\nabla_{\xi}\hat{\mathbf{n}}\right\}, \tag{10.160}$$

$$= -\text{tr}\left\{\frac{\mathcal{C}^T}{w^2}\nabla_{\xi}\hat{\mathbf{n}}\right\}, \tag{10.161}$$

$$= \frac{1}{w^2}\text{tr}\left\{\left(\frac{\mathcal{C}^T\mathcal{C}}{\sqrt{g}}\right)\mathcal{W}\right\}, \tag{10.162}$$

$$= \frac{\sqrt{g}}{w^2}\text{tr}\{\mathcal{G}^{-1}\mathcal{W}\}, \tag{10.163}$$

and the relation (10.93) finishes the proof. §

LEMMA 10.4 The projection of the second term in (10.154) onto the unit surface normal is

$$\hat{\mathbf{n}} \cdot \text{div}_{\xi}\left(\frac{\mathcal{J}\mathcal{B}}{w^2}\right) = \frac{1}{w^2}\text{tr}\{\mathcal{B}\mathcal{W}\}. \text{ §} \tag{10.164}$$

Proof.

$$\hat{\mathbf{n}} \cdot \text{div}_{\xi}\left(\frac{\mathcal{J}\mathcal{B}}{w^2}\right) = \text{div}_{\xi} \cdot \left(\frac{\mathcal{J}\mathcal{B}}{w^2}\right)^T\hat{\mathbf{n}} - \text{tr}\left\{\left(\frac{\mathcal{J}\mathcal{B}}{w^2}\right)^T\nabla_{\xi}\hat{\mathbf{n}}\right\}, \tag{10.165}$$

$$= \text{div}_{\xi} \cdot \left(\frac{\mathcal{B}\mathcal{J}^T\hat{\mathbf{n}}}{w^2}\right) - \frac{1}{w^2}\text{tr}\{\mathcal{B}\mathcal{J}^T\nabla_{\xi}\hat{\mathbf{n}}\}, \tag{10.166}$$

$$= \frac{1}{w^2}\text{tr}\{\mathcal{B}\mathcal{W}\}. \text{ §} \tag{10.167}$$

Pulling these lemmas together, one has

THEOREM 10.5 The Lagrange Multiplier in (10.154) is

$$\lambda = -\frac{1}{w^2}\text{tr}\{\mathcal{B}\mathcal{W}\}. \text{ §} \tag{10.168}$$

Note that λ depends on the curvature tensor, the partial derivatives of H, and the weight function. If the surface has zero curvature, the multiplier is zero. Finally, the Euler-Lagrange equation (10.154) in conservative form is

$$\frac{H}{\sqrt{g}}\text{div}_{\xi}\left(\frac{\mathcal{C}}{w^2}\right) - \text{div}_{\xi}\left\{\frac{\mathcal{J}\mathcal{B}}{w^2}\right\} = \frac{1}{w^2}[\text{tr}\{\mathcal{G}^{-1}(H\mathcal{I} - \mathcal{G}\mathcal{B})\mathcal{W}\}]\hat{\mathbf{n}}. \tag{10.169}$$

Exercise 10.2.14 Evaluate the surface Euler-Lagrange equations (10.169) for $H = \sqrt{g}$ with $w = 1$. Relate the results to the surface metric identity. §

Exercise 10.2.15 Find the surface Euler-Lagrange equations for the *area* functional and show $\sqrt{g} = \beta w$ is a generic property of the solution. §

Numerical approaches to (10.169) are not recommended because the Lagrange Multiplier is difficult to evaluate for arbitrary functionals H, it is not an invariant of the parameterization, and it would be difficult to invert the parameterization to evaluate the right-hand side at the correct location. Excessive off-surface truncation error is also a possibility.

10.2.5 The Surface Covariant Projections

The Euler-Lagrange equation (10.169) is projected onto the tangent plane. As shown previously, this is equivalent to multiplying the Euler-Lagrange equation by the matrix $\mathcal{J}^T$. Thus, the projected form of (10.169) is

$$\left(\frac{H}{\sqrt{g}}\right)\mathcal{J}^T \operatorname{div}_\xi\left(\frac{C}{w^2}\right) - \mathcal{J}^T \operatorname{div}_\xi\left(\frac{\mathcal{J}\mathcal{B}}{w^2}\right) = \mathbf{0}. \qquad (10.170)$$

Using (10.115), the first term is

$$\left(\frac{H}{\sqrt{g}}\right)\mathcal{J}^T \operatorname{div}_\xi\left(\frac{C}{w^2}\right) = \left(\frac{H}{\sqrt{g}}\right)\mathcal{J}^T\left\{C\,\nabla_\xi\left(\frac{1}{w^2}\right)\right. \qquad (10.171)$$

$$\left. + \frac{2}{w^2}\sqrt{g}\,\bar{\kappa}\,\hat{\mathbf{n}}\right\}, \qquad (10.172)$$

$$= \left(\frac{H}{\sqrt{g}}\right)\mathcal{J}^T\left\{C\,\nabla_\xi\left(\frac{1}{w^2}\right)\right\}, \qquad (10.173)$$

$$= H\,\nabla_\xi\left(\frac{1}{w^2}\right). \qquad (10.174)$$

To treat the second term, one needs the planar result (8.55),

$$[\nabla_\xi \mathcal{J}^T](\mathcal{J}\mathcal{B}) = \nabla_\xi H \qquad (10.175)$$

generalized to surfaces.

Exercise 10.2.16 Verify (10.175). §

The second term is

$$\mathcal{J}^T \operatorname{div}_\xi\left(\frac{\mathcal{J}\mathcal{B}}{w^2}\right) = \operatorname{div}_\xi\left(\frac{\mathcal{J}^T\mathcal{J}\mathcal{B}}{w^2}\right) - [\nabla_\xi \mathcal{J}^T]\left(\frac{\mathcal{J}\mathcal{B}}{w^2}\right), \qquad (10.176)$$

$$= \operatorname{div}_\xi\left(\frac{\mathcal{G}\mathcal{B}}{w^2}\right) - \frac{1}{w^2}\nabla_\xi H. \qquad (10.177)$$

Observing that $\nabla_\xi H = \operatorname{div}_\xi(H\mathcal{I})$, one has

THEOREM 10.6 The covariant projection of the Lagrange Multiplier form of the surface Euler-Lagrange equations is

$$\operatorname{div}_\xi\left\{\frac{H\mathcal{I} - \mathcal{G}\mathcal{B}}{w^2}\right\} = \mathbf{0}. \qquad (10.178)$$

Proof. The proof readily follows from the lemmas established in this section. §

One can solve (10.178) in a Newton iteration (Knupp [113]), but since this form reverts to (10.148) when the parameter space is introduced, the numerical approach suggested in Section 10.2.3 may be more convenient.

Chapter 11

Contravariant Functionals: Alignment, Diagonalization, and Rotation

A weakness of the variational methods described in previous chapters is that there is no clear interpretation of the weight functions (except in the case of the area functional); the weights are merely hooks that permit one to tinker with the grid to achieve a desired result. The variational method of Steinberg and Roache (see Chapter 6) uses a reference grid to construct the weight function in an intuitive manner. However, a clear geometric meaning to the weight can only be found for the area functional (and this results in a nonelliptic generator). The variational method presented in Chapter 8 does not remedy this problem of the weights. The nonvariational, weighted elliptic grid generator (Section 5.5) may also be criticized, again because there is no direct geometric connection between the weight functions and the interior tangents (this partly accounts for the rise of interactive grid generation as a workable substitute for automatic grid generation).

Ideally, one would like to specify an arbitrary *local condition* that gives the local relationship between the two covariant tangent vectors; for example, relation (5.28) for orthogonality and cell-aspect ratio control. This is done in this chapter using functionals that depend on a quadratic integrand (expressed in terms of the elliptic norm). Matrix functions of position serve as physical-space weights that describe the local condition. Physical-space weights are handled more conveniently using the Brackbill-Saltzman (or contravariant) variational approach. The main advantage of the variational approach of this chapter, then, is that the resulting Euler-Lagrange equations provide a least-squares fit to a given local condition. Another advantage is that the second variation of the functional depends only on the physical-space weight functions (i.e., the matrix functions). It is thus straightforward to show that the second variation is nonnegative. The approach has strong connections to the theory of harmonic grid generation (see Dvinsky [54, 55], Liao [126], and Sritharan [185] for a complete discussion of harmonic grid generation) and, in particular, to Winslow's variable diffusion generator.

The remaining difficulty is, of course, that the desired local condition does not necessarily result from solving the proposed Euler-Lagrange equations. This is due to the fact that the equations give a least-squares fit to the condition; the resulting grid is critically dependent upon the imposed boundary conditions. Furthermore, a grid having the desired local conditions everywhere may simply not exist (this is the beauty of the least-squares approach: if the grid exists, the method will produce it if the proper boundary conditions are given; if the grid does not exist, a least-squares approximation will be produced).

For simplicity, the present discussion is confined to the planar problem (described in Chapter 5); generalizations to the surface and volume cases require additional work. In Section 11.1 two basic classes of functionals are introduced, based on the elliptic norm and the rank of two matrices, which lead to alignment and diagonalization grid generators. A nonsymmetric form of the diagonalization equations is derived in Section 11.2 for numerical purposes, while Section 11.3 shows how to construct weight functions for the diagonalization equations to attain least-square solutions to a given local condition on the tangent vectors. Section 11.4 gives a weighted grid generator whose local condition is stated in terms of rotation matrices. The chapter closes with Section 11.5, in which it is shown how to combine ellipticity and alignment into a single functional.

11.1 FUNCTIONALS BASED ON THE ELLIPTIC NORM

Let $\tilde{\mathcal{A}}$ and $\tilde{\mathcal{B}}$ be 2×2 real, positive semidefinite matrices. Define the nonnegative function $G \mid R^2 \times R^2 \to R$ by

$$G(\mathbf{x}_1, \mathbf{x}_2) = \mathbf{x}_1^T \cdot \tilde{\mathcal{A}}\,\mathbf{x}_1 + \mathbf{x}_2^T \cdot \tilde{\mathcal{B}}\,\mathbf{x}_2. \tag{11.1}$$

It is easy to show that the function G is convex when the two matrices are positive-definite, so it is believed that grid-generation functionals based on G (such as the ones to follow) will turn out to be convex (or perhaps polyconvex). The proof of this is left as a future research topic.

The matrices $\tilde{\mathcal{A}}$ and $\tilde{\mathcal{B}}$ are to serve as physical-space weighting functions, so assume further that the two matrices are functions of $\mathbf{x}$, the position in the plane. For physical-space weighting, contravariant functionals result in simpler Euler-Lagrange equations than do covariant functionals. Therefore, consider the contravariant functional form

$$I[\,\xi(\mathbf{x}),\, \eta(\mathbf{x})\,] = \frac{1}{2}\int_\Omega G(\,\nabla_\mathbf{x}\xi,\, \nabla_\mathbf{x}\eta\,)\,dx\,dy, \tag{11.2}$$

to be minimized subject to the constraint of the boundary data (the functional is clearly nonnegative). The second variation can be shown to also be nonnegative due to the assumption that the matrices are positive semidefinite.

The Euler-Lagrange equations for this functional are examined to see what kinds of solution properties are possible for various assumptions on the matrices $\tilde{\mathcal{A}}$ and $\tilde{\mathcal{B}}$. The Euler-Lagrange equations read

$$\mathrm{div}_\mathbf{x}\,[\,\mathcal{A}\,\nabla_\mathbf{x}\xi,\, \mathcal{B}\,\nabla_\mathbf{x}\eta\,]^T = \mathbf{0} \tag{11.3}$$

with

$$\mathcal{A} = \frac{1}{2}(\tilde{\mathcal{A}} + \tilde{\mathcal{A}}^T), \tag{11.4}$$

$$\mathcal{B} = \frac{1}{2}(\tilde{\mathcal{B}} + \tilde{\mathcal{B}}^T), \tag{11.5}$$

i.e., the matrices that appear in the Euler-Lagrange equations are the *symmetric* part of the matrices $\tilde{\mathcal{A}}$ and $\tilde{\mathcal{B}}$ that appear in the functional. This means that the skew-symmetric part of the matrices appearing in the functional have no effect on the grid, nor on the second variation of the functional, i.e., one may as well assume that $\tilde{\mathcal{A}}$ and $\tilde{\mathcal{B}}$ are symmetric to begin with.

Observe that the matrix $\mathcal{A}$ in the Euler-Lagrange equations is real, symmetric (and thus has real eigenvalues), and positive semidefinite; the same statement holds for $\mathcal{B}$. There are two cases to be considered that have a major impact on the type of grids that can be generated. These cases lead to *alignment* and *diagonalization* functionals; they are based on the ranks of the matrices.

11.1.1 The Alignment Functional

Suppose both $\mathcal{A}$ and $\mathcal{B}$ have rank 1. Then the matrices are singular and have only one nonzero eigenvalue each. The null space of each matrix has dimension one. Functionals based on matrices with rank 1 are referred to as *alignment functionals* because they can be used to align the coordinate system tangents with prescibed vector fields.

The following lemma is used to find the geometric condition that leads to potential solutions to the Euler-Lagrange equation (11.3); many other solutions are, of course, possible.

LEMMA 11.1 Let $\mathcal{M}$ be any real, symmetric matrix of rank 1. Then $\mathcal{M}$ obeys the proportionality

$$\mathcal{M} \propto \mathbf{v} \otimes \mathbf{v} \tag{11.6}$$

with $\mathbf{v}$ an arbitrary nonzero vector. The null space of $\mathcal{M}$ is the span of the vector $\mathbf{v}^{\perp}$, while the eigenspace is the span of $\mathbf{v}$, with eigenvalue $\lambda \propto \mathbf{v} \cdot \mathbf{v}$. The proof depends on the fact that the colums of a rank-1 matrix are linearly dependent. §

Now that the form of the two matrices has been established, it is easy to constuct functionals that align the tangents with two given (possibly nonorthogonal) vector fields $\mathbf{v}_1$ and $\mathbf{v}_2$. Take

$$\mathcal{A} = \mathbf{v}_2 \otimes \mathbf{v}_2, \tag{11.7}$$

$$\mathcal{B} = \mathbf{v}_1 \otimes \mathbf{v}_1. \tag{11.8}$$

Then the null space of $\mathcal{A}$ is the span of $\mathbf{v}_2^{\perp}$ and the null space of $\mathcal{B}$ is the span of $\mathbf{v}_1^{\perp}$, so the contravariant pair $\nabla_{\mathbf{x}}\xi \propto \mathbf{v}_2^{\perp}$ and $\nabla_{\mathbf{x}}\eta \propto \mathbf{v}_1^{\perp}$ is a potential solution of the Euler-Lagrange equation; this is equivalent to the covariant pair $\mathbf{x}_\xi \propto \mathbf{v}_1$ and $\mathbf{x}_\eta \propto \mathbf{v}_2$.

If only one vector field, $\mathbf{v}$, is given, one may strive for alignment and orthogonality by setting

$$\mathcal{A} = \mathbf{v}^{\perp} \otimes \mathbf{v}^{\perp}, \tag{11.9}$$

$$\mathcal{B} = \mathbf{v} \otimes \mathbf{v}, \tag{11.10}$$

to align $\mathbf{x}_\xi$ with $\mathbf{v}$ and $\mathbf{x}_\eta$ with $\mathbf{v}^{\perp}$.

The matrices (11.7)-(11.8) give

$$G = \nabla_{\mathbf{x}}\xi \cdot (\mathbf{v}_2 \otimes \mathbf{v}_2)\, \nabla_{\mathbf{x}}\xi + \nabla_{\mathbf{x}}\eta \cdot (\mathbf{v}_1 \otimes \mathbf{v}_1)\, \nabla_{\mathbf{x}}\eta, \tag{11.11}$$

$$= (\mathbf{v}_2 \cdot \nabla_{\mathbf{x}}\xi)^2 + (\mathbf{v}_1 \cdot \nabla_{\mathbf{x}}\eta)^2, \tag{11.12}$$

which shows $G = 0$ (and thus the functional is minimized) if the alignment solution holds. It is possible to show that the Euler-Lagrange equations for the alignment functional based on (11.12) are nonelliptic (aligment being an inherently nonelliptic task).

It is readily shown that the alignment functional proposed here is essentially equivalent to the *directional control* functionals of Giannakopoulos and Engel [81]. Applying various vector and matrix identities:

$$G = \nabla_{\mathbf{x}}\xi \cdot (\mathbf{v}_2 \otimes \mathbf{v}_2)\, \nabla_{\mathbf{x}}\xi + \nabla_{\mathbf{x}}\eta \cdot (\mathbf{v}_1 \otimes \mathbf{v}_1)\, \nabla_{\mathbf{x}}\eta, \tag{11.13}$$

$$\begin{aligned} &= \nabla_{\mathbf{x}}\xi \cdot \{(\mathbf{v}_2^{\perp} \cdot \mathbf{v}_2^{\perp})\mathcal{I} - (\mathbf{v}_2^{\perp} \otimes \mathbf{v}_2^{\perp})\}\, \nabla_{\mathbf{x}}\xi \\ &\quad + \nabla_{\mathbf{x}}\eta \cdot \{(\mathbf{v}_1^{\perp} \cdot \mathbf{v}_1^{\perp})\mathcal{I} - (\mathbf{v}_1^{\perp} \otimes \mathbf{v}_1^{\perp})\}\, \nabla_{\mathbf{x}}\eta, \end{aligned} \tag{11.14}$$

$$\begin{aligned} &= (\mathbf{v}_2^{\perp} \cdot \mathbf{v}_2^{\perp})\,(\nabla_{\mathbf{x}}\xi \cdot \nabla_{\mathbf{x}}\xi) - (\mathbf{v}_2^{\perp} \cdot \nabla_{\mathbf{x}}\xi)^2 \\ &\quad + (\mathbf{v}_1^{\perp} \cdot \mathbf{v}_1^{\perp})\,(\nabla_{\mathbf{x}}\eta \cdot \nabla_{\mathbf{x}}\eta) - (\mathbf{v}_1^{\perp} \cdot \nabla_{\mathbf{x}}\eta)^2, \end{aligned} \tag{11.15}$$

$$= \|\, \mathbf{v}_2^{\perp} \times \nabla_{\mathbf{x}}\xi \,\|^2 + \|\, \mathbf{v}_1^{\perp} \times \nabla_{\mathbf{x}}\eta \,\|^2, \tag{11.16}$$

which is recognized to be of the form of Giannakopoulos and Engel. The main drawback to the alignment approach given in this section is the lack of ellipticity of the governing equations; this problem is addressed in section 11.5.

11.1.2 The Diagonalization Functionals

Suppose both $\mathcal{A}$ and $\mathcal{B}$ in (11.3) have rank 2. Then their inverses exist, their null spaces are just the zero vector, and both matrices are symmetric and positive definite, so $G > 0$. The Euler-Langrange equation now cannot possess the previously-constructed alignment solution because this would force both tangent vectors to be zero since the null spaces have dimension zero. Another possible solution is based on the metric identity $\mathrm{div}_{\mathbf{x}} C^{-1} = \mathbf{0}$. Since $C^{-T} = [-(\nabla_{\mathbf{x}}\eta)^{\perp}, (\nabla_{\mathbf{x}}\xi)^{\perp}]$, one has the pair of local conditions:

$$\mathcal{A}\, \nabla_{\mathbf{x}}\xi = -(\nabla_{\mathbf{x}}\eta)^{\perp}, \tag{11.17}$$

$$\mathcal{B}\, \nabla_{\mathbf{x}}\eta = +(\nabla_{\mathbf{x}}\xi)^{\perp}. \tag{11.18}$$

Each relation in (11.17)-(11.18) makes a local statement relating the contravariant tangents. For arbitrary matrices $\mathcal{A}$ and $\mathcal{B}$, there is no reason to expect the two relationships to be consistent with (i.e., derivable from) one another. In the most general case, in which the two rank-2 matrices are unrelated, *two* local conditions on the tangents are implied—not a very useful situation. To obtain consistency between the two local conditions, the two matrices must be related.

LEMMA 11.2 The two local conditions (11.17)-(11.18) are consistent with one another if

$$\mathcal{B} = \frac{1}{\alpha}\mathcal{A} \tag{11.19}$$

(where $\alpha = \det \mathcal{A}$). §

Equation (11.19) will be referred to as the *consistency* condition. The proof of the lemma uses the fact that $\mathcal{A}$ and $\mathcal{B}$ are invertible and symmetric; also useful is the matrix

$$\mathcal{P} = \begin{pmatrix} 0 & -1 \\ 1 & 0 \end{pmatrix}. \tag{11.20}$$

$\mathcal{P}$ has many interesting properties. It is orthogonal, skew-symmetric, and

$$\det \mathcal{P} = 1\,, \quad \mathcal{P}^T = \mathcal{P}^{-1}\,, \quad \mathcal{P}^{-1} = -\mathcal{P}\,, \quad \mathcal{P}^2 = -\mathcal{I}, \tag{11.21}$$

$$\mathcal{P}\mathbf{v} = \mathbf{v}^{\perp}\,, \quad \mathcal{P}^T\mathbf{v} = -\mathbf{v}^{\perp}, \tag{11.22}$$

$$\mathcal{P}\mathcal{A}^T\mathcal{P}^T = \text{cofactor } \mathcal{A}\,, \quad \mathcal{P}^T\mathcal{A}^T\mathcal{P} = \text{cofactor } \mathcal{A} \tag{11.23}$$

with arbitrary $\mathbf{v} \in R^2$ and $\mathcal{A}$ an arbitrary 2×2 matrix. The lemma is now easily proved.

Proof. Rewrite conditions (11.17)-(11.18) as

$$\mathcal{A}\,\nabla_{\mathbf{x}}\xi = \mathcal{P}^T\,\nabla_{\mathbf{x}}\eta, \tag{11.24}$$

$$\mathcal{B}\,\nabla_{\mathbf{x}}\eta = \mathcal{P}\,\nabla_{\mathbf{x}}\xi. \tag{11.25}$$

If (11.24) holds, then

$$\mathcal{P}\mathcal{A}\,\nabla_{\mathbf{x}}\xi = \nabla_{\mathbf{x}}\eta, \tag{11.26}$$

$$\mathcal{B}\mathcal{P}\mathcal{A}\,\nabla_{\mathbf{x}}\xi = \mathcal{B}\,\nabla_{\mathbf{x}}\eta, \tag{11.27}$$

thus, (11.25) is derivable from (11.24) provided $\mathcal{P}^T\mathcal{B}\mathcal{P}\mathcal{A} = \mathcal{I}$. In the same manner, one can show that (11.24) is derivable from (11.25) provided $\mathcal{P}\mathcal{A}\mathcal{P}^T\mathcal{B} = \mathcal{I}$. These two conditions turn out to be the same since $\mathcal{A}$ and $\mathcal{B}$ are invertible. Applying (11.23) gives (cofactor $\mathcal{A})\mathcal{B} = \mathcal{I}$, from which (11.19) follows. §

If the consistency condition on the matrices holds, then the function G becomes

$$G = \nabla_{\mathbf{x}}\xi \cdot \mathcal{A}\,\nabla_{\mathbf{x}}\xi + \nabla_{\mathbf{x}}\eta \cdot \frac{1}{\alpha}\mathcal{A}\,\nabla_{\mathbf{x}}\eta, \tag{11.28}$$

and the Euler-Lagrange equation reduces to

$$\text{div}_{\mathbf{x}}\mathcal{D}^{-1}\mathcal{J}^{-1}\mathcal{A} = \mathbf{0} \tag{11.29}$$

with $\mathcal{D} = \text{diag}(1, \alpha)$. A potential solution is $\mathcal{D}^{-1}\mathcal{J}^{-1}\mathcal{A} = C^{-1}$, from which one obtains the following equivalent contravariant and covariant local conditions:

$$\mathcal{J}^{-1} = \mathcal{D}\, C^{-1}\mathcal{A}^{-1}, \tag{11.30}$$

$$\mathcal{J} = \mathcal{A}\, C\, \mathcal{D}^{-1}. \tag{11.31}$$

Explicitly, (11.31) can be stated as the consistent pair of local conditions

$$\mathbf{x}_\xi = -\mathcal{A}\,\mathbf{x}_\eta^\perp, \tag{11.32}$$

$$\mathbf{x}_\eta = \frac{1}{\alpha}\mathcal{A}\,\mathbf{x}_\xi^\perp. \tag{11.33}$$

The local relation (11.33) can be shown to comprise a first-order elliptic system (e.g., using the classification in [229]); the second-order system (11.29) can also be shown to be elliptic using an ellipticity test similar to the one in Chapter 6. Rearranging the local condition (11.30) or (11.31) yet again, one obtains the *diagonalization* condition

$$\frac{C^T \mathcal{A}\, C}{\sqrt{g}} = \mathcal{D}. \tag{11.34}$$

This relation is interpreted to mean that the coordinate system is a least-squares fit to the problem of diagonalizing the matrix $\mathcal{A}$. One satisfies the *consistency* condition (11.19) if and only if the *diagonalization* condition (11.34) holds.

Some special cases of interest:

- $\mathcal{D}^{-1} = \mathcal{I}$. Then $\alpha = 1$ so $\mathcal{A}$ has the form $\mathcal{A} = \mathcal{T}/\sqrt{\tau}$ with $\mathcal{T}$ symmetric, positive definite and $\tau = \det \mathcal{T}$. This results in the Winslow variable diffusion generator (see Section 11.3 for further discussion of this generator).
- $\mathcal{A} = \mathcal{I}$ (then $\mathcal{D}^{-1} = \mathcal{I}$). The Winslow (homogeneous Thompson-Thames-Mastin) grid generator results.
- If $\mathcal{A}$ has the form $\mathcal{A} = (1/f)\,\mathcal{I}$, then $\alpha = 1/f^2$ and $\mathcal{D} = \text{diag}(1, 1/f^2)$, from which one obtains the local condition $\mathcal{J} = C\mathcal{F}$ with $\mathcal{F} = \text{diag}(1/f, f)$. This is recognized as the covariant statement for an orthogonal grid, with cell-aspect ratio proportional to f (see equation (5.28)). The corresponding Euler-Lagrange equation, $\text{div}_{\mathbf{x}}\mathcal{F}\mathcal{J}^{-1} = \mathbf{0}$, is not the same as the scaled Laplacian (equation (5.30)), but is a contravariant analog. Solutions to the scaled Laplacian potentially satisfy the same local condition as given here, but cannot be said to be a least-squares fit since the scaled-Laplacian is not derivable from a variational principle when f depends on $\mathbf{x}$. Note this *cell-apsect* case is not a special case of the Winslow Variable Diffusion generator.

11.2 NONSYMMETRIC FORM OF THE DIAGONALIZATION EQUATIONS

Numerical solutions to (11.29) are most easily obtained by inverting and computing a nonsymmetric form in a manner analogous to the procedure given in Section 7.3.5. Inversion is readily achieved using the relationship (7.73):

$$\text{div}_{\boldsymbol{\xi}}\mathcal{D}^{-1}\mathcal{J}^{-1}\mathcal{A}\, C = \mathbf{0}. \tag{11.35}$$

To obtain a nonsymmetric form similar to the Winslow equations, project the transformed equation using the matrix product $\mathcal{J}\mathcal{D}$, giving

$$\mathcal{J}\mathcal{D}\,\mathrm{div}_{\boldsymbol{\xi}}\mathcal{D}^{-1}\mathcal{J}^{-1}\mathcal{A}\,C = \mathbf{0} \tag{11.36}$$

and expand to obtain

$$[\nabla_{\boldsymbol{\xi}}\mathcal{J}\mathcal{D}]\mathcal{D}^{-1}C^T\mathcal{A}\,C = \sqrt{g}[\nabla_{\boldsymbol{\xi}}\mathcal{A}]C. \tag{11.37}$$

Putting this equation into the form

$$\mathcal{Q}_{\text{diag}}\mathbf{x} = \mathbf{R}, \tag{11.38}$$

the left-hand side explicitly reads

$$\mathcal{Q}_{\text{diag}}\mathbf{x} = \alpha_{22}\,\mathbf{x}_{\xi\xi} - 2\alpha_{12}\,\mathbf{x}_{\xi\eta} + \alpha_{11}\,\mathbf{x}_{\eta\eta} + \frac{1}{\alpha}\,(\alpha_{11}\,\alpha_{\eta} - \alpha_{12}\,\alpha_{\xi})\,\mathbf{x}_{\eta} \tag{11.39}$$

where A_{ij} are the elements of $\mathcal{A}$ and

$$\alpha_{22} = y_{\eta}^2\,A_{11} - 2\,x_{\eta}\,y_{\eta}\,A_{12} + x_{\eta}^2\,A_{22}\,, \tag{11.40}$$

$$\alpha_{12} = y_{\xi}\,y_{\eta}\,A_{11} - (x_{\xi}\,y_{\eta} + x_{\eta}\,y_{\xi})\,A_{12} + x_{\xi}\,x_{\eta}\,A_{22}\,, \tag{11.41}$$

$$\alpha_{11} = y_{\xi}^2\,A_{11} - 2\,x_{\xi}\,y_{\xi}\,A_{12} + x_{\xi}^2\,A_{22}\,, \tag{11.42}$$

while the right-hand side reads

$$R_1 = \sqrt{g}\,\{(A_{11})_{\xi}\,y_{\eta} - (A_{12})_{\xi}\,x_{\eta} - (A_{11})_{\eta}\,y_{\xi} + (A_{12})_{\eta}\,x_{\xi}\}\,, \tag{11.43}$$

$$R_2 = \sqrt{g}\,\{(A_{12})_{\xi}\,y_{\eta} - (A_{22})_{\xi}\,x_{\eta} - (A_{12})_{\eta}\,y_{\xi} + (A_{22})_{\eta}\,x_{\xi}\}\,. \tag{11.44}$$

As an example, the *cell-aspect* ratio equations, with $\mathcal{A} = \mathcal{I}/f$, are

$$\mathcal{Q}_w\,\mathbf{x} = \frac{1}{f}\,\{-(g_{22}\,f_{\xi} - g_{12}\,f_{\eta})\,\mathbf{x}_{\xi} + (g_{11}\,f_{\eta} - g_{12}\,f_{\xi})\,\mathbf{x}_{\eta}\} \tag{11.45}$$

with $\mathcal{Q}_w\,\mathbf{x}$ defined as in (5.52).

11.3 WINSLOW'S VARIABLE DIFFUSION GENERATOR

The case $\mathcal{A} = \mathcal{T}/\sqrt{\tau}$, leading to Winslow's variable diffusion grid generator, is reviewed. The functional to be minimized is

$$I[\xi(x,y),\eta(x,y)] = \int_{\Omega}\left\{\nabla_{\mathbf{x}}\xi\cdot\frac{\mathcal{T}}{\sqrt{\tau}}\nabla_{\mathbf{x}}\xi + \nabla_{\mathbf{x}}\eta\cdot\frac{\mathcal{T}}{\sqrt{\tau}}\nabla_{\mathbf{x}}\eta\right\}\,dx\,dy\,, \tag{11.46}$$

where it is required that $\mathcal{T}(\mathbf{x})$ be a symmetric, positive-definite matrix.

The Euler-Lagrange equation is readily calculated from (11.29) to be

$$\mathrm{div}_{\mathbf{x}} \mathcal{J}^{-1} \frac{\mathcal{T}}{\sqrt{\mathcal{T}}} = \mathbf{0}, \tag{11.47}$$

or more explicitly,

$$\nabla_{\mathbf{x}} \cdot \frac{\mathcal{T}}{\sqrt{\mathcal{T}}} \nabla_{\mathbf{x}} \xi = 0, \tag{11.48}$$

$$\nabla_{\mathbf{x}} \cdot \frac{\mathcal{T}}{\sqrt{\mathcal{T}}} \nabla_{\mathbf{x}} \eta = 0. \tag{11.49}$$

Equations (11.48)-(11.49) are recognized as similar to Winslow's variable diffusion model (Winslow [232]); the only difference is that Winslow's tensor was a function of ξ and η, i.e., it was a logical-space weight. Various forms of the generator have also been considered by Anderson [8] (who attempted to relate it to area grid generation within a solution-adaptive context) and by Brackbill [22].

The original part of this discussion is the interpretation of the weight tensor using the theory developed in this book. As demonstrated, a potential solution to equation (11.47) is

$$\mathcal{J}^{-1} \frac{\mathcal{T}}{\sqrt{\mathcal{T}}} = C^{-1}, \tag{11.50}$$

i.e., the Jacobian matrix at each point of the grid satisfies

$$\mathcal{J} = \frac{\mathcal{T}}{\sqrt{\mathcal{T}}} C. \tag{11.51}$$

Explicitly, equation (11.51) reads

$$-\frac{\mathcal{T}}{\sqrt{\mathcal{T}}} \mathbf{x}_\eta^\perp = \mathbf{x}_\xi, \tag{11.52}$$

$$\frac{\mathcal{T}}{\sqrt{\mathcal{T}}} \mathbf{x}_\xi^\perp = \mathbf{x}_\eta. \tag{11.53}$$

These equations show that the potential solution is a quasiconformal mapping (see, for example, Thompson [212]).

The weight function derived from (11.51) is related to the Jacobian matrix by

$$\frac{\mathcal{T}}{\sqrt{\mathcal{T}}} = \frac{1}{\sqrt{g}} \mathcal{J}\mathcal{J}^T. \tag{11.54}$$

In principle, then, if the Jacobian matrix of the desired transformation is specified over the domain Ω, (11.54) can be used to construct the weight tensor, and (11.47) can be solved to obtain a least-squares solution potentially having the desired Jacobian matrix (numerically, one would solve the inverted equation (11.38) with $\mathcal{A} = \mathcal{T} / \sqrt{\mathcal{T}}$). The matrix on the right-hand side of (11.54) is sometimes referred to as the left-Cauchy-Green-Tensor (see Jacquotte [101]).

One may construct the weight function $\mathcal{T}$ from (11.54) by specifying a desired local condition. Suppose, for example, one desires both coordinate alignment with a given vector field and to control the aspect ratio through the function f. The most succinct statement of this is

$$\mathbf{x}_\xi = a\,\mathbf{v}, \tag{11.55}$$

$$\mathbf{x}_\eta = f\,\mathbf{x}_\xi^\perp \tag{11.56}$$

with a an unspecified nonzero constant. The first relation, of course, gives the direction of the tangent $\mathbf{x}_\xi$, while one can deduce from both relations that

$$g_{12} = 0, \tag{11.57}$$

$$\sqrt{\frac{g_{22}}{g_{11}}} = f, \tag{11.58}$$

i.e., the grid is orthogonal and the aspect ratio is controlled by the function f. One also finds $\sqrt{g} = a^2|\mathbf{v}|^2 f$ and that

$$\mathcal{J} = a\,[\,\mathbf{v}, \mathbf{v}^\perp\,]\,\mathrm{diag}(1, f), \tag{11.59}$$

so that

$$\frac{1}{\sqrt{g}}\mathcal{J}\mathcal{J}^T = \mathcal{M}\,\mathcal{F}\,\mathcal{M}^T \tag{11.60}$$

with $\mathcal{M} = [\mathbf{v}, \mathbf{v}^\perp]\,/\,|\mathbf{v}|$ and $\mathcal{F} = \mathrm{diag}(1/f, f)$. It is believed that solving (11.47) with the weight $\mathcal{T}/\sqrt{\mathcal{T}} = \mathcal{M}\,\mathcal{F}\,\mathcal{M}^T$ will generate a grid that is a least-squares fit to the requirement (11.60).

11.4 LOCAL CONDITION WITH NONSYMMETRIC MATRICES

Equations (11.32) and (11.33) specify a local condition between the covariant tangents via a symmetric, positive-definite matrix. Alternatively, it would seem natural to give the local condition using a rotation matrix. Such matrices are generally nonsymmetric, so the variational approach given in the first part of this chapter is precluded. In this section, it is shown that it is nonetheless possible to devise a variational principle leading to this alternate local condition. Although the resulting grid-generation equations are quite similar to those in the preceding sections of this chapter, two differences stand out; first, the new weight function is proportional to the metric tensor $\mathcal{G}$ (instead of the left-Cauchy-Green-Tensor), and second, the weights appear on the right-hand side of the generator instead of being mixed in with the operator. The grid generation equations are thus similar in form to the standard inhomogeneous Thompson-Thames-Mastin equations (but better motivated than the latter).

Define the two vectors

$$\boldsymbol{\xi}_x = \begin{pmatrix} \xi_x \\ \eta_x \end{pmatrix}, \tag{11.61}$$

$$\boldsymbol{\xi}_y = \begin{pmatrix} \xi_y \\ \eta_y \end{pmatrix}, \tag{11.62}$$

and let $\mathcal{A}$ and $\mathcal{B}$ again be 2×2 symmetric, positive-definite real matrices. Let G be as in (11.1) and consider the functional

$$I[\,\xi(\mathbf{x}),\,\eta(\mathbf{x})\,] = \frac{1}{2}\int_\Omega G\,(\,\xi_x,\,\xi_y\,)\,dx\,dy, \tag{11.63}$$

to be minimized with the boundary data as constraint (compare to (11.2)).

Proceeding as usual, the Euler-Lagrange equation is

$$\mathrm{div}_x[\mathcal{A}\,\xi_x, \mathcal{B}\,\xi_y] = \mathbf{0}. \tag{11.64}$$

A potential solution to (11.64) is

$$\mathcal{A}\,\xi_x = -\xi_y^{\perp}, \tag{11.65}$$

$$\mathcal{B}\,\xi_y = +\xi_x^{\perp}. \tag{11.66}$$

The requirement of consistency between these two point conditions again results in condition (11.19) on the matrices, so consider the special case

$$G = \xi_x \cdot \mathcal{A}\,\xi_x + \xi_y \cdot \frac{1}{\alpha}\mathcal{A}\,\xi_y. \tag{11.67}$$

When (11.19) holds, the integrand G can also be expressed in terms of the contravariant vectors as

$$\begin{aligned} G = {} & A_{11}\nabla_{\mathbf{x}}\xi \cdot \mathcal{D}^{-1}\nabla_{\mathbf{x}}\xi \\ & + 2A_{12}\nabla_{\mathbf{x}}\xi \cdot \mathcal{D}^{-1}\nabla_{\mathbf{x}}\eta \\ & + A_{22}\nabla_{\mathbf{x}}\eta \cdot \mathcal{D}^{-1}\nabla_{\mathbf{x}}\eta \end{aligned} \tag{11.68}$$

with $\mathcal{D} = \mathrm{diag}(1, \alpha)$, as before. The Euler-Lagrange equation for this case reads

$$\mathrm{div}_{\mathbf{x}}\mathcal{A}\,\mathcal{J}^{-1}\mathcal{D}^{-1} = \mathbf{0}. \tag{11.69}$$

Compare the Euler-Lagrange equations (11.69) and (11.29); the reversal of order in (11.69) is a consequence of using the vectors (11.61)-(11.62) instead of the usual contravariant tangents in the functional. This reversal of order carries through in the analogs to (11.30), (11.31), and (11.34):

$$\mathcal{J}^{-1} = \mathcal{A}^{-1}\mathcal{C}^{-1}\mathcal{D}, \tag{11.70}$$

$$\mathcal{J} = \mathcal{D}^{-1}\mathcal{C}\mathcal{A}, \tag{11.71}$$

$$\frac{\mathcal{C}\mathcal{A}\,\mathcal{C}^T}{\sqrt{g}} = \mathcal{D}. \tag{11.72}$$

A significant difference in the explicit local condition arises from the reversal of order; writing these in terms of the covariant tangents:

$$\mathbf{x}_\xi = -\hat{\mathcal{A}}\,\mathbf{x}_\eta^{\perp}, \tag{11.73}$$

$$\mathbf{x}_\eta = \frac{1}{\hat{\alpha}}\hat{\mathcal{A}}^T\mathbf{x}_\xi^{\perp}, \tag{11.74}$$

where $\hat{\mathcal{A}}$ is the *nonsymmetric*, positive-definite matrix

$$\hat{\mathcal{A}} = \frac{1}{A_{22}} \begin{pmatrix} \alpha & -A_{12} \\ A_{12} & 1 \end{pmatrix}. \tag{11.75}$$

and $\hat{\alpha} = \det(\hat{\mathcal{A}}) = A_{11}/A_{22}$. Comparing (11.73)-(11.74) to (11.32)-(11.33), one sees that the form is the same, but the matrix is now nonsymmetric, whereas before it was symmetric. In both cases they are positive definite (which guarantees $\sqrt{g} > 0$, locally). The first-order equations (11.73)-(11.74) are elliptic, as is (11.69).

The polar decomposition theorem [87] may be applied to express $\hat{\mathcal{A}}$ as the product of a symmetric, positive-definite matrix $\mathcal{V}$ and a rotation matrix $\mathcal{R}$, i.e., $\hat{\mathcal{A}} = \mathcal{V}\mathcal{R}$, where

$$\mathcal{V} = \frac{\rho}{A_{22}} \begin{pmatrix} \alpha(1+\alpha) + 2A_{12}^2 & (\alpha-1)A_{12} \\ (\alpha-1)A_{12} & 1+\alpha+2A_{12}^2 \end{pmatrix}, \tag{11.76}$$

$$\mathcal{R} = \rho \begin{pmatrix} 1+\alpha & -2A_{12} \\ 2A_{12} & 1+\alpha \end{pmatrix}, \tag{11.77}$$

and

$$\rho = \frac{1}{\sqrt{(1+\alpha)^2 + 4A_{12}^2}}. \tag{11.78}$$

Since the matrix in the present local condition is nonsymmetric, it is possible to obtain an orthogonal matrix from (11.75); this requires $\alpha = 1$, so let $\mathcal{A} = \mathcal{T}/\sqrt{\tau}$ with $\mathcal{T}$ symmetric and positive definite. Then, from (11.68), the integrand in the variational principle is

$$G = \frac{1}{\sqrt{\tau}} \{ T_{11} (\nabla_{\mathbf{x}}\xi \cdot \nabla_{\mathbf{x}}\xi) + 2\,T_{12} (\nabla_{\mathbf{x}}\xi \cdot \nabla_{\mathbf{x}}\eta) + T_{22} (\nabla_{\mathbf{x}}\eta \cdot \nabla_{\mathbf{x}}\eta) \} \tag{11.79}$$

and the Euler-Lagrange equation reduces to

$$\mathrm{div}_{\mathbf{x}} \frac{\mathcal{T}}{\sqrt{\tau}} \mathcal{J}^{-1} = \mathbf{0} \tag{11.80}$$

(compare to (11.47)). The potential solution has the properties

$$\mathcal{J}^{-1} = \left(\frac{\mathcal{T}}{\sqrt{\tau}}\right)^{-1} C^{-1}, \tag{11.81}$$

$$\mathcal{J} = C \frac{\mathcal{T}}{\sqrt{\tau}}, \tag{11.82}$$

$$\frac{C\mathcal{T}C^T}{\sqrt{g}} = \sqrt{\tau}\,\mathcal{I}. \tag{11.83}$$

By rearranging (11.82), one observes that the weight function in the local solution is

$$\frac{\mathcal{T}}{\sqrt{\tau}} = \frac{\mathcal{G}}{\sqrt{g}}, \tag{11.84}$$

i.e., the weight is proportional to the metric tensor (compare to (11.54)).

When $\alpha = 1$, $\mathcal{V} = \sqrt{T_{11}/T_{22}}\,\mathcal{I}$; the explicit local condition is thus given in terms of a rotation and a stretch:

$$\mathbf{x}_\xi = -\sqrt{\frac{T_{11}}{T_{22}}}\mathcal{R}\,\mathbf{x}_\eta^\perp, \tag{11.85}$$

$$\mathbf{x}_\eta = +\sqrt{\frac{T_{22}}{T_{11}}}\mathcal{R}^T\mathbf{x}_\xi^\perp \tag{11.86}$$

where $\mathcal{R}$ is the *orthogonal* matrix

$$\mathcal{R} = \frac{1}{\sqrt{T_{11}T_{22}}}\begin{pmatrix} \sqrt{\tau} & -T_{12} \\ T_{12} & \sqrt{\tau} \end{pmatrix}. \tag{11.87}$$

One can obtain the usual stretch by the factor $r > 0$ and rotation through an angle θ by choosing

$$\mathcal{T} = \begin{pmatrix} 1/r & \sin\theta \\ \sin\theta & r \end{pmatrix}, \tag{11.88}$$

which gives

$$\mathcal{R} = \begin{pmatrix} \cos\theta & -\sin\theta \\ \sin\theta & \cos\theta \end{pmatrix}, \tag{11.89}$$

and

$$\mathbf{x}_\xi = -\frac{1}{r}\mathcal{R}\,\mathbf{x}_\eta^\perp, \tag{11.90}$$

$$\mathbf{x}_\eta = +r\mathcal{R}^T\mathbf{x}_\xi^\perp. \tag{11.91}$$

The nonsymmetric (inverted) form of equation (11.80) is readily shown (by the usual technique) to be

$$\mathcal{Q}_w\mathbf{x} = -\mathcal{J}\left[\nabla_\xi\left(\frac{\mathcal{T}}{\sqrt{\tau}}\right)^{-1}\right]\frac{\mathcal{T}}{\sqrt{\tau}}g\mathcal{G}^{-1}, \tag{11.92}$$

where the left-hand side is just the Winslow operator defined in (5.52). Observe that the weights in (11.92) lie entirely upon the right-hand side, whereas in (11.38) they are mixed into the second-order part of the operator. Equation (11.92) is thus closer in form to the standard inhomogenous grid-generation equations of Section 5.5 than is (11.38). The generator (11.92) may be more useful than (11.38) in some instances; for example, it is difficult to devise a Steger-Sorenson-like algorithm with (11.38) while (11.92) preserves the approach given in Section 5.6.2. Finally, it is observed that the *cell-aspect-ratio* generator (11.45) is obtained from (11.92) when $\mathcal{T}/\sqrt{\tau} = \text{diag}(1/f, f)$.

The inhomogeneous equations in Section 5.5 can be put into the form

$$\text{div}_{\mathbf{x}} \mathcal{J}^{-1} = \frac{1}{g} S\mathbf{P} \tag{11.93}$$

with $\mathbf{P} = (P, Q)$ and $S = g\mathcal{I}$ in the original weighted formulation and $S = \text{diag}(g_{22}, g_{11})$ in Warsi's modification. Comparing (11.93) to (11.80), there appears to be no direct connection between the standard inhomogeneous generators and the rotation/stretch approach given here. Although the practical value of the present approach remains to be established, the generator (11.92) gives the first geometrically motivated version of the inhomogenous grid generators.

11.5 ALIGNMENT WITH AN ELLIPTIC GENERATOR

This section considers the question: is it possible to construct a smooth coordinate system that is aligned with a given a pair of vector fields $\{\mathbf{v}_1 \mid \mathbf{x} \in \Omega \to E^2\}$ and $\{\mathbf{v}_2 \mid \mathbf{x} \in \Omega \to E^2\}$? To ensure that the vector field is properly oriented, assume that $v = \hat{k} \cdot (\mathbf{v}_1 \times \mathbf{v}_2) > 0$. A nonelliptic alignment scheme based on rank 1 matrices was described in subsection 11.1.1. In this section, it is shown that it is possible to achieve both alignment and ellipticity within a single generator. As usual, the solution to the Euler-Lagrange equations is not guaranteed to be the desired local condition, but merely a least-squares fit to the condition.

Let $\mathcal{A}$ and $\mathcal{B}$ be 2×2 symmetric positive-definite matrices with determinants $\alpha > 0$ and $\beta > 0$, respectively. Solutions to the Euler-Lagrange equation (11.3) are sought, based on the metric identity. Therefore, consider solutions of the form $[\mathcal{A}\nabla_{\mathbf{x}}\xi, \mathcal{B}\nabla_{\mathbf{x}}\eta]^T = \mu C^{-1}$, where μ is a positive constant independent of $\mathbf{x}$ and $C^{-1} = \mathcal{J}^T/\sqrt{g}$ with $\mathcal{J}$ the Jacobian matrix. This condition is the same as the following pair of local conditions on the vectors $\nabla_{\mathbf{x}}\xi$ and $\nabla_{\mathbf{x}}\eta$:

$$\mathcal{A}\nabla_{\mathbf{x}}\xi = -\mu\,(\nabla_{\mathbf{x}}\eta)^{\perp}, \tag{11.94}$$

$$\mathcal{B}\nabla_{\mathbf{x}}\eta = +\mu\,(\nabla_{\mathbf{x}}\xi)^{\perp}. \tag{11.95}$$

In Lemma 11.2, a relationship called the *consistency condition* was derived; this relationship gave the condition between the two weight matrices if one of the local conditions was to be derive-able from the other. With the constant μ introduced, the consistency condition generalizes to $\mathcal{B} = (\mu^2/\alpha)\mathcal{A}$; the consistency condition implies $\alpha\beta = \mu^4$. The consistency condition leads to the diagonalization generator described in subsection 11.1.2. Assumption of the consistency condition is not necessary to obtain useful grid generators, however.

The following lemma gives the general form of the weight matrices in the local condition if consistency is not required.

LEMMA 11.3 The local condition (11.94)-(11.95) holds for arbitrary contravariant tangent vectors $\nabla_{\mathbf{x}}\xi$ and $\nabla_{\mathbf{x}}\eta$ if and only if the matrices $\mathcal{A}$ and $\mathcal{B}$ have the form

$$\mathcal{A} = \frac{\mathcal{J}\,\mathcal{D}_{\mathcal{A}}\,\mathcal{J}^T}{\sqrt{g}}, \tag{11.96}$$

$$\mathcal{B} = \frac{\mathcal{J}\,\mathcal{D}_{\mathcal{B}}\,\mathcal{J}^T}{\sqrt{g}}, \tag{11.97}$$

where $\mathcal{D}_{\mathcal{A}} = \operatorname{diag}(\mu, \alpha/\mu)$ and $\mathcal{D}_{\mathcal{B}} = \operatorname{diag}(\beta/\mu, \mu)$. An algebraic proof of this is readily constructed. §

From (11.96)-(11.97), the weight matrices are symmetric, positive definite provided α, β, and $\sqrt{g}$ are postive. Observe that the local condition is insufficient to constrain the determinants α and β of the two matrices, so these remain aribitrary.

The expressions (11.96)-(11.97) are equivalent to the following useful outer-product forms:

$$\mathcal{A} = \sqrt{g}\left\{\mu\,(\nabla_{\mathbf{x}}\eta)^{\perp}\otimes(\nabla_{\mathbf{x}}\eta)^{\perp} + \frac{\alpha}{\mu}\,(\nabla_{\mathbf{x}}\xi)^{\perp}\otimes(\nabla_{\mathbf{x}}\xi)^{\perp}\right\}, \tag{11.98}$$

$$\mathcal{B} = \sqrt{g}\left\{\frac{\beta}{\mu}\,(\nabla_{\mathbf{x}}\eta)^{\perp}\otimes(\nabla_{\mathbf{x}}\eta)^{\perp} + \mu\,(\nabla_{\mathbf{x}}\xi)^{\perp}\otimes(\nabla_{\mathbf{x}}\xi)^{\perp}\right\}. \tag{11.99}$$

where $1\,/\,\sqrt{g} = (\nabla_{\mathbf{x}}\xi)^{\perp}\cdot(\nabla_{\mathbf{x}}\eta)$.

To generate a grid whose covariant tangents are aligned with the given vector field (i.e., $\mathbf{x}_\xi \propto \mathbf{v}_1$ and $\mathbf{x}_\eta \propto \mathbf{v}_2$) and having cell-aspect ratio

$$\sqrt{\frac{g_{22}}{g_{11}}} = \sqrt{\frac{\nabla_{\mathbf{x}}\xi\cdot\nabla_{\mathbf{x}}\xi}{\nabla_{\mathbf{x}}\eta\cdot\nabla_{\mathbf{x}}\eta}} = f, \tag{11.100}$$

the contravariant tangents must satisfy

$$\nabla_{\mathbf{x}}\xi = -\,a\,\mathbf{v}_2^{\perp}\,/|\mathbf{v}_2|, \tag{11.101}$$

$$\nabla_{\mathbf{x}}\eta = +\,\frac{a}{f}\,\mathbf{v}_1^{\perp}\,/|\mathbf{v}_1|, \tag{11.102}$$

with "a" an arbitrary scale factor. Substitution of (11.101)-(11.102) into (11.98)-(11.99) gives the necessary form of the weight matrices:

$$\mathcal{A} = \frac{1}{\nu}\,\{\,\frac{\mu r}{f}\,(\mathbf{v}_1\otimes\mathbf{v}_1) + \alpha\,\frac{f}{\mu r}\,(\mathbf{v}_2\otimes\mathbf{v}_2)\,\}, \tag{11.103}$$

$$\mathcal{B} = \frac{1}{\nu}\,\{\,\beta\,\frac{r}{\mu f}\,(\mathbf{v}_1\otimes\mathbf{v}_1) + \frac{\mu f}{r}\,(\mathbf{v}_2\otimes\mathbf{v}_2)\,\}. \tag{11.104}$$

where $r = |\mathbf{v}_2|\,/\,|\mathbf{v}_1|$. It may be shown that the weight matrices $\mathcal{A}$ and $\mathcal{B}$ remain well-defined in the limit that the length of either velocity-vector goes to zero. The weight matrices in (11.103)-(11.104) may also be expressed as

$$\mathcal{A} = \frac{\mathcal{V}\,\hat{\mathcal{D}}_{\mathcal{A}}\,\mathcal{V}^T}{\nu}, \tag{11.105}$$

$$\mathcal{B} = \frac{\mathcal{V}\,\hat{\mathcal{D}}_{\mathcal{B}}\,\mathcal{V}^T}{\nu}, \tag{11.106}$$

where $\mathcal{V} = [\mathbf{v}_1, \mathbf{v}_2]$, $\hat{\mathcal{D}}_{\mathcal{A}} = \mathrm{diag}(\mu r/f\,, \alpha f/\mu r)$, and $\hat{\mathcal{D}}_{\mathcal{B}} = \mathrm{diag}(\beta r/\mu f\,, \mu f/r)$.

Assuming weight matrices as constructed in (11.103)-(11.104), it is natural to consider whether or not the solution (11.101)-(11.102) to the local condition (11.94)-(11.95) is unique. Using the matrix $\mathcal{P}$ defined in (11.20), the local condition implies that the contravariant tangents must satisfy the auxilliary system:

$$\mathcal{Z}\,\nabla_{\mathbf{x}}\xi = \mu^2\,\nabla_{\mathbf{x}}\xi, \tag{11.107}$$

$$\mathcal{Z}\,\nabla_{\mathbf{x}}\eta = \frac{\alpha\beta}{\mu^2}\,\nabla_{\mathbf{x}}\eta \tag{11.108}$$

with $\mathcal{Z} = \beta\mathcal{B}^{-1}\mathcal{A}$. $\mathcal{Z}$ is easily calculated from (11.105)-(11.106): $\mathcal{Z} = \mathcal{V}^{-T}\,\mathcal{D}_{\mathcal{Z}}\,\mathcal{V}^T$ where $\mathcal{D}_{\mathcal{Z}} = \mathrm{diag}(\mu^2, \alpha\beta/\mu^2)$. The eigenpairs of $\mathcal{Z}$ are $(\mu^2, \mathbf{v}_2^{\perp})$ and $(\alpha\beta/\mu^2, \mathbf{v}_1^{\perp})$; thus, the contravariant tangents are aligned with the eigenvectors of $\mathcal{Z}$. In particular, if $\alpha\beta \neq \mu^4$, $\nabla_{\mathbf{x}}\xi$ must align with $\mathbf{v}_2^{\perp}$ and not with $\mathbf{v}_1^{\perp}$. Similarly, $\nabla_{\mathbf{x}}\eta$ must align with $\mathbf{v}_1^{\perp}$ and not $\mathbf{v}_2^{\perp}$. Using this result, it may be shown that the solution (11.101)-(11.102) is unique up-to the scalar multiplier "a," provided the consistency condition is not enforced.

The case $\mathbf{v}_1 = \mathbf{v}$, $\mathbf{v}_2 = \mathbf{v}^{\perp}$ is of particular interest since it corresponds to alignment with a single vector field. In this case, the weight matrices reduce to

$$\mathcal{A} = \frac{1}{|\mathbf{v}|^2}\left\{\frac{\mu}{f}\,(\mathbf{v}\otimes\mathbf{v}) + \alpha\,\frac{f}{\mu}\,(\mathbf{v}^{\perp}\otimes\mathbf{v}^{\perp})\right\}, \tag{11.109}$$

$$\mathcal{B} = \frac{1}{|\mathbf{v}|^2}\left\{\frac{\beta}{\mu f}\,(\mathbf{v}\otimes\mathbf{v}) + \mu f\,(\mathbf{v}^{\perp}\otimes\mathbf{v}^{\perp})\right\}. \tag{11.110}$$

These can also be expressed as

$$\mathcal{A} = \frac{\mu}{f}\left\{\phi\,\mathcal{I} + (1-\phi)\,\frac{(\mathbf{v}\otimes\mathbf{v})}{|\mathbf{v}|^2}\right\}, \tag{11.111}$$

$$\mathcal{B} = \mu f\left\{\psi\,\mathcal{I} + (1-\psi)\,\frac{(\mathbf{v}^{\perp}\otimes\mathbf{v}^{\perp})}{|\mathbf{v}|^2}\right\}, \tag{11.112}$$

where $\phi = \alpha f^2/\mu^2$ and $\psi = \beta/f^2\mu^2$, showing that the determinants of the matrices can be interpreted as smoothing parameters.

A derivation of the computationally-convenient inverted, nonsymmetric form of the Euler-Lagrange equations for the elliptic-alignment case is now given. When $\mathcal{B} \neq \mathcal{A}$, the inversion techniques derived in earlier chapters fail on (11.3) if the usual approach is taken. The following identity is useful in giving the inverted, nonsymmetric form of the Euler-Lagrange equations

$$\mathrm{div}_{\mathbf{x}}\,[\,\mathcal{A}\,\nabla_{\mathbf{x}}\xi,\ \mathcal{B}\,\nabla_{\mathbf{x}}\eta\,]^T = \begin{bmatrix} \nabla_{\mathbf{x}}\xi\cdot(\,\mathrm{div}_{\mathbf{x}}\mathcal{A} - \mathcal{Q}_{\mathcal{A}}\,\mathbf{x}) \\ \nabla_{\mathbf{x}}\eta\cdot(\,\mathrm{div}_{\mathbf{x}}\mathcal{B} - \mathcal{Q}_{\mathcal{B}}\,\mathbf{x}) \end{bmatrix}, \tag{11.113}$$

with the generalized Winslow operator $\mathcal{Q}_{\mathcal{A}}\,\mathbf{x}$ defined in terms of the matrix $\mathcal{A}$ by

$$\mathcal{Q}_{\mathcal{A}}\,\mathbf{x} = (\nabla_{\mathbf{x}}\xi \cdot \mathcal{A}\,\nabla_{\mathbf{x}}\xi)\,\mathbf{x}_{\xi\xi} + 2\,(\nabla_{\mathbf{x}}\xi \cdot \mathcal{A}\,\nabla_{\mathbf{x}}\eta)\,\mathbf{x}_{\xi\eta} + (\nabla_{\mathbf{x}}\eta \cdot \mathcal{A}\,\nabla_{\mathbf{x}}\eta)\,\mathbf{x}_{\eta\eta}\,. \quad (11.114)$$

The derivation of (11.113) goes as follows:

$$\mathcal{Q}_{\mathcal{A}}\,\mathbf{x} = [\,\nabla_{\boldsymbol{\xi}}\mathcal{J}\,](\mathcal{J}^{-1}\mathcal{A}\,\mathcal{J}^{-T})\,, \quad (11.115)$$

$$= \frac{1}{\sqrt{g}}[\,\nabla_{\boldsymbol{\xi}}\mathcal{J}\,](\mathcal{J}^{-1}\mathcal{A}\,C)\,, \quad (11.116)$$

$$= \frac{1}{\sqrt{g}}\{\,\mathrm{div}_{\boldsymbol{\xi}}\mathcal{A}\,C - \mathcal{J}\mathrm{div}_{\boldsymbol{\xi}}\mathcal{J}^{-1}\mathcal{A}\,C\,\}\,, \quad (11.117)$$

$$= \mathrm{div}_{\mathbf{x}}\mathcal{A} - \mathcal{J}\mathrm{div}_{\mathbf{x}}\mathcal{J}^{-1}\mathcal{A}. \quad (11.118)$$

The result (11.118) readily leads to the identity

$$\mathrm{div}_{\mathbf{x}} \cdot \mathcal{A}\,\nabla_{\mathbf{x}}\xi = \nabla_{\mathbf{x}}\xi \cdot (\mathrm{div}_{\mathbf{x}}\mathcal{A} - \mathcal{Q}_{\mathcal{A}}\,\mathbf{x})\,, \quad (11.119)$$

from which (11.113) follows. Note that (11.113) is stated in terms of projections onto the contravariant tangents (the Euler-Lagrange equation can only be stated as a vector relationship when the consistency condition is satisfied). The equation is conveniently solved computationally in the projected form:

$$\nabla_{\mathbf{x}}\xi \cdot \mathcal{Q}_{\mathcal{A}}\,\mathbf{x} = \frac{1}{\sqrt{g}}\,\nabla_{\mathbf{x}}\xi \cdot (\,[\,\nabla_{\boldsymbol{\xi}}\mathcal{A}\,]\,C)\,, \quad (11.120)$$

$$\nabla_{\mathbf{x}}\eta \cdot \mathcal{Q}_{\mathcal{B}}\,\mathbf{x} = \frac{1}{\sqrt{g}}\,\nabla_{\mathbf{x}}\eta \cdot (\,[\,\nabla_{\boldsymbol{\xi}}\mathcal{B}\,]\,C)\,, \quad (11.121)$$

which puts all second-derivatives on the left-hand side.

Bibliography

[1] Abrahamsson, L. 1991. Orthogonal grid generation for two- dimensional ducts. *J. Comp. and Appl. Math.* **34**, 305–314.

[2] Allievi, A., and Calisal, S. M. 1992. Application of Bubnov-Galerkin formulation to orthogonal grid generation. *J. Comp. Physics.* **98**, 163–173.

[3] Amsden, A. A., and Hirt, C. W. 1973. A simple scheme for generating general curvilinear grids. *J. Comp. Physics* **11**, 348–359.

[4] Anderson, D. A., and Rai, M. M. 1982. The use of solution-adaptive grids in solving partial differential equations. In *Numerical Grid Generation*, J. F. Thompson, ed., pp. 317-338 North-Holland, New York.

[5] Anderson, D. A. 1983. Adaptive grid methods for partial differential equations. In *Advances in Grid Generation*, K. N. Ghia and U. Ghia, eds., pp. 1–15. ASME, Houston.

[6] Anderson, D. A. 1986. Constructing adaptive grids with Poisson grid generators. In *Numerical Grid Generation in Computational Fluid Dynamics*, J. Haüser and C. Taylor, eds., pp. 125–136. Pineridge Press, Swansea, UK.

[7] Anderson, D. A. 1987. Equidistribution Schemes, Poisson Generators, and Adaptive Grids. *Appl. Math. and Comp.* **24**, 211–227.

[8] Anderson, D. A. 1990. Grid Cell Volume Control with an Adaptive Grid Generator. *Appl. Math. and Comp.* **35**, 209–217.

[9] Arcilla, A. S., Häuser, J., Eiseman, P. R., and Thompson, J. F. 1991. Numerical grid generation in computational fluid dynamics. In *Proceedings of the Third International Conference on Numerical Grid Generation in Computational Fluid Dynamics and Related Fields*, Barcelona, Spain. North-Holland, Amsterdam.

[10] Arina, R. 1986. Orthogonal grids with adaptive control. In *Numerical Grid Generation in Computational Fluid Dynamics*, J. Haüser and C. Taylor, eds., pp. 113–124. Pineridge Press, Swansea, UK.

[11] Arina, R. 1988. Adaptive orthogonal surface coordinates. In *Numerical Grid Generation in Computational Fluid Mechanics '88*, S. Sengupta, J. Haüser, P. R. Eisemen, and J. F. Thompson, eds., pp. 351–359. Pineridge Press, Swansea, UK.

[12] Arina, R., and Casella, M. 1991. A harmonic grid generation technique for surfaces and three-dimensional regions. In *Numerical Grid Generation in Computational Fluid Dynamics*, A. S.-Arcilla, J. Häuser, P. R. Eiseman, J. F. Thompson, eds., pp. 935–945.

[13] Armstrong, M. A. 1987. *Basic Topology*. Springer-Verlag, New York.

[14] Ascoli, E. P., Dandy, D. S., and Leal, L. G. 1987. On distortion functions for the strong constraint method of numerically generating orthogonal coordinate grids. *J. Comp. Physics* **72**, 513–519.

[15] Barfield, W. D. 1970. An optimal mesh generator for Lagrangian hydrodynamics calculations in two space dimensions. *J. Comp. Physics* **6**, 417–429.

[16] Barrera, P., Perez, A., and Castellanos, L. 1992. *Curvilinear Coordinate System Generation over Plane Irregular Regions*. Facultad de Ciencias, U.N.A.M., Mexico.

[17] Bell, J. B., Shubin, G. R., and Stephens, A. B. 1982. A segmentation approach to grid generation using biharmonics. *J. Comp. Physics* **47**, 463–472.

[18] Bers, L., John, F., and Schecter, M. 1964. *Partial Differential Equations*. Interscience, New York.

[19] Birkhoff, G., and Lynch, R. E. 1984. *Numerical Solution of Elliptic Problems*. SIAM, Philadelphia.

[20] Brackbill, J. U. 1982a. Coordinate system control: Adaptive meshes. In *Numerical Grid Generation*, J. F. Thompson, ed., pp. 277–294. North-Holland, New York.

[21] Brackbill, J. U., and Saltzman, J. S. 1982b. Adaptive zoning for singular problems in two dimensions. *J. Comp. Physics* **46**, 342–368.

[22] Brackbill, J. U. *An Adaptive Grid with Directional Control*. Preprint.

[23] Camarero, R., and Younis, M. 1980. Efficient generation of boundary-fitted coordinates for cascades using multigrid. *AIAA J.* **18**, 487–488.

[24] Carey, G. F. 1993. *Grid Generation, Refinement, and Redistribution*. Wiley, New York.

[25] Castillo, J. E. 1986. *Mathematical Aspects of Grid Generation I*. In *Numerical Grid Generation in Computational Fluid Dynamics*, J. Haüser and C. Taylor, eds., pp. 35–43. Pineridge Press, Swansea, UK.

[26] Castillo, J. E. 1987a. *On Variational Grid Generation*. Ph.D. thesis, University of New Mexico.

[27] Castillo, J. E., Steinberg, S., and Roache, P. J. 1987b. Mathematical aspects of variational grid generation II. *J. Comp. and Appl. Math.* **20**, 127–135.

[28] Castillo, J. E. 1988a. A direct variational grid generation method: Orthogonality control. In *Numerical Grid Generation in Computational Fluid Me-*

chanics '88, S. Sengupta, J. Haüser, P. R. Eiseman, and J. F. Thompson, eds., pp. 247–256. Pineridge Press, Swansea, UK.

[29] Castillo, J. E., Steinberg, S., and Roache, P. J. 1988b. Parameter estimation in variational grid generation. *Appl. Math. and Comp.* **28**(2), 1–23.

[30] Castillo, J. E., Steinberg, S., and Roache, P. J. 1988c. On the folding of numerically generated grids: Use of a reference grid. *Comm. Applied Num. Methods* **4**, 471–481.

[31] Castillo, J. E. 1990. An adaptive direct variational grid generation method. *Computers Math. Applic.* **4**(1), 1–9.

[32] Castillo, J. E., ed. 1991a. *Mathematical Aspects of Numerical Grid Generation.* SIAM, Philidelphia.

[33] Castillo, J. E. 1991b. The discrete grid generation method on curves and surfaces. In *Numerical Grid Generation in Computational Fluid Dynamics*, A. S.-Arcilla, J. Häuser, P. R. Eiseman, J. F. Thompson, eds., pp. 915–924.

[34] Castillo, J. E. 1991a. A discrete variational grid generation method. *SIAM J. Sci. Stat. Comp.* **12**(2), 454–468.

[35] Celia, M. A., and Gray, W. G. 1992. *Numerical Methods for Differential Equations.* Prentice-Hall, Engelwood Cliffs, NJ.

[36] Chakravarthy, S., and Anderson, D. 1979. Numerical conformal mapping. *Math. Comp.* **33**, 953–969.

[37] Challis, N. V., and Burley, D. M. 1982. Numerical method for conformal mapping. *IMA J. Numer. Anal.* **2**, 169–181.

[38] Chawner, J. R., and Anderson, D. A. 1991. Development of an algebraic grid generation method with orthogonality and clustering control. In *Numerical Grid Generation in Computational Fluid Dynamics*, A. S.-Arcilla, J. Häuser, P. R. Eiseman, J. F. Thompson, eds., pp. 107–117.

[39] Chikhliwala, E. D., and Yortsos, Y. C. 1985. Application of orthogonal mapping to some two-dimensional domains. *J. Comp. Physics* **57**, 391–402.

[40] Christdoulou, K. N., and Soriven, L. E. 1993. Discretization of free surface flows and other moving boundary problems. *J. Comp. Physics* **99**, 39–55.

[41] Chu, W. H. 1971. Development of a general finite difference approximation for a general domain. Part 1: Machine transformation. *J. Comp. Physics* **8**, 392–408.

[42] Corwin, L. J., and Szczarba, R. H. 1982. *Multivariable Calculus.* Marcel Dekker, New York.

[43] Dacorogna, B. 1989. *Direct Methods in the Calculus of Variations.* Springer-Verlag, Berlin.

[44] Dannenhoffer, J. F. 1988. A comparison of two adaptive grid techniques. In *Numerical Grid Generation in Computational Fluid Mechanics '88*, S. Sengupta, J. Haüser, P. R. Eiseman, and J. F. Thompson, eds., pp. 319–328. Pineridge Press, Swansea, UK.

[45] Daripa, P. 1991a. A new theory for one-dimensional adaptive grid generation and its applications. *SIAM J. Numer. Anal.* **28**, 1635–1660.

[46] Daripa, P. 1991b. On a numerical method for quasiconformal grid generation. *J. Comp. Phys.* **96**, 229–236.

[47] Daripa, P. 1992a. Iterative schemes and algorithms for adaptive grid generation in one dimension. *J. Comp. Phys.* **100**, 284–293.

[48] Daripa, P. 1992b. A fast algorithm to solve non-homogeneous Cauchy Riemann equations in the complex plane. *SIAM J. Sci. Stat. Comput.* Preprint.

[49] Daripa, P. 1992c. *A Fast Algorithm to Solve Beltrami Equation with Applications to Quasiconformal Mappings.* Preprint.

[50] Davies, C. W. 1981. An initial value approach to the production of discrete orthogonal coordinates. *J. Comp. Physics* **39**, 164–178.

[51] Dembo, R. S., and Steihaug, T. 1983. Truncated Newton algorithms for large-scale uncontrained optimization. *Math. Prog.* **26**, 190–212.

[52] Desbois, F., and Jacquotte O.-P. 1991. Surface mesh generation and optimization. In *Numerical Grid Generation in Computational Fluid Dynamics,* A. S.-Arcilla, J. Häuser, P. R. Eiseman, J. F. Thompson, eds., pp. 131–141.

[53] Dougherty, R. L., and Hyman, J. M. 1990. *A Divide and Conquer Algorithm for Grid Generation.* Tech. Rep. LA-UR-90-2128, Los Alamos National Laboratory, Los Alamos, NM.

[54] Duraiswami, R., and Prosperetti, A. 1992. Orthogonal mapping in two dimensions. *J. Comp. Phys.* **98**, 254–268.

[55] Dvinsky, A. S. 1988. Adaptive grid generation from harmonic maps. In *Numerical Grid Generation in Computational Fluid Mechanics '88,* S. Sengupta, J. Haüser, P. R. Eiseman, and J. F. Thompson, eds., pp. 299–308. Pineridge Press, Swansea, UK.

[56] Dvinsky, A. S. 1991. Adaptive grid generation from harmonic maps on Riemannian manifolds. *J. Comp. Phys.* **95**, 450–476.

[57] Dwyer, H. A., Kee, R., and Sanders, B. 1980. Adaptive grid method for problems in fluid mechanics and heat transfer. *AIAA Journal* **18**, 1205–1212.

[58] Dwyer, H. A., Smooke, M. D., and Kee, R. J. 1982. Adaptive gridding for finite difference solutions to heat and mass transfer problems. In *Numerical Grid Generation,* J. F. Thompson, ed., pp. 339–356. North-Holland, New York.

[59] Eiseman, P. R. 1978. A coordinate system for a viscous transonic cascade analysis. *J. Comp. Physics* **26**, 307–338.

[60] Eiseman, P. R. 1979. A multi-surface method of coordinate generation. *J. Comp. Physics* **33**, 118–150.

[61] Eiseman, P. R. 1982a. Coordinate generation with precise controls over mesh properties. *J. Comp. Physics* **47**, 331–351.

[62] Eiseman, P. R. 1982b. High level continuity for coordinate generation with precise controls. *J. Comp. Physics* **47**, 352–374.

[63] Eiseman, P. R. 1982c. Orthogonal grid generation. In *Numerical Grid Generation,* J. F. Thompson, ed., pp. 193–226. North-Holland, New York.

[64] Eiseman, P. R. 1980. *Alternating Direction Adaptive Grid Generation for Three-Dimensional Regions.* Preprint.

[65] Eiseman, P. R. 1985a. Alternating Direction Adaptive Grid Generation. *AIAA Journal* **23**(4), 551–560.

[66] Eiseman, P. R. 1985b. Adaptive grid generation by mean value relaxation. *J. Fluids Eng.* **107**, 477–483.

[67] Eiseman, P. R. 1985c. Grid generation for fluid mechanics computations. *Ann. Rev. Fluid Mech.* **17**, 487–522.

[68] Eiseman, P. R. 1987. Adaptive grid generation. *Comp. Meth. in Appl. Mech. and Engr.* **64**(1–3), 321–376.

[69] Eiseman, P. R. 1988. A control point form of algebraic grid generation. *Int. J. for Num. Methods in Fluids* **8**, 1165–1181.

[70] Eiseman, P. R. 1992. Control point grid generation. *Computers Math. Applic.* **24**(5–6), 57–67.

[71] Epstein, B. 1962. *Partial Differential Equations, An Introduction.* Robert E. Krieger Pub. Co., Malarbar, Florida.

[72] Eyler, L. L., and White, M. D. 1988. Surface constrained grid generation with Lagrange multipliers. In *Numerical Grid Generation in Computational Fluid Mechanics '88*, S. Sengupta, J. Haüser, P. R. Eiseman, and J. F. Thompson, eds., pp. 125–136. Pineridge Press, Swansea, UK.

[73] Fletcher, C. A. J. 1988. *Computational Techniques for Fluid Dynamics.* Springer-Verlag, Berlin.

[74] Fletcher, R. 1980. *Practical Methods of Optimization*, Vol. 1, *Unconstrained Optimization.* Wiley, New York.

[75] Fornberg, B. 1980. A numerical method for conformal mapping. *SIAM J. Sci. Stat. Comput.* **1**, 386–400.

[76] Forsythe, G. E., and Wasow, W. R. 1960. *Finite Difference Methods for Partial Differential Equations.* Wiley, New York.

[77] Garon, A., and Camarero, R. 1983. Generation of surface-fitted coordinate grids. In *Advances in Grid Generation*, K. N. Ghia and U. Ghia, eds., pp. 117–122. ASME, Houston, TX.

[78] Gelfand, I. M., and Fomin, S. V. 1963. *Calculus of Variations.* Prentice-Hall, Englewood Cliffs, NJ.

[79] George, P. L. 1991. *Automatic Mesh Generation: Applicatons to Finite Element Methods.* Wiley, New York.

[80] Ghia, K. N., and Ghia, U., eds. 1983a. *Advances in Grid Generation*, FED-Vol. 5, ASME Applied Mechanics, Bioengineering, and Fluids Engineering Conference, Houston, TX.

[81] Ghia, K. N., Ghia, U., and Shin, C. T. 1983b. Adaptive grid generation for flows with local high gradient regions. In *Advances in Grid Generation*, K. N. Ghia and U. Ghia, eds., pp. 35–47. ASME, Houston, TX.

[82] Giannakopoulos, A. E., and Engel, A. J. 1988. Directional control in grid generation. *J. Comp. Physics* **74**, 422–439.

[83] Gilding, B. H. 1988. A numerical grid generation technique. *Computers and Fluids* **16**(1), 47–58.

[84] Godunov, S. K., and Prokopov, G. P. 1972. The use of moving meshes in gas-dynamical computations. *USSR Comp. Math. Math. Phys.* **12**, 182.

[85] Golub, G. H., and Ortega, J. M. 1991. *Scientific Computing and Differential Equations.* Academic Press, Boston.

[86] Gordon, W. J., and Hall, C. A. 1973. Construction of curvilinear coordinate systems and applications to mesh generation. *Inst. J. Num. Meth. Engr.* **7**, 461–477.

[87] Gordon, W. J., and Thiel, L. C. 1982. Transfinite mappings and their applications to grid generation. In *Numerical Grid Generation,* J. F. Thompson, ed., pp. 171–192. North-Holland, New York.

[88] Gurtin, M. E. 1981. *An Introduction to Continuum Mechanics.* Academic Press, New York.

[89] Hall, C. A., and Porsching, T. A. 1990. *Numerical Analysis of Partial Differential Equations.* Prentice-Hall, Englewood Cliffs, NJ.

[90] Haüser, J., and Taylor, C., eds. 1986. *Numerical Grid Generation in Computational Fluid Dynamics.* Pineridge Press, Swansea, UK.

[91] Haussling, H. J., and Coleman, R. M. 1981. A method for generation of orthogonal and nearly orthogonal boundary-fitted coordinate systems. *J. Comp. Physics* **43**, 373–381.

[92] Hawken, D. F. 1987. Review of adaptive-grid techniques for solution of partial differential equations. *Prog. Aerospace Sci.* **24**, 29–49.

[93] Hawken, D. F., Gottlieb, J. J., and Hansen, J. S. 1991. Review of some adaptive node-movement techniques in finite-element and finite-difference solutions of partial differential equations. *J. Comp. Phys.* **95**, 254–302.

[94] Hayes, J. K., Kahaner, D. K., and Kellner, R. 1972. An improved method for numerical conformal mapping. *Math. Comp.* **26**, 327–334.

[95] Heinrich, B. 1987. *Finite Difference Methods on Irregular Networks, A Generalized Approach to Second Order Elliptic Problems.* Birkäuser, Basel.

[96] Henrici, P. 1974. *Applied and Computational Complex Analysis,* 2 volumes. John Wiley and Sons, New York.

[97] Horn, R. A., and Johnson, C. R. 1985. *Matrix Analysis.* Cambridge University Press, Cambridge.

[98] Hsu, K. I., and Lee, S. L. 1991. A numerical technique for two-dimensional grid generation with grid control at all of the boundaries. *J. Comp. Phys.* **96**, 451–469.

[99] Ives, D. C. 1982. Conformal grid generation. In *Numerical Grid Generation,* J. F. Thompson, ed., pp. 107–137. North-Holland, New York.

[100] Jacquotte, O. P., and Cabello, J. 1988a. A variational method for the optimization and adaption of grids in computational fluid dynamics. In *Numerical Grid Generation in Computational Fluid Mechanics '88,* S. Sengupta, J. Haüser, P. R. Eiseman, and J. F. Thompson, eds., pp. 405–413. Pineridge Press, Swansea, UK.

[101] Jacquotte, O.-P. 1988b. A mechanical model for a new grid generation method in computational fluid dynamics. *Comp. Meth. Appl. Mech. and Engr.* **66**, 323–338.

[102] Jacquotte, O-P. 1991. Recent progress on mesh optimization. In *Numerical Grid Generation in Computational Fluid Dynamics,* A. S.-Arcilla, J. Häuser, P. R. Eiseman, J. F. Thompson, eds., pp. 581–596.

[103] Jain, R. K. 1986. Generation of body-fitted grids around airfoils using multigrid method. In *Numerical Grid Generation in CFD,* J. Haüser and C. Taylor, eds., pp. 305–317. Pineridge Press, Swansea, UK.

[104] Jain, R. K. 1986. Grid generation about airfoils using multigrid method in solution of elliptic partial differential equations. GMD Arbeitspapiere No. 208.

[105] Jain, R. K. 1987. Generation of body-fitted grids for two dimensional interior domains by the solution of a system of Poisson equations using multigrid techniques. GMD Arbeitspapiere No. 246.

[106] Jeng, Y. N., and Liou, Y. C. 1992. Two modified versions of Hsu-Lee's elliptic solver of grid generation. *Num. Heat Trans.,* Part B, Vol. 22, pp. 125–140.

[107] Jeng, Y. N., and Liou, Y. C. 1993. Adaptive grid generation by elliptic equations with grid control at all of the boundaries. *Num. Heat Trans.,* Part B, Vol. 23, pp. 135–151.

[108] Kang, I. S., and Leal, L. G. 1992. Orthogonal grid generation in a 2D domain via the boundary integral technique. *J. Comp. Phys.* **102**, 78–87.

[109] Kennon, S. R., and Dulikravich, G. S. 1986. Generation of computational grids using optimization. *AIAA Journal* **24**(7), 1069–1073.

[110] Knupp, P. M. 1989. *Robust Grid Generation on Curves and Surfaces.* Ph.D. thesis, University of New Mexico.

[111] Knupp, P. M. 1990. On the invertibility of the isoparametric map. *Comp. Meth. Appl. Mech. Engr.* **78**, 313–329.

[112] Knupp, P. M. 1991a. Intrinsic Algebraic grid generation. In *Mathematical Aspects of Numerical Grid Generation,* J. E. Castillo, ed. SIAM, Philadelphia.

[113] Knupp, P. M. 1991b. The direct variational grid generation method extended to curves. *Appl. Math. Comp.* **43**, 65–78.

[114] Knupp, P. M. 1992a. A robust elliptic grid generator. *J. Comp. Phys.* **100**, 409–418.

[115] Knupp, P. M. 1992b. *Surface Grid Generation and the Tangent Plane Methodology.* Preprint.

[116] Knupp, P. M. *Truncaton Error in Grid Generation: A Case Study.* Preprint.

[117] Knupp, P. M. 1993. *Alignment of Boundary-Fitted Coorinates with a Vector-Field Pair: A New Variational/Elliptic Generator.* Preprint.

[118] Kreis, B. 1983. Construction of a curvilinear grid. *SIAM J. Sci. Stat. Comput.* **4**(2).

[119] Kreis, R. I., Thames, F. C., and Hassan, H. A. 1986. Application of a variational method for generating adaptive grids. *AIAA Journal* **24**(3), 404–410.

[120] Kumar, A., and Kumar, N. S. 1988. A new approach to grid generation based on local optimisation. In *Numerical Grid Generation in Computational Fluid Mechanics '88*, S. Sengupta, J. Haüser, P. R. Eiseman, and J. F. Thompson, eds., pp. 177–184. Pineridge Press, Swansea, UK.

[121] Lancaster, P., and Salkauskas, K. 1990. *Curve and Surface Fitting.* Academic Press, San Diego, CA.

[122] Lee, D., and Tsuei, Y. M. 1992. A modified adaptive grid method for recirculating flows. *Intl. J. Num. Meth. Fluids* **14**, 775–791.

[123] Lehto, O., and Virtaanen, K. I. 1973. *Quasiconformal Mappings in the Plane.* Springer-Verlag, Berlin.

[124] Liao, G., and Liu, H. 1993. Existence and $C^{(0,\alpha)}$ regularity of a minimum of a functional related to grid generation problem. *Num. Math. PDE's* **9**(3).

[125] Liao, G., and Su, J. 1993. A direct method in Dacorogna-Moser's approach of grid generation problems. *Applicable Analysis*, to appear.

[126] Liao, G., and Su, J. 1992. Grid generation via deformation. *Applied Math. Letters* **5**(3), 27–29.

[127] Liao, G. 1992. Variational approach to grid generation. *Num. P.D.E.'s* **8**, 143–147.

[128] Liao, G., and Anderson, D. A. 1993. A new approach to grid generation. *Applicable Analysis*, to appear.

[129] Liao, G. 1991. On harmonic maps. In *Mathematical Aspects of Numerical Grid Generation*, J. E. Castillo, ed. SIAM, Philadelphia.

[130] Mastin, C. W., and Thompson, J. F. 1978a. Transformation of three-dimensional regions onto rectangular regions by elliptic systems. *Numer. Math.* **29**, 397–407.

[131] Mastin, C. W., and Thompson, J. F. 1978b. Elliptic systems and numerical transformations. *J. Math. Anal. and Appl.* **62**, 52–62.

[132] Mastin, C. W. 1981. Coordinate generation by conformal and quasi-conformal mappings. *Elliptic Problem Solvers*, M. Shultz, ed. Academic Press, New York.

[133] Mastin, C. W. 1982. Error induced by coordinate systems. In *Numerical Grid Generation*, J. F. Thompson, ed., pp. 31–40. North-Holland, New York.

[134] Mastin, C. W., and Thompson, J. F. 1984. Quasiconformal mappings and grid generation. *SIAM J. Sci. Stat. Comput.* **5**(2), 305–310.

[135] Mastin, C. W. 1992. Linear variational methods and adaptive grids. *Computers Math. Applic.* **24**(5–6), 51–56.

[136] Matsuno, K., and Dwyer, H. A. 1988. Adaptive methods for elliptic grid generation. *J. Comp. Physics* **77**, 40–52.

[137] Mavriplis, D. J. 1992. *Unstructured Mesh Algorithms for Aerodynamic Calculations.* ICASE Report No. 92-35, NASA-Langley Research Center, Hampton, VA.

[138] Melas, A. D. 1991. *An Example of a Harmonic Map between Euclidean Balls.* Preprint, University of California at Los Angeles.

[139] Menikoff, R., and Zemach, C. 1980. Methods for numerical conformal mapping. *J. Comp. Phys.* **36**, 366–410.

[140] Minoux, M. 1986. *Mathematical Programming.* John Wiley and Sons, New York.

[141] Mobley, C. D., and Stewart, R. J. 1980. On the numerical generation of boundary-fitted orthogonal curvilinear coordinate systems. *J. Comp. Physics* **34**, 124–135.

[142] Moretti, G. 1980. Grid generation using classical techniques. In *Numerical Grid Generation Techniques*, R. E. Smith, ed., pp. 1–35. NASA CP 2166, NASA Langley Research Center, Hampton VA.

[143] Morice, P. 1983. Numerical generation of boundary-fitted coordinate systems with optimal control of orthogonality. In *Advances in Grid Generation*, K. N. Ghia and U. Ghia, eds., pp. 71–78. ASME, Houston, TX.

[144] Nakahashi, K., and Deiwert, G. S. 1986. Three-dimensional adaptive grid method. *AIAA Journal* **24**(6), 948–954.

[145] Nakamura, S. 1982. Marching grid generation using parabolic partial differential equations. In *Numerical Grid Generation*, J. F. Thompson, ed., pp. 775–786. North-Holland, New York.

[146] Nakamura, S. 1983. Adaptive grid relocation algorithm for transonic full potential calculation using one-dimensional and two-dimensional diffusion equations. In *Advances in Grid Generation*, K. N. Ghia and U. Ghia, eds., pp. 49–58. ASME, Houston, TX.

[147] Niederdrenk, P. 1991. Solution-adaptive grid generation by hyperbolic/parabolized hyperbolic P.D.E.s. In *Numerical Grid Generation in Computational Fluid Dynamics*, A. S.-Arcilla, J. Häuser, P. R. Eiseman, J. F. Thompson, eds., pp. 173–181.

[148] Noak, R. W., and Anderson, D. A. 1990. Solution-adaptive grid generation using parabolic partial differential equations. *AIAA Journal* **28**(6), 1016–1023.

[149] Oh, H. J., and Kang, I. S. 1993. *Orthogonal Grid Generation in a 2D Domain with Specified Boundary Correspondence at Three Sides.* Pohang Institute of Science and Tech., Korea.

[150] Pardhanani, A., and Carey, G. F. 1988. Optimization of computational grids. *Num. Meth. for P.D.E.'s* **4**(2), 95–117.

[151] Peyret, R., and Taylor, T. D. 1983. *Computational Methods for Fluid Flow.* Springer-Verlag, New York.

[152] Potter, D. E., and Tuttle, G. H. 1973. Construction of discrete orthogonal coordinates. *J. Comp. Physics* **13**, 483–501.

[153] Pulliam, T. H., and Steger, J. L. 1980. Implicit finite-difference simulations of three-dimensional compressible flow. *AIAA J.* **18**, 159–167.

[154] Quirk, J. J. 1992. *An Alternative to Unstructured Grids for Computing Gas Dynamic Flows Around Arbitrary Complex Two-Dimensional Bodies.* ICASE Report No. 92-7, NASA Langley, Hampton, VA.

[155] Rai, M. M., and Anderson, D. A. 1981. Grid evolution in time assymptotic problems. *J. Comp. Physics* **43**, 327–344.

[156] Renelt, H. 1988. *Elliptic Systems and Quasiconformal Mappings.* Wiley, New York.

[157] Roache, P. J. 1978. Marching methods for elliptic problems, Part 1. *Num. Heat Transfer* **1**, 1–25.

[158] Roache, P. J., and Steinberg, S. 1985. *A New Approach to Grid Generation Using a Variational Formulation.* AIAA 7th C.F.D. Conference, paper no. 85-1527, pp. 360–370.

[159] Roache, P. J., Salari, K., and Steinberg, S. 1991. Hybrid adaptive Poisson grid generation and grid smoothness. *Commun. in Applied Num. Meth.* **7**, 345–354.

[160] Ryskin, G., and Leal, L. G. 1983. Orthogonal mapping. *J. Comp. Physics* **50**, 71–100.

[161] Saha, S., and Basu, B. C. 1991. Simple algebraic technique for nearly orthogonal grid generation. *AIAA Journal* **29**, 1340–1341.

[162] Saitou, M., and Hirata, A. 1992. Two-dimensional unsteady solidification problem calculated by using the boundary-fitted coordinate system. *J. Comp. Phys.* **100**, 188–196.

[163] Salari, K. 1989. Personal communication.

[164] Saltzman, J. S., and Brackbill, J. U. 1982. Applications and generalizations of variational methods for generating adaptive meshes. In *Numerical Grid Generation*, J. F. Thompson, ed., pp. 865–884. North-Holland, New York.

[165] Saltzman, J. S. 1986. Variational methods for generating meshes on surfaces in three dimensions. *J. Comp. Physics* **63**, 1–19.

[166] Samarskii, A. A., Tishkin, V. F., Favorskii, A. P., and Shashkov, M. Yu. 1981. Operational finite-difference schemes. *Diff. Eqns.* **17**, 863–885.

[167] Samarskii, A. A., Tishkin, V. F., Favorskii, A. P., and Shashkov, M. Yu. 1982. Employment of the reference-operator method in the construction of finite difference analogs of tensor operations. *Diff. Eqns.* **18**, 881–885.

[168] Schwartz, A. L., and Connett, W. E. 1986. Evaluating algebraic adaptive grid strategies. In *Numerical Grid Generation in Computational Fluid Dynamics*, J. Haüser and C. Taylor, eds., pp. 63–174. Pineridge Press, Swansea, UK.

[169] Seidl, A., and Klose, H. 1985. Numerical conformal mapping of a towel-shaped region onto a rectangle. *SIAM J. Sci. Stat. Comput.* **6**(4), 833–842.

[170] Sengupta, S., Haüser, J., Eiseman, P. R., and Thompson, J. F., eds. 1988. *Numerical Grid Generation in Computational Fluid Mechanics '88.* Pineridge Press, Swansea, UK.

[171] Shanno, D. F. 1978. Conjugate gradient methods with inexact searches. *Math. Oper. Res.* **3**, 244–256.

[172] Shih, T. I-P., Bailey, R. T., Nguyen, H. L., and Roelke, R. J. 1991. Algebraic grid generation for complex geometries. *Intl. J. for Num. Meth. in Fluids* **13**(1), 1–31.

[173] Showalter, R. E. 1977. *Hilbert Space Methods for Partial Differential Equations.* Pitman, London.

[174] Shubin, G. R., Stephens, A. B., and Bell, J. B. 1982. Three dimensional grid generation using biharmonics. In *Numerical Grid Generation,* J. F. Thompson, ed., pp. 761–774. North-Holland, New York.

[175] Smarch-Abolhassani, J., and Smith, R. E. 1992. A practical approach to algebraic grid generation. *Computers Math. Applic.* **24**(5–6), 69–81.

[176] Smith, P. W., and Sritharan, S. S. 1988. Theory of harmonic grid generation. *Complex Variables* **10**, 359–369.

[177] Smith, R. E., ed. 1980. *Numerical Grid Generation Techniques.* NASA Conference Publication 2166, NASA Langley Research Center, Hampton, VA.

[178] Smith, R. E. 1981. *Two-Boundary Grid Generation for the Solution of the Three Dimensional Navier-Stokes Equations.* NASA TM-83123, Langley Research Center, Hampton, VA.

[179] Smith, R. E. 1982. Algebraic grid generation. In *Numerical Grid Generation,* J. F. Thompson, ed., pp. 137–170. North-Holland, New York.

[180] Sod, G. A. 1985. *Numerical Methods in Fluid Dynamics.* Cambridge University Press, Cambridge.

[181] Soni, B. K., and Wang, T.-S. *Goodness of Grid: Quantitative Measures.* Preprint.

[182] Soni, B. K., and Mastin, C. W. 1990. *Variational Methods for Grid Optimization.* Preprint.

[183] Soni, B. K. 1992. Grid generation for internal flow configuration. *Computers Math. Applic.* **24**(5–6), 191–201.

[184] Sorenson, R. L., and Steger, J. L. 1980. Numerical generation of two dimensional grids by the use of Poisson equations with grid control. In *Numerical Grid Generation Techniques,* R. E. Smith, ed., pp. 449–461. NASA CP 2166, NASA Langley Research Center, Hampton, VA.

[185] Sorenson, R. L., and Steger, J. L. 1981. *Grid Generation in Three Dimensions by Poisson Equations with Control of Cell Size and Skewness at Boundary Surfaces.* Preprint.

[186] Sparis, P. D. 1985. A method for generating boundary-orthogonal curvilinear coordinate systems using the biharmonic equation. *J. Comp. Physics* **61**, 445–462.

[187] Sparis, P. D., Karkanis, A., and Pergantis, S. 1992. Conjugate method of solutions of the biharmonic equation for the generation of boundary orthogonal grids. *Comp. Meth. Appl. Mech. and Engr.* **98**, 273–290.

[188] Sritharan, S. S. 1991. Mathematical aspects of harmonic grid generation. In *Mathematical Aspects of Numerical Grid Generation,* J. Castillo, ed., 131–145.

[189] Starius, G. 1977. Constructing orthogonal curvilinear meshes by solving initial value problems. *Numer. Math.* **28**, 25–48.

[190] Steger, J. L., and Sorenson, R. L. 1979. Automatic mesh-point clustering near a boundary in grid generation with elliptic partial differential equations. *J. Comp. Physics* **33**, 405–410.

[191] Steger, J. L., and Chausee, D. S. 1980. Generation of body-fitted coordinates using hyperbolic partial differential equations. *SIAM J. Sci. Stat. Comput.* **1**(4), 431–437.

[192] Steger, J. L. 1991. Grid generation with hyperbolic partial differential equations for application to complex configurations. In *Numerical Grid Generation in Computational Fluid Dynamics*, A. S.-Arcilla, J. Häuser, P. R. Eiseman, J. F. Thompson, eds., pp. 871–886.

[193] Steinberg, S., and Roache, P. J. 1985. Symbolic manipulation and computational fluid dynamics. *J. Comp. Physics* **57**, 251–284.

[194] Steinberg, S., and Roache, P. J. 1986. Variational grid generation. *Num. Meth. for P.D.E.s* **2**, 71–96.

[195] Steinberg, S. 1988. Overview of symbol manipulation. *CWI Quarterly* **1**, 65–72.

[196] Steinberg, S., and Roache, P. J. 1990. Anomalies in grid generation on curves. *J. Comp. Physics* **91**, 255–277.

[197] Steinberg, S., and Roache, P. J. 1991. *Discretizing Symmetric Operators in General Coordinates.* Preprint.

[198] Steinberg, S., and Roache, P. J. 1992. Variational curve and surface grid generation. *J. Comp. Phys.* **100**, 163–178.

[199] Steinbrenner, J. P., and Anderson, D. A. 1990. Grid-generation methodology in applied aerodynamics. In *Applied Computational Aerodynamics*, Vol. 125, *Progress in Astronautics and Aeronautics*, P. A. Henne, ed., pp. 91–132. AIAA, Washington, D.C.

[200] Steinthorsson, E., Shih, T. I-P., and Roelke, R. J. 1992. Enhancing control of grid distribution in algebraic grid generation. *Intl. J. Num. Meth. Fluids* **15**, 297–311.

[201] Stoker, J. J. 1969. *Differential Geometry.* Wiley, New York.

[202] Strikwerda, J. C. 1989. *Finite Difference Schemes and Partial Differential Equations.* Wadsworth & Brooks Cole, Pacific Grove, CA.

[203] Struik, D. J. 1988. *Lectures on Classical Differential Geometry.* Dover, New York.

[204] Symm, G. T. 1966. An integral equation method in conformal mapping. *Numer. Math.* **9**, 250–258.

[205] Tamamidis, P., and Assanis, D. N. 1991. Generation of orthogonal grids with control of spacing. *J.C.P.* **94**, 437–453.

[206] Theodoropoulos, T., and Bergeles, G. C. 1989. A Laplacian equation method for numerical generation of boundary-fitted 3D orthogonal grids. *J. Comp. Physics* **82**, 269–288.

[207] Thinking 1992. *Getting Started in CM Fortran.* Thinking Machines Corporation, 245 First Street, Cambridge, MA 02142-1214, USA.

[208] Thomas, P. D., and Lombard, C. K. 1979. Geometric conservation law and its application to flow computations on movings grids. *AIAA Journal* **17**, 1030–1037.

[209] Thomas, P., and Middlecoff, J. 1980. Direct control of the grid point distribution in meshes generated by elliptic equations. *AIAA Journal* **18**(6), 652–656.

[210] Thomas, P. D. 1982. Numerical generation of composite three dimensional grids by quasilinear elliptic systems. In *Numerical Grid Generation*, J. F. Thompson, ed., pp. 667–686.

[211] Thomas, P. D. 1984. Stationary interior grids and computation of moving interfaces. In *Advances in Computational Methods for Boundary and Interior Layers*. Boole Press.

[212] Thompson, J. F., Thames, F. C., and Mastin, C. W. 1974. Automatic numerical generation of body-fitted curvilinear coordinate system for field containing any number of arbitrary two-dimensional bodies. *J. Comp. Physics* **15**, 299–319.

[213] Thompson, J. F., Thames, F. C., and Mastin, C. W. 1977. TOMCAT—A code for numerical generation of boundary-fitted curvilinear cordinate systems on fields containing any number of arbitrary two-dimensional bodies. *J. Comp. Physics* **24**, 274–302.

[214] Thompson, J. F., ed. 1982a. *Numerical Grid Generation*. North-Holland, New York. (Also published as Vols. 10 and 11 of *Applied Mathematics and Computation*, 1982.)

[215] Thompson, J. F. 1982b. Elliptic grid generation. In *Numerical Grid Generation*, J. F. Thompson, ed., pp. 79–105. North-Holland, New York.

[216] Thompson, J. F., Warsi, Z. U. A., and Mastin, C. W. 1982c. Boundary-fitted coordinate systems for numerical solution of partial differential equations—A review. *J. Comp. Physics* **47**, 1–108.

[217] Thompson, J. F., and Mastin, C. W. 1983. Order of difference expressions in curvilinear coordinate systems. In *Advances in Grid Generation*, Ghia and Ghia, eds., pp. 17–28. ASME.

[218] Thompson, J. F. 1984. Grid generation techniques in computational fluid dynamics. *AIAA Journal* **22**(11), 1505–1523.

[219] Thompson, J. F., Warsi, Z. U. A., and Mastin, C. W. 1985a. *Numerical Grid Generation: Foundations and Applications*. North-Holland, Elsevier, New York.

[220] Thompson, J. F. 1985b. A survey of dynamically-adaptive grids in the numerical solution of partial differential equations. *Applied Numerical Mathematics* **1**, 3–27.

[221] Thompson, J. F., Martinez, A., and Mounts, J. S. 1987. *Program EAGLE Numerical Grid Generation System Users Manual*, Vols. I–III. AFATL-TR-87-15, Eglin AFB.

[222] Vinokur, M. 1983. On one-dimensional stretching functions for finite difference calculations. *J. Comp. Phys.* **50**, 215–234.

[223] Vinokur, M. 1989. An analysis of finite-difference and finite-volume formulations of conservation laws. *J. Comp. Phys.* **81**, 1–52.

[224] Vreugdenhil, C. B. 1991. Grid control for non-orthogonal elliptic grids. *Comm. Appl. Num. Meth.* **7**, 633–637.

[225] Warsi, Z. U. A., and Thompson, J. F. 1980a. Application of variational methods in the fixed and adaptive grid generation. *Computers Math. Applic.* **19**(8/9), 31–41.

[226] Warsi, Z. U. A., and Thompson, J. F. 1980b. A non-iterative method for the generation of two-dimensional orthogonal curvilinear coordinates in an euclidean space. In *Numerical Grid Generation Techniques*, R. E. Smith, ed., pp. 516. NASA CP 2166, NASA Langley Research Center, Hampton, VA.

[227] Warsi, Z. U. A., and Thompson, J. F. 1982a. A noniterative method for the generation of orthogonal coordinates in doubly-connected regions. *Math. of Comp.* **38**(158), 501–516.

[228] Warsi, Z. U. A. 1982b. Basic differential models for coordinate generation. In *Numerical Grid Generation*, J. F. Thompson, ed., pp. 41–78. North-Holland, New York.

[229] Warsi, Z. U. A., and Ziebarth, J. P. 1982c. Numerical generation of three-dimensional coordinates between bodies of arbitrary shapes. In *Numerical Grid Generation*, J. F. Thompson, ed., pp. 717–728.

[230] Warsi, Z. U. A. 1986. Numerical grid generation in arbitrary surfaces through a second-order differential model. *J. Comp. Physics* **64**, 82–96.

[231] Warsi, Z. U. A. 1990. Theoretical foundation of the equations for the generation of surface coordinates. *AIAA J.* **28**(6), 1140–1142.

[232] Warsi, Z. U. A., and Koomullil, G. P. 1991. Application of spectral techniques in surface grid generation. In *Numerical Grid Generation in Computational Fluid Dynamics*, A. S.-Arcilla, J. Häuser, P. R. Eiseman, J. F. Thompson, eds., pp. 955–963.

[233] Wendland, W. L. 1979. *Elliptic Systems in the Plane.* Pitman, London.

[234] Whitney, A. K., and Thomas, P. D. 1983. Construction of grids on curved surfaces described by generalized coordinates through the use of an elliptic system. In *Advances in Grid Generation.* K. N. Ghia and U. Ghia, eds., pp. 173–179. ASME, Houston, TX.

[235] Winslow, A. 1967. Numerical solution of the quasilinear Poisson equations in a nonuniform triangle mesh. *J. Comp. Physics* **2**, 149–172.

[236] Winslow, A. 1981. *Adaptive Mesh Zoning by the Equipotential Method.* UCID-19062.

[237] Yeung, R. W., and Vaidhyanathan, M. 1992. Non-linear interaction of water waves with submerged obstacles. *Intl. J. Num. Meth. Fluids* **14**, 1111–1130.

[238] Zegeling, P. A., Verwer, J. G., and Van Eijkeren, J. C. H. 1992. Application of a moving-grid method to a class of 1D brine transport problems in porous media. *Intl. J. Num. Meth. Fluids* **15**, 175–191.

Appendix A

Tensor Coefficients

This appendix gives expressions for the tensor coefficients in the planar (8.71) and volume (9.59) nonconservative forms of the Euler-Lagrange equations arising from functionals of the form $H = H(g_{ij}, \sqrt{g})$.

A.1 PLANAR TENSOR COEFFICIENTS

A.1.1 Covariant Form

$$\begin{aligned}
\mathcal{T}_{11} &= \frac{\partial H}{\partial g_{11}} \mathcal{I} \\
&+ \left\{ 2\frac{\partial^2 H}{\partial g_{11}^2} + 2\frac{g_{22}}{\sqrt{g}} \frac{\partial^2 H}{\partial g_{11} \partial \sqrt{g}} + \frac{g_{22}^2}{2g} \frac{\partial^2 H}{\partial \sqrt{g}^2} \right\} (\mathbf{x}_\xi \otimes \mathbf{x}_\xi) \\
&+ \left\{ \frac{\partial^2 H}{\partial g_{11} \partial g_{12}} - \frac{g_{12}}{\sqrt{g}} \frac{\partial^2 H}{\partial g_{11} \partial \sqrt{g}} + \frac{g_{22}}{2\sqrt{g}} \frac{\partial^2 H}{\partial g_{12} \partial \sqrt{g}} - \frac{g_{12} g_{22}}{2g} \frac{\partial^2 H}{\partial \sqrt{g}^2} \right\} \\
&\times [(\mathbf{x}_\xi \otimes \mathbf{x}_\eta) + (\mathbf{x}_\eta \otimes \mathbf{x}_\xi)] \\
&+ \left\{ \frac{1}{2} \frac{\partial^2 H}{\partial g_{12}^2} - \frac{g_{12}}{\sqrt{g}} \frac{\partial^2 H}{\partial g_{12} \partial \sqrt{g}} + \frac{g_{12}^2}{2g} \frac{\partial^2 H}{\partial \sqrt{g}^2} \right\} (\mathbf{x}_\eta \otimes \mathbf{x}_\eta)
\end{aligned} \tag{A.1}$$

and

$$\begin{aligned}
\mathcal{T}_{12} + \mathcal{T}_{12}^T &= \left\{ \frac{\partial H}{\partial g_{12}} + \sqrt{g} \frac{\partial^2 H}{\partial g_{12} \partial \sqrt{g}} \right\} \mathcal{I} \\
&+ \left\{ 2\frac{\partial^2 H}{\partial g_{11} \partial g_{12}} - 2\frac{g_{12}}{\sqrt{g}} \frac{\partial^2 H}{\partial g_{11} \partial \sqrt{g}} - \frac{g_{12} g_{22}}{g} \frac{\partial^2 H}{\partial \sqrt{g}^2} \right\} (\mathbf{x}_\xi \otimes \mathbf{x}_\xi) \\
&+ \left\{ 2\frac{\partial^2 H}{\partial g_{11} \partial g_{22}} + \frac{g_{11}}{\sqrt{g}} \frac{\partial^2 H}{\partial g_{11} \partial \sqrt{g}} + \frac{1}{2} \frac{\partial^2 H}{\partial g_{12}^2} + \frac{g_{22}}{\sqrt{g}} \frac{\partial^2 H}{\partial g_{22} \partial \sqrt{g}} \right. \\
&+ \left. \frac{g_{11} g_{22} + g_{12}^2}{2g} \frac{\partial^2 H}{\partial \sqrt{g}^2} \right\} [(\mathbf{x}_\xi \otimes \mathbf{x}_\eta) + (\mathbf{x}_\eta \otimes \mathbf{x}_\xi)]
\end{aligned}$$

$$+ \left\{ 2\frac{\partial^2 H}{\partial g_{12}\partial g_{22}} - 2\frac{g_{12}}{\sqrt{g}}\frac{\partial^2 H}{\partial g_{22}\partial\sqrt{g}} - \frac{g_{11}g_{12}}{g}\frac{\partial^2 H}{\partial\sqrt{g}^2} \right\}(\mathbf{x}_\eta \otimes \mathbf{x}_\eta) \tag{A.2}$$

and

$$\begin{aligned}
\mathcal{T}_{22} &= \frac{\partial H}{\partial g_{22}}\mathcal{I} \\
&+ \left\{ \frac{1}{2}\frac{\partial^2 H}{\partial g_{12}^2} - \frac{g_{12}}{\sqrt{g}}\frac{\partial^2 H}{\partial g_{12}\partial\sqrt{g}} + \frac{g_{12}^2}{2g}\frac{\partial^2 H}{\partial\sqrt{g}^2} \right\}(\mathbf{x}_\xi \otimes \mathbf{x}_\xi) \\
&+ \left\{ \frac{\partial^2 H}{\partial g_{22}\partial g_{12}} - \frac{g_{12}}{\sqrt{g}}\frac{\partial^2 H}{\partial g_{22}\partial\sqrt{g}} + \frac{g_{11}}{2\sqrt{g}}\frac{\partial^2 H}{\partial g_{12}\partial\sqrt{g}} - \frac{g_{12}g_{11}}{2g}\frac{\partial^2 H}{\partial\sqrt{g}^2} \right\} \\
&\times [(\mathbf{x}_\xi \otimes \mathbf{x}_\eta) + (\mathbf{x}_\eta \otimes \mathbf{x}_\xi)] \\
&+ \left\{ 2\frac{\partial^2 H}{\partial g_{22}^2} + 2\frac{g_{11}}{\sqrt{g}}\frac{\partial^2 H}{\partial g_{22}\partial\sqrt{g}} + \frac{g_{11}^2}{2g}\frac{\partial^2 H}{\partial\sqrt{g}^2} \right\}(\mathbf{x}_\eta \otimes \mathbf{x}_\eta).
\end{aligned} \tag{A.3}$$

A.1.2 Contravariant Form

$$\begin{aligned}
g\mathcal{T}_{11} &= \frac{\partial H}{\partial g_{11}}g\mathcal{I} + \left\{ 2g_{11}^2\frac{\partial^2 H}{\partial g_{11}^2} + 2g_{11}g_{12}\frac{\partial^2 H}{\partial g_{11}\partial g_{12}} \right. \\
&\left. + \frac{g_{12}^2}{2}\frac{\partial^2 H}{\partial g_{12}^2} + 2g_{11}\sqrt{g}\frac{\partial^2 H}{\partial g_{11}\partial\sqrt{g}} + g_{12}\sqrt{g}\frac{\partial^2 H}{\partial g_{12}\partial\sqrt{g}} + \frac{g}{2}\frac{\partial^2 H}{\partial\sqrt{g}^2} \right\}(\mathbf{x}_\eta^\perp \otimes \mathbf{x}_\eta^\perp) \\
&- \left\{ 2g_{11}g_{12}\frac{\partial^2 H}{\partial g_{11}^2} + (g_{11}g_{22} + g_{12}^2)\frac{\partial^2 H}{\partial g_{11}\partial g_{12}} + \frac{g_{12}g_{22}}{2}\frac{\partial^2 H}{\partial g_{12}^2} \right. \\
&\left. + g_{12}\sqrt{g}\frac{\partial^2 H}{\partial g_{11}\partial\sqrt{g}} + \frac{g_{22}\sqrt{g}}{2}\frac{\partial^2 H}{\partial g_{12}\partial\sqrt{g}} \right\}[(\mathbf{x}_\xi^\perp \otimes \mathbf{x}_\eta^\perp) + (\mathbf{x}_\eta^\perp \otimes \mathbf{x}_\xi^\perp)] \\
&+ \left\{ 2g_{12}^2\frac{\partial^2 H}{\partial g_{11}^2} + 2g_{12}g_{22}\frac{\partial^2 H}{\partial g_{11}\partial g_{12}} + \frac{g_{22}^2}{2}\frac{\partial^2 H}{\partial g_{12}^2} \right\}(\mathbf{x}_\xi^\perp \otimes \mathbf{x}_\xi^\perp)
\end{aligned} \tag{A.4}$$

and

$$\begin{aligned}
g(\mathcal{T}_{12} + \mathcal{T}_{12}^T) &= \left\{ \frac{\partial H}{\partial g_{12}} + \sqrt{g}\frac{\partial^2 H}{\partial g_{12}\partial\sqrt{g}} \right\} g\mathcal{I} \\
&+ \left\{ 2g_{11}^2\frac{\partial^2 H}{\partial g_{11}\partial g_{12}} + 4g_{11}g_{12}\frac{\partial^2 H}{\partial g_{11}\partial g_{22}} + g_{11}g_{12}\frac{\partial^2 H}{\partial g_{12}^2} \right. \\
&\left. + 2g_{12}^2\frac{\partial^2 H}{\partial g_{12}\partial g_{22}} + 2\sqrt{g}g_{12}\frac{\partial^2 H}{\partial g_{22}\partial\sqrt{g}} \right\}(\mathbf{x}_\eta^\perp \otimes \mathbf{x}_\eta^\perp) \\
&- \left\{ 2g_{11}g_{12}\frac{\partial^2 H}{\partial g_{11}\partial g_{12}} + 2(g_{11}g_{22} + g_{12}^2)\frac{\partial^2 H}{\partial g_{11}\partial g_{22}} \right. \\
&+ g_{11}\sqrt{g}\frac{\partial^2 H}{\partial g_{11}\partial\sqrt{g}} + \frac{1}{2}(g_{11}g_{22} + g_{12}^2)\frac{\partial^2 H}{\partial g_{12}^2} \\
&+ 2g_{12}g_{22}\frac{\partial^2 H}{\partial g_{12}\partial g_{22}} + g_{22}\sqrt{g}\frac{\partial^2 H}{\partial g_{22}\partial\sqrt{g}}
\end{aligned}$$

$$
\begin{aligned}
&+ \left. \frac{g}{2} \frac{\partial^2 H}{\partial \sqrt{g}^2} \right\} [(\mathbf{x}_\xi^\perp \otimes \mathbf{x}_\eta^\perp) + (\mathbf{x}_\eta^\perp \otimes \mathbf{x}_\xi^\perp)] \\
&+ \left\{ 2 g_{12}^2 \frac{\partial^2 H}{\partial g_{12} \partial g_{11}} + 4 g_{12} g_{22} \frac{\partial^2 H}{\partial g_{11} \partial g_{22}} \right. \\
&+ 2 g_{12} \sqrt{g} \frac{\partial^2 H}{\partial g_{11} \sqrt{g}} + g_{12} g_{22} \frac{\partial^2 H}{\partial g_{12}^2} \\
&+ \left. 2 g_{22}^2 \frac{\partial^2 H}{\partial g_{12} \partial g_{22}} \right\} (\mathbf{x}_\xi^\perp \otimes \mathbf{x}_\xi^\perp)
\end{aligned} \tag{A.5}
$$

$$
\begin{aligned}
g \mathcal{T}_{22} = {} & \frac{\partial H}{\partial g_{22}} g \mathcal{I} + \left\{ 2 g_{12}^2 \frac{\partial^2 H}{\partial g_{22}^2} + 2 g_{12} g_{11} \frac{\partial^2 H}{\partial g_{22} \partial g_{12}} + \frac{g_{11}^2}{2} \frac{\partial^2 H}{\partial g_{12}^2} \right\} (\mathbf{x}_\eta^\perp \otimes \mathbf{x}_\eta^\perp) \\
& - \left\{ 2 g_{22} g_{12} \frac{\partial^2 H}{\partial g_{22}^2} + (g_{11} g_{22} + g_{12}^2) \frac{\partial^2 H}{\partial g_{22} \partial g_{12}} + \frac{g_{12} g_{11}}{2} \frac{\partial^2 H}{\partial g_{12}^2} \right. \\
& + \left. g_{12} \sqrt{g} \frac{\partial^2 H}{\partial g_{22} \partial \sqrt{g}} + \frac{g_{11} \sqrt{g}}{2} \frac{\partial^2 H}{\partial g_{12} \partial \sqrt{g}} \right\} [(\mathbf{x}_\xi^\perp \otimes \mathbf{x}_\eta^\perp) + (\mathbf{x}_\eta^\perp \otimes \mathbf{x}_\xi^\perp)] \\
& + \left\{ 2 g_{22}^2 \frac{\partial^2 H}{\partial g_{22}^2} + 2 g_{22} g_{12} \frac{\partial^2 H}{\partial g_{22} \partial g_{12}} \right. \\
& + \frac{g_{12}^2}{2} \frac{\partial^2 H}{\partial g_{12}^2} + 2 g_{22} \sqrt{g} \frac{\partial^2 H}{\partial g_{22} \partial \sqrt{g}} + g_{12} \sqrt{g} \frac{\partial^2 H}{\partial g_{12} \partial \sqrt{g}} \\
& + \left. \frac{g}{2} \frac{\partial^2 H}{\partial \sqrt{g}^2} \right\} (\mathbf{x}_\xi^\perp \otimes \mathbf{x}_\xi^\perp)
\end{aligned} \tag{A.6}
$$

A.2 VOLUME TENSOR COEFFICIENTS

A.2.1 Mixed Covariant and Contravariant Form

$$
\begin{aligned}
\mathcal{T}_{11} = {} & \left(\frac{\partial H}{\partial g_{11}} \right) \mathcal{I} + 2 \frac{\partial^2 H}{\partial g_{11}^2} (\mathbf{x}_\xi \otimes \mathbf{x}_\xi) \\
& + \frac{\partial^2 H}{\partial g_{11} \partial g_{12}} [(\mathbf{x}_\xi \otimes \mathbf{x}_\eta) + (\mathbf{x}_\eta \otimes \mathbf{x}_\xi)] + \frac{\partial^2 H}{\partial g_{11} \partial g_{13}} [(\mathbf{x}_\xi \otimes \mathbf{x}_\zeta) + (\mathbf{x}_\zeta \otimes \mathbf{x}_\xi)] \\
& + \frac{1}{2} \frac{\partial^2 H}{\partial g_{12}^2} (\mathbf{x}_\eta \otimes \mathbf{x}_\eta) \\
& + \frac{1}{2} \frac{\partial^2 H}{\partial g_{13}^2} (\mathbf{x}_\zeta \otimes \mathbf{x}_\zeta) + \frac{1}{2} \frac{\partial^2 H}{\partial g_{12} \partial g_{13}} [(\mathbf{x}_\eta \otimes \mathbf{x}_\zeta) + (\mathbf{x}_\zeta \otimes \mathbf{x}_\eta)] \\
& + \sqrt{g} \frac{\partial^2 H}{\partial g_{11} \partial \sqrt{g}} [(\mathbf{x}_\xi \otimes \nabla \xi) + (\nabla \xi \otimes \mathbf{x}_\xi)] \\
& + \frac{\sqrt{g}}{2} \frac{\partial^2 H}{\partial g_{12} \partial \sqrt{g}} [(\mathbf{x}_\eta \otimes \nabla \xi) + (\nabla \xi \otimes \mathbf{x}_\eta)] \\
& + \frac{\sqrt{g}}{2} \frac{\partial^2 H}{\partial g_{13} \partial \sqrt{g}} [(\mathbf{x}_\zeta \otimes \nabla \xi) + (\nabla \xi \otimes \mathbf{x}_\zeta)] + \frac{g}{2} \frac{\partial^2 H}{\partial \sqrt{g}^2} (\nabla \xi \otimes \nabla \xi)
\end{aligned} \tag{A.7}
$$

$$
\begin{aligned}
\mathcal{T}_{12} + \mathcal{T}_{12}^{T} = {} & \frac{\partial H}{\partial g_{12}} \mathcal{I} \\
& + 2 \frac{\partial^2 H}{\partial g_{11} \partial g_{12}} (\mathbf{x}_\xi \otimes \mathbf{x}_\xi) + 2 \frac{\partial^2 H}{\partial g_{11} \partial g_{22}} [(\mathbf{x}_\xi \otimes \mathbf{x}_\eta) + (\mathbf{x}_\eta \otimes \mathbf{x}_\xi)] \\
& + \frac{\partial^2 H}{\partial g_{11} \partial g_{23}} [(\mathbf{x}_\xi \otimes \mathbf{x}_\zeta) + (\mathbf{x}_\zeta \otimes \mathbf{x}_\xi)] \\
& + \sqrt{g} \frac{\partial^2 H}{\partial g_{11} \partial \sqrt{g}} [(\mathbf{x}_\xi \otimes \nabla \eta) + (\nabla \eta \otimes \mathbf{x}_\xi)] \\
& + \frac{1}{2} \frac{\partial^2 H}{\partial g_{12}^2} [(\mathbf{x}_\xi \otimes \mathbf{x}_\eta) + (\mathbf{x}_\eta \otimes \mathbf{x}_\xi)] + \frac{1}{2} \frac{\partial^2 H}{\partial g_{12} \partial g_{13}} [(\mathbf{x}_\xi \otimes \mathbf{x}_\zeta) + (\mathbf{x}_\zeta \otimes \mathbf{x}_\xi)] \\
& + 2 \frac{\partial^2 H}{\partial g_{12} \partial g_{22}} (\mathbf{x}_\eta \otimes \mathbf{x}_\eta) + \frac{1}{2} \frac{\partial^2 H}{\partial g_{12} \partial g_{23}} [(\mathbf{x}_\eta \otimes \mathbf{x}_\zeta) + (\mathbf{x}_\zeta \otimes \mathbf{x}_\eta)] \\
& + \frac{\partial^2 H}{\partial g_{13} \partial g_{23}} (\mathbf{x}_\zeta \otimes \mathbf{x}_\zeta) \\
& + \frac{\sqrt{g}}{2} \frac{\partial^2 H}{\partial g_{12} \partial \sqrt{g}} [(\mathbf{x}_\xi \otimes \nabla \xi) + (\nabla \xi \otimes \mathbf{x}_\xi) + (\mathbf{x}_\eta \otimes \nabla \eta) + (\nabla \eta \otimes \mathbf{x}_\eta)] \\
& + \frac{\partial^2 H}{\partial g_{13} \partial g_{22}} [(\mathbf{x}_\eta \otimes \mathbf{x}_\zeta) + (\mathbf{x}_\zeta \otimes \mathbf{x}_\eta)] \\
& + \frac{\sqrt{g}}{2} \frac{\partial^2 H}{\partial g_{13} \partial \sqrt{g}} [(\mathbf{x}_\zeta \otimes \nabla \eta) + (\nabla \eta \otimes \mathbf{x}_\zeta)] \\
& + \sqrt{g} \frac{\partial^2 H}{\partial g_{22} \partial \sqrt{g}} [(\mathbf{x}_\eta \otimes \nabla \xi) + (\nabla \xi \otimes \mathbf{x}_\eta)] \\
& + \frac{\sqrt{g}}{2} \frac{\partial^2 H}{\partial g_{23} \partial \sqrt{g}} [(\mathbf{x}_\zeta \otimes \nabla \xi) + (\nabla \xi \otimes \mathbf{x}_\zeta)] \\
& + \frac{g}{2} \frac{\partial^2 H}{\partial \sqrt{g}^2} [(\nabla \xi \otimes \nabla \eta) + (\nabla \eta \otimes \nabla \xi)]. \qquad \text{(A.8)}
\end{aligned}
$$

The other four matrices can be obtained from these two by cyclic permutations.

Appendix **B**

Fortran Code Directory

This appendix provides a description of the codes that are on the disk that comes with this book. The first section in this appendix provides programs for generating grids from analytic formulas in one and two dimensions. Note that these codes were written by different people at different times, so their programming styles are not consistent. Makefiles are provided for compiling the codes under UNIX. On other systems, the Makefiles exhibit explicitly all the source code needed to compile any given program. Both ".f" and ".for" endings are used on the source files to help *make* distinguish different programming styles.

B.1 ANALYTIC GRID GENERATORS

B.1.1 Program linear

Program linear uses formula (1.13)

$$x = (1 - \xi)\, x_0 + \xi x_1\,, \quad 0 \le \xi \le 1\,, \tag{B.1}$$

to produce a linear grid on the interval $[x_0, x_1]$. The program is in

- fortran/analytic/linear.f

B.1.2 Program oned

For one dimensional grids, it is assumed that a formula $x = x(\xi)$ is given and that the grid is given by

$$x_i = x\left(\frac{i}{M}\right), \quad 0 \le i \le M\,, \tag{B.2}$$

where M is a given positive integer. There are subroutines for creating a general grid, the identity grid, an exponentially compressed grid, and a tangent grid:

- fortran/analytic/oned.f
- fortran/analytic/identity1.f
- fortran/analytic/compress.f
- fortran/analytic/tangent.f

B.1.3 Program bilinear

To generate a bilinear map in two dimensions, the physical region must be a quadrilateral defined by four points: $\mathbf{x}_{0,0}$, $\mathbf{x}_{1,0}$, $\mathbf{x}_{0,1}$, $\mathbf{x}_{1,1}$. The biliner map is given by (1.14), which when written is coordinate notation gives

$$\begin{aligned} x(\xi,\eta) &= (1-\xi)(1-\eta)\,x_{0,0} + (1-\xi)\,\eta\,x_{0,1} \\ &\quad + \xi\,(1-\eta)\,x_{1,0} + \xi\,\eta\,x_{1,1}\,, \\ y(\xi,\eta) &= (1-\xi)(1-\eta)\,y_{0,0} + (1-\xi)\,\eta\,y_{0,1} \\ &\quad + \xi\,(1-\eta)\,y_{1,0} + \xi\,\eta\,y_{1,1}\,. \end{aligned} \tag{B.3}$$

The program is in

- fortran/analytic/bilinear.f

B.1.4 Program twod

For two-dimensional grids, it is assumed that formulas $x = x(\xi,\eta)$ and $y = y(\xi,\eta)$ are given and that the grid is given by

$$x_{i,j} = x\left(\frac{i}{M},\frac{j}{N}\right),\quad y_{i,j} = y\left(\frac{i}{M},\frac{j}{N}\right),\qquad 0 \le i \le M\,,\ 0 \le j \le N\,. \tag{B.4}$$

There are programs for creating general grids given by formulas, the identity grid, polar grids, parabolic grids, elliptic grids, horseshoe grids, modified horseshoe grids, and bipolar grids:

- fortran/analytic/twod.f
- fortran/analytic/identity2.f
- fortran/analytic/polar.f
- fortran/analytic/parabolic.f
- fortran/analytic/elliptic.f
- fortran/analytic/horse.f
- fortran/analytic/modified.f
- fortran/analytic/bipolar.f

B.2 TRANSFINITE INTERPOLATION (TFI)

See Section 1.5 for a description of the algorithm. The main program is in the **transf.for** file. The subroutines are in

- fortran/analytic/transf.for
- fortran/analytic/b_parm.for
- fortran/analytic/bset.for
- fortran/analytic/discrtz.for
- fortran/analytic/select.for
- fortran/analytic/tfi.for
- fortran/analytic/ts.for

B.3 HOSTED EQUATIONS

These are codes for solving one- and two-dimensional scalar hosted equations with Dirichlet boundary conditions.

B.3.1 One-D-Programs

The one-dimensional main program is in **hosted_1d.f**. The subroutines are in

- fortran/hosted/one/hosted_1d.f
- fortran/hosted/one/alp.f
- fortran/hosted/one/bndry.f
- fortran/hosted/one/error.f
- fortran/hosted/one/exact.f
- fortran/hosted/one/grid_gen.f
- fortran/hosted/one/sor.f
- fortran/hosted/one/stencils.f
- fortran/hosted/one/trans.f

B.3.2 Two-D Programs

The two-dimensional main program is in **hosted_2d.f**. The subroutines are in

- fortran/hosted/two/hosted_2d.f
- fortran/hosted/two/alp.f
- fortran/hosted/two/bet.f
- fortran/hosted/two/bndry.f
- fortran/hosted/two/error.f
- fortran/hosted/two/exact.f
- fortran/hosted/two/gam.f

- fortran/hosted/two/grid_gen.f
- fortran/hosted/two/sor_9pt.f
- fortran/hosted/two/stencils.f
- fortran/hosted/two/trans.f

B.4 LINE GENERATORS

Codes for Chapter 3 Projects:

- fortran/line/proj345.f
- fortran/line/proj346.f
- fortran/line/proj347.f

B.5 HYPERBOLIC GRID GENERATOR

Codes for the hyperbolic grid generator are in

- fortran/planar/hyperbolic/boundary.f
- fortran/planar/hyperbolic/hyperbolic.f
- fortran/planar/hyperbolic/weight.f

B.6 TTM GENERATOR

The main code for the TTM generator is in **ittm.f**. Here is a listing of all of the files:

- fortran/planar/ttm/b_parm.for
- fortran/planar/ttm/bset.for
- fortran/planar/ttm/coeff_ittm.f
- fortran/planar/ttm/discrtz.for
- fortran/planar/ttm/ittm.f
- fortran/planar/ttm/mets.f
- fortran/planar/ttm/pq.f
- fortran/planar/ttm/select.for
- fortran/planar/ttm/sor_9pt.f
- fortran/planar/ttm/tfi.for
- fortran/planar/ttm/tngts.f
- fortran/planar/ttm/ts.for

B.7 VARIATIONAL GENERATORS

B.7.1 Length Functional

This code numerically solves the unweighted Length equations (5.42).

- fortran/planar/length/length.f

B.7.2 Weighted Combination Functional

This code numerically solves the weighted Euler-Lagrange equations based on equation (6.69). The main code is in **comb.f**. Here is a list of all of the files:

- fortran/planar/comb/coeff_cmb.f
- fortran/planar/comb/comb.f
- fortran/planar/comb/sor_9pt_xy.f

The code also uses some of the subroutines in B.6.

B.7.3 Metric Elements Functional

This code numerically solves equation (8.35) with H an arbitrary function of the elements of the metric tensor. The main code is in **tensor.f**. Here is a list of all of the files:

- fortran/planar/vari/coeff.f
- fortran/planar/vari/grd_met.f
- fortran/planar/vari/hessian.f
- fortran/planar/vari/partials.f
- fortran/planar/vari/tensor.f
- fortran/planar/vari/tnsr.f
- fortran/planar/vari/t11.f
- fortran/planar/vari/t12.f
- fortran/planar/vari/t22.f

The code also uses some of the subroutines in B.6.

Appendix C

A Rogue's Gallery of Grids

The appendix provides a large number of examples on some regions that are nontrivial to grid.

Figure C.1 *Unit square grids*

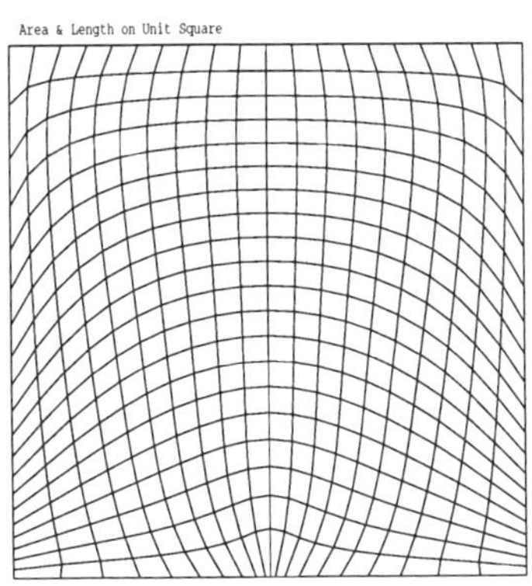

Area & Length

Figure C.1 *Continued*

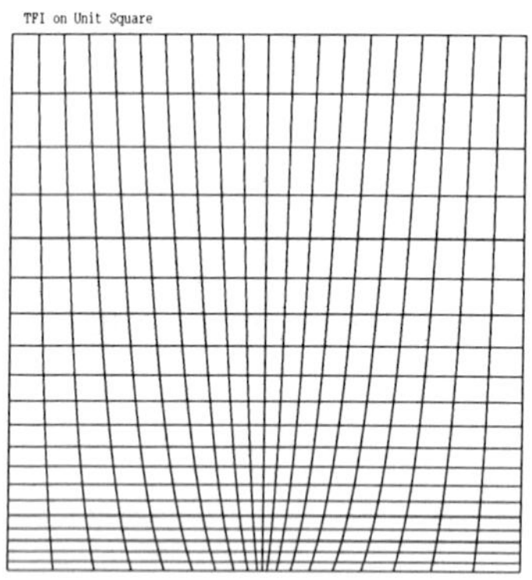

TFI

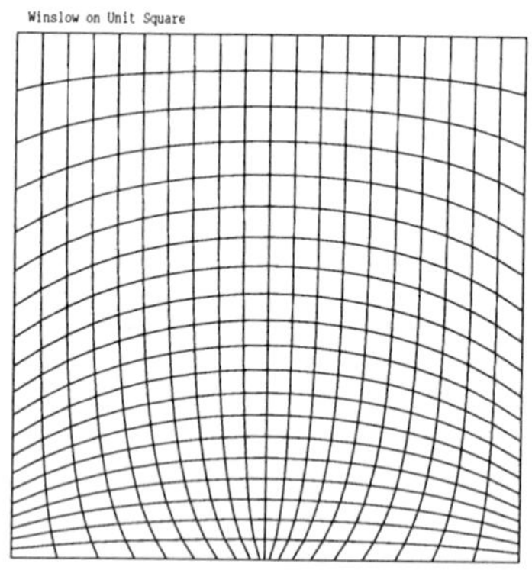

Winslow

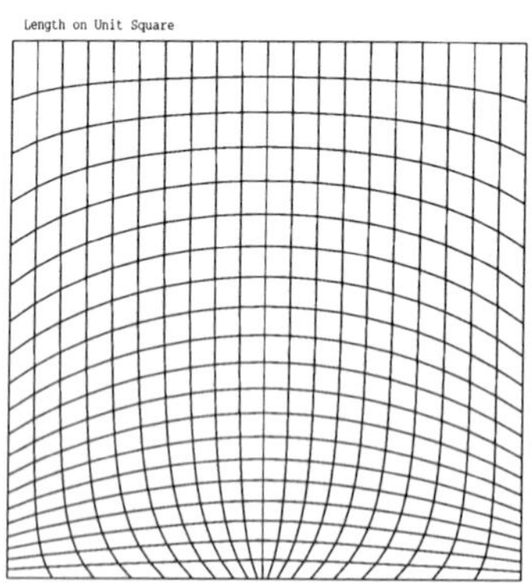

Length

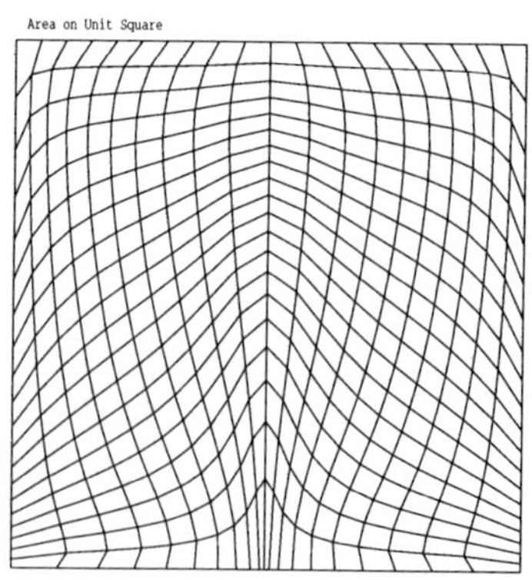

Area

Figure C.1 *Continued*

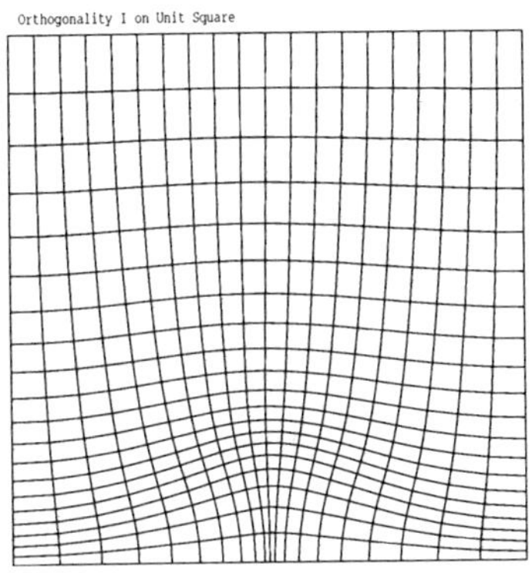

Orthogonality I

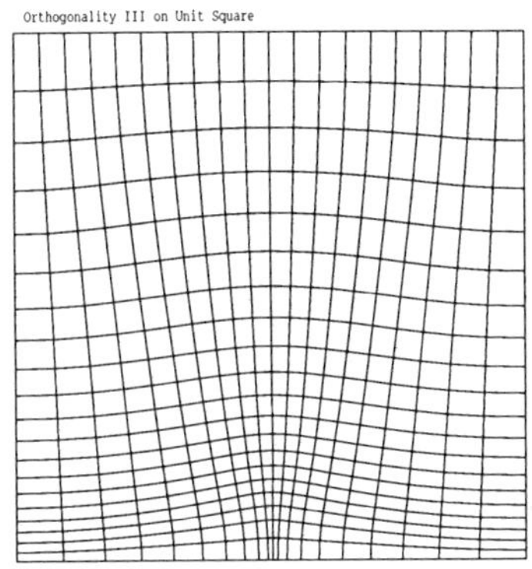

Orthogonality III

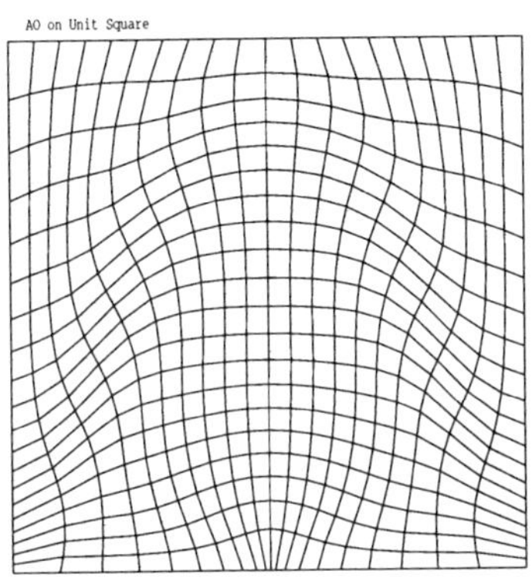

AO

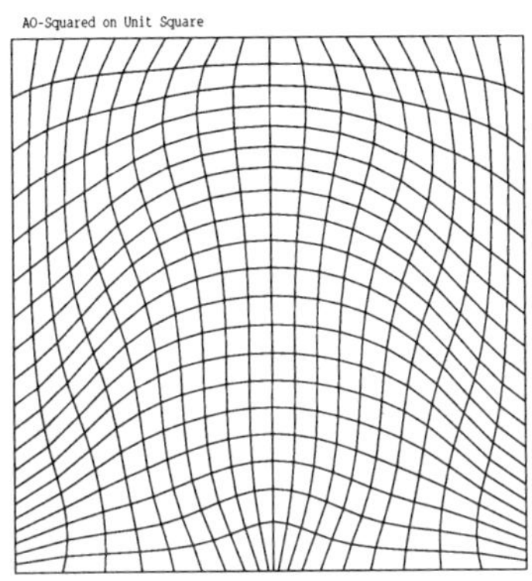

AO-Squared

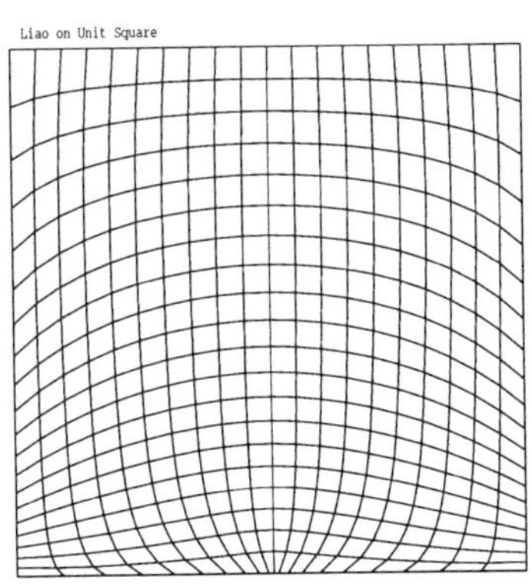

Liao

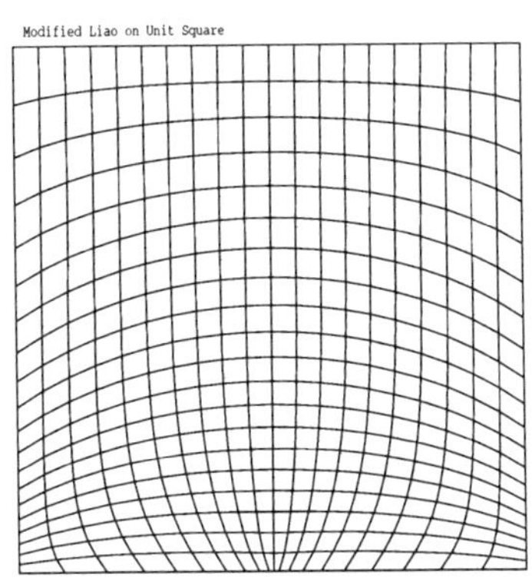

Modified Liao

Figure C.2 *Trapezoid grids*

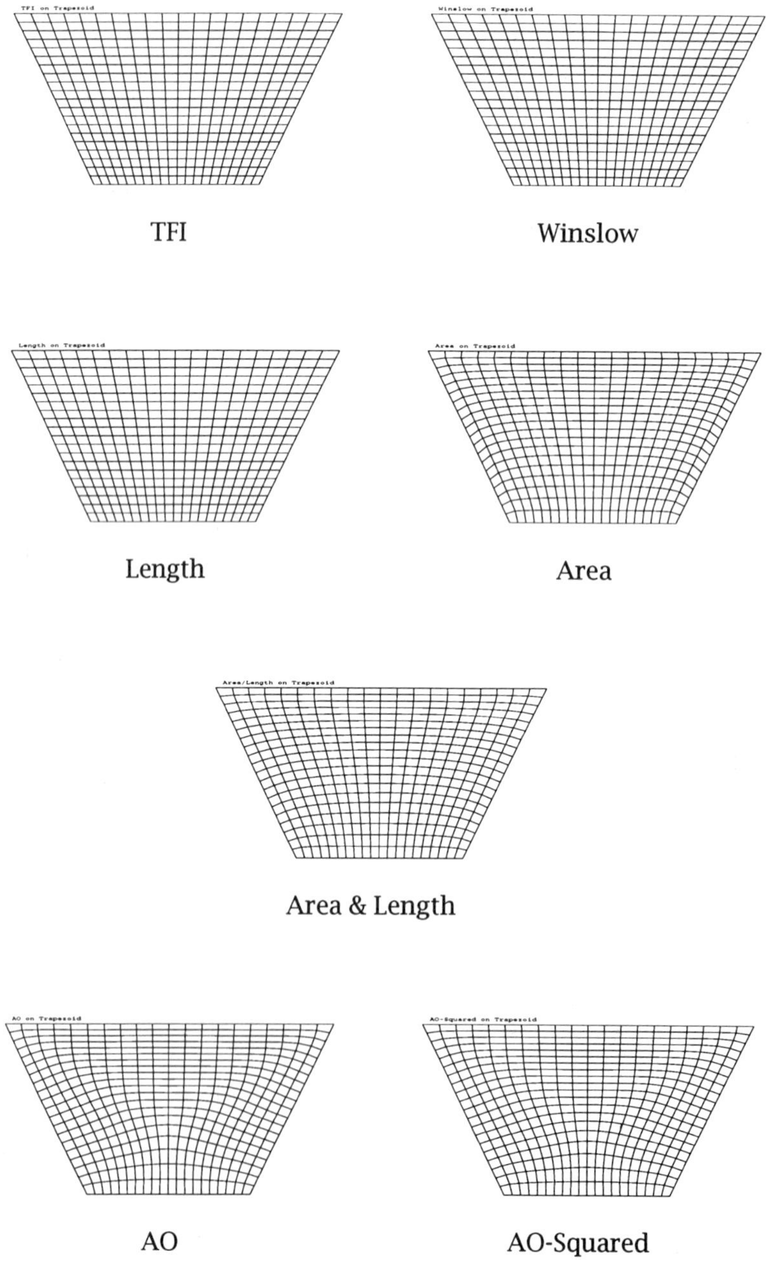

Figure C.2 *Continued*

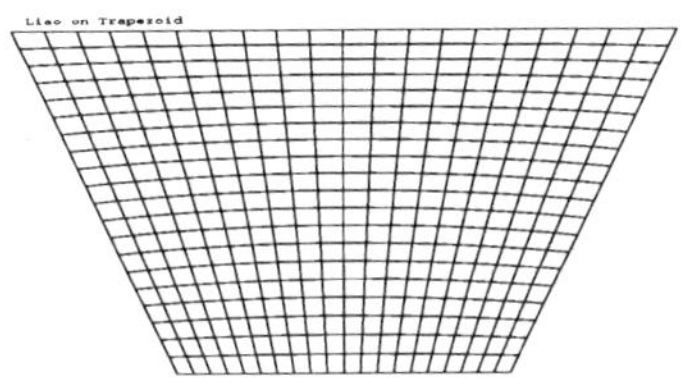

Liao

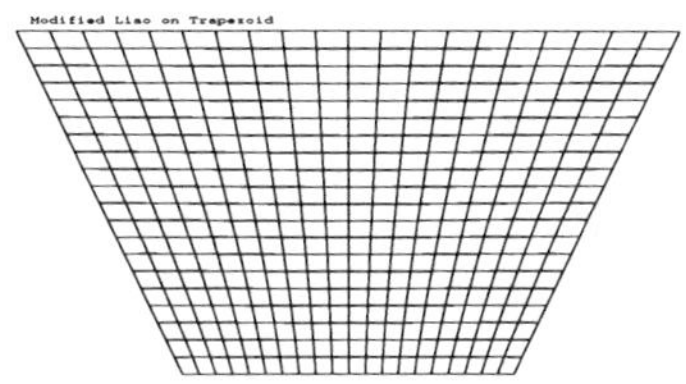

Modified Liao

Figure C.3 *Annulus grids*

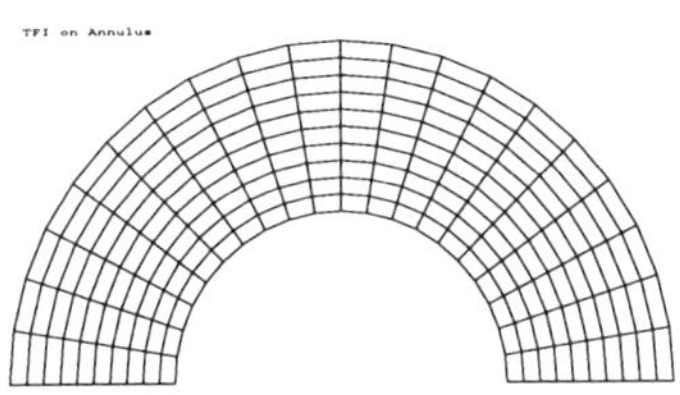

TFI

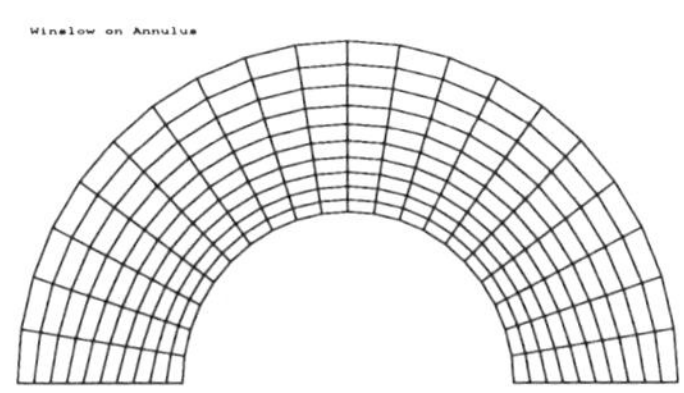

Winslow

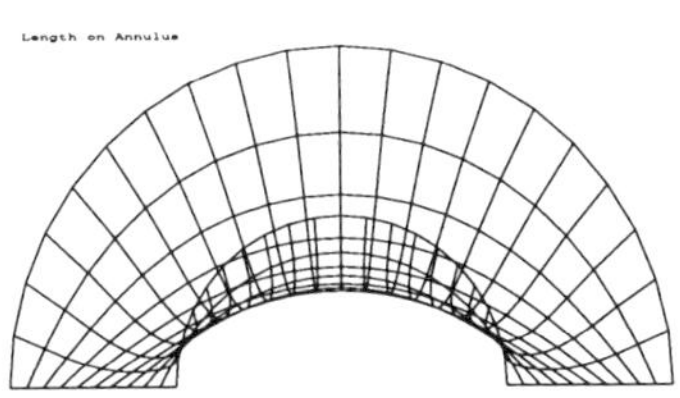

Length

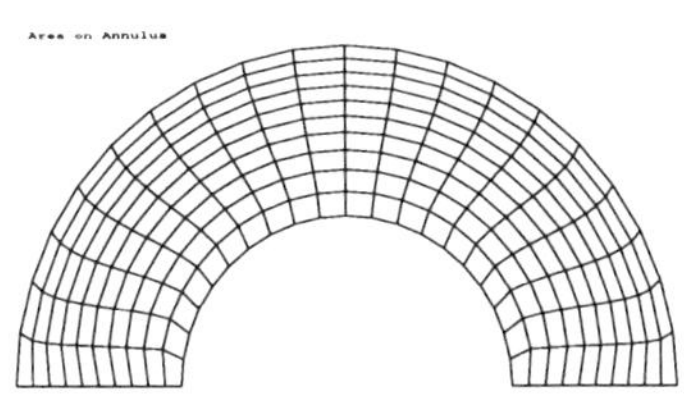

Area

Figure C.3 *Continued*

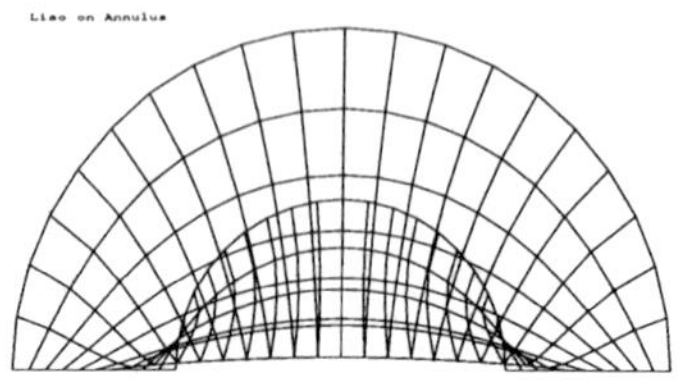

Liao

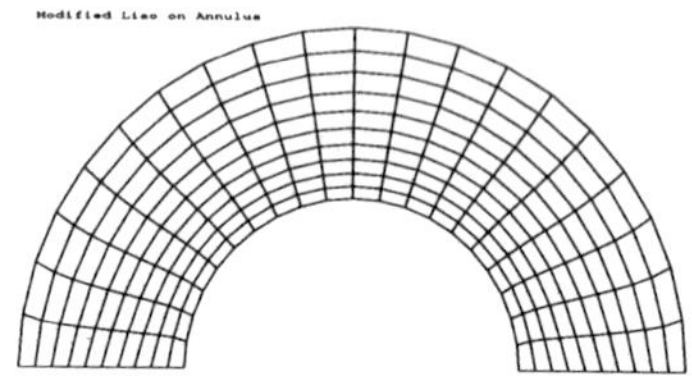

Modified Liao

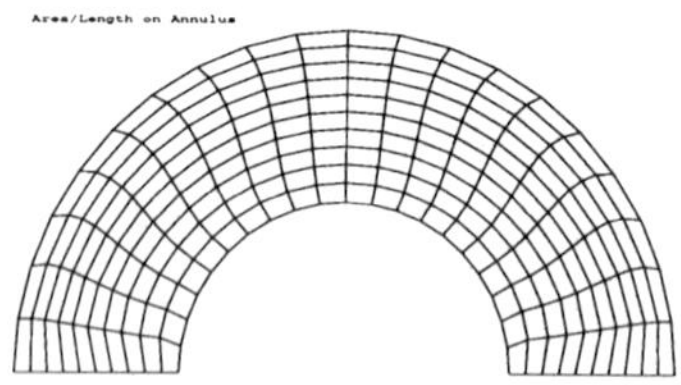

Area & Length

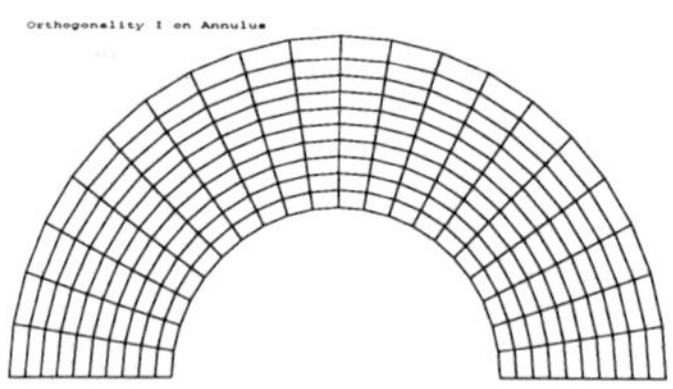

Orthogonality I

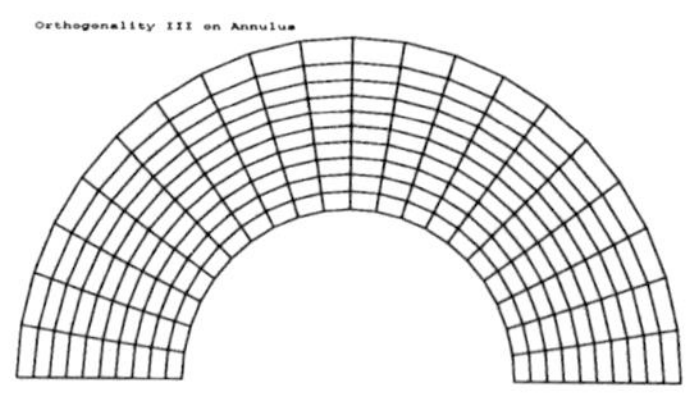

Orthogonality III

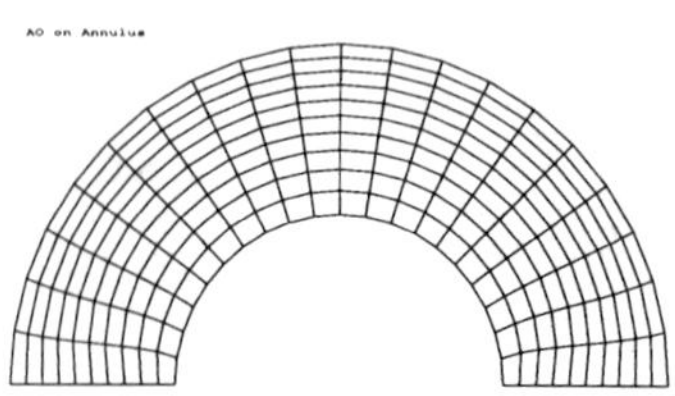

AO

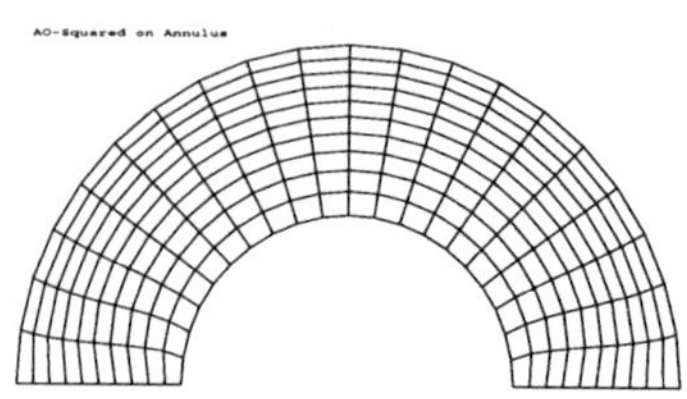

AO-Squared

Figure C.4 *Horseshoe grids*

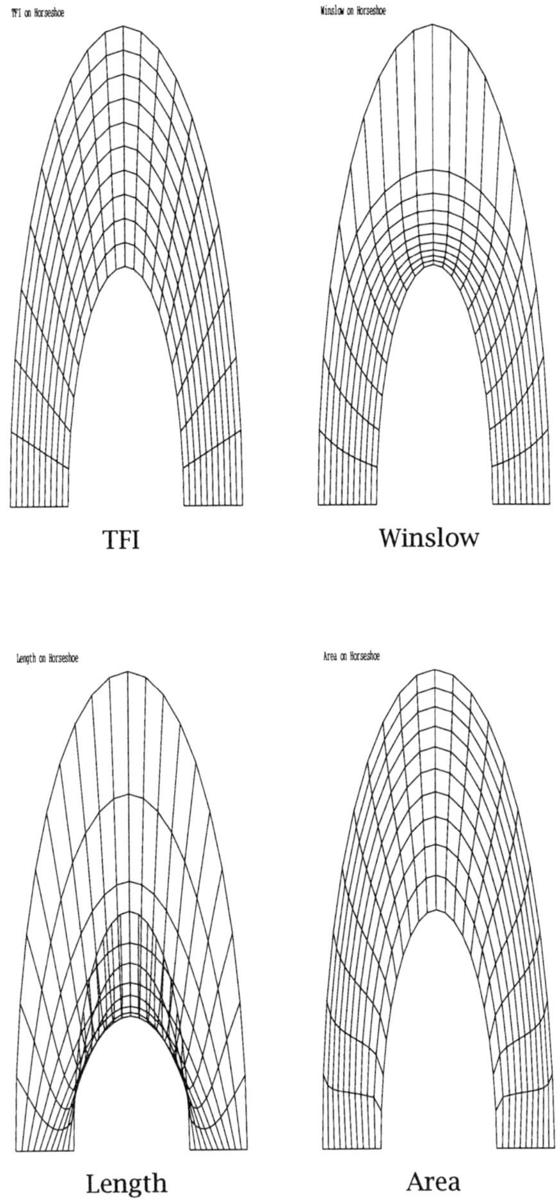

Figure C.4 *Continued*

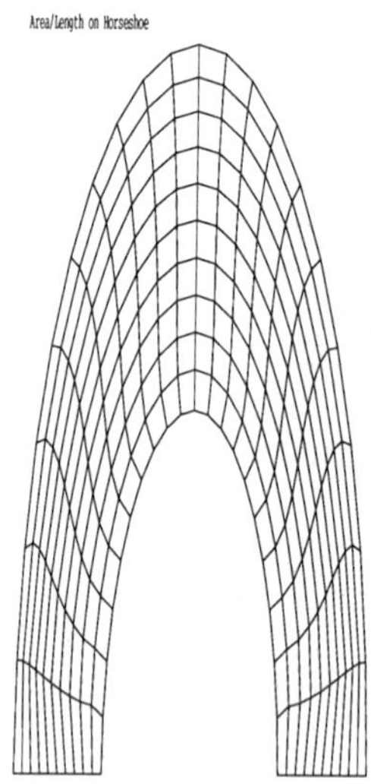

Area & Length

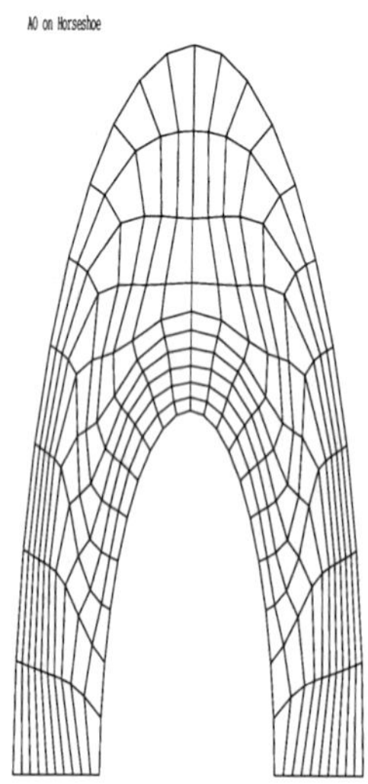

AO

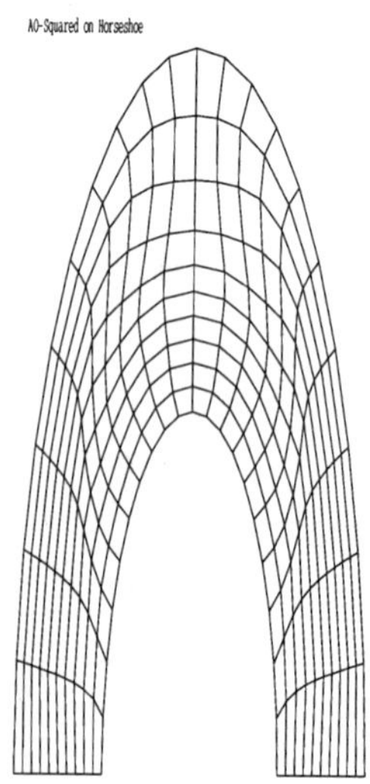

AO-Squared

Figure C.4 *Continued*

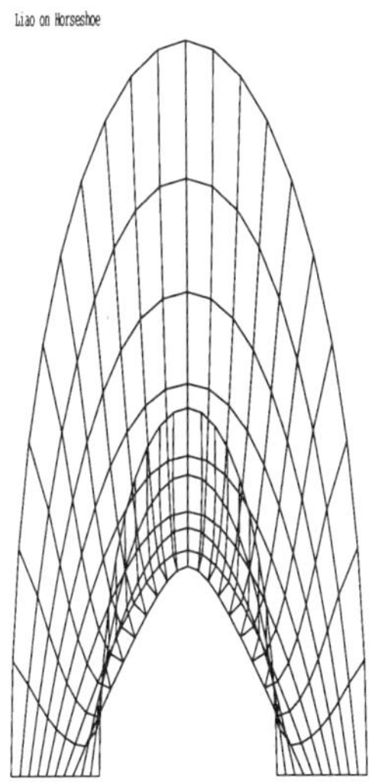

Liao

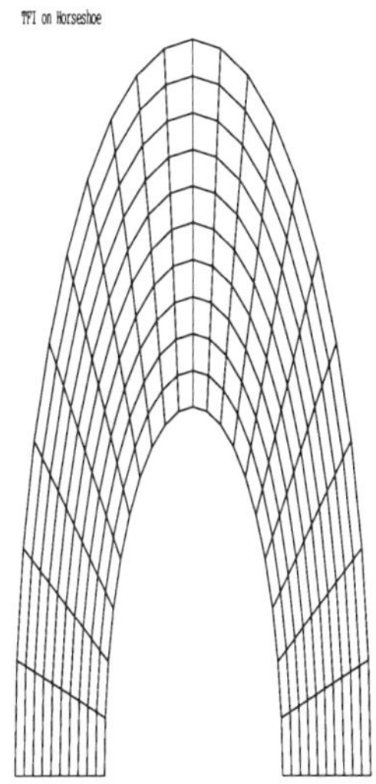

Modified Liao

Figure C.5 *Swan grids*

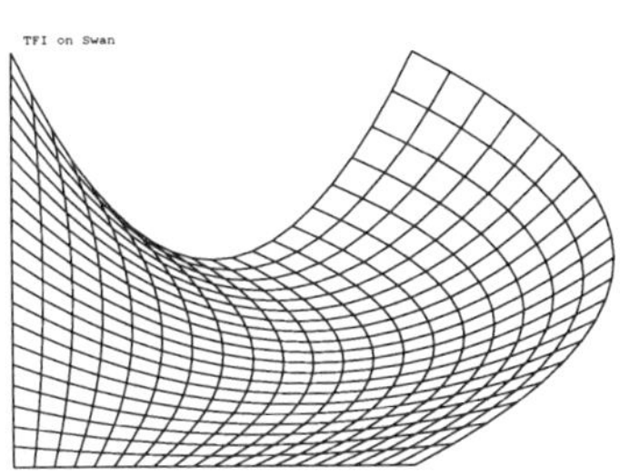

TFI

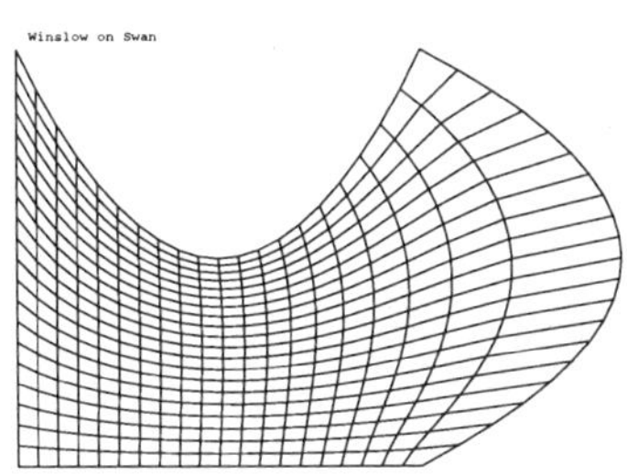

Winslow

Figure C.5 *Continued*

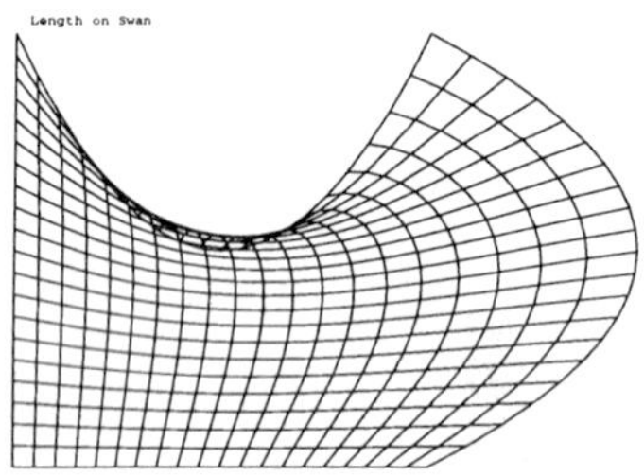

Length

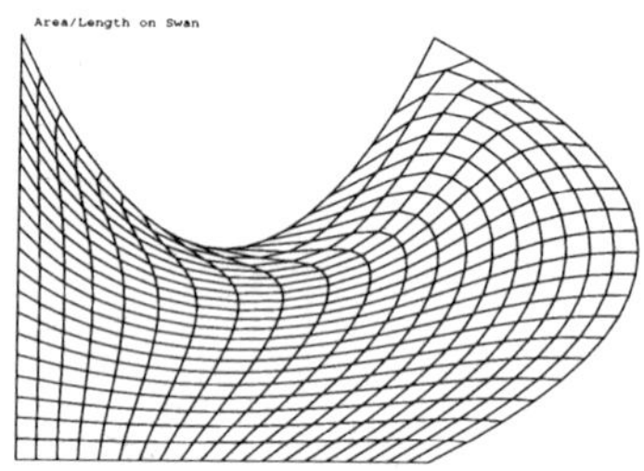

Area & Length

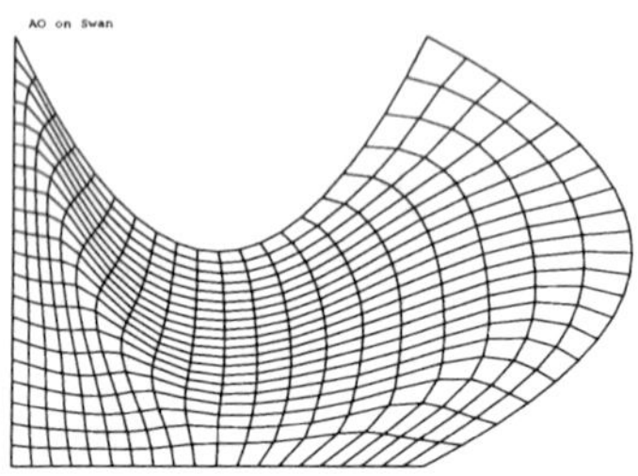

AO

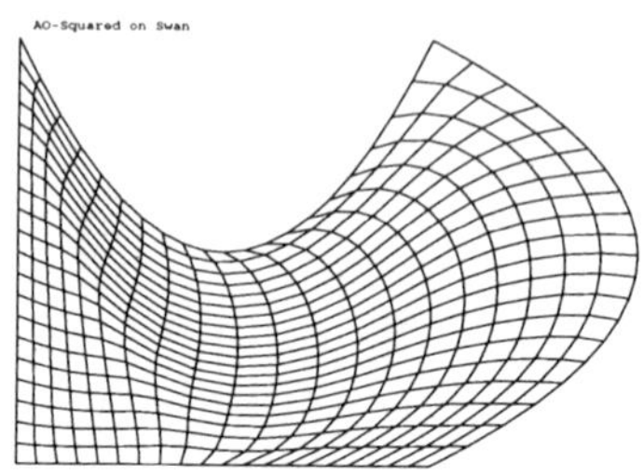

AO-Squared

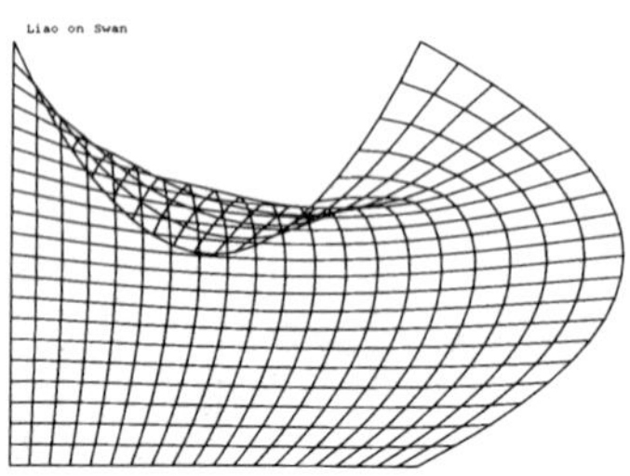

Liao

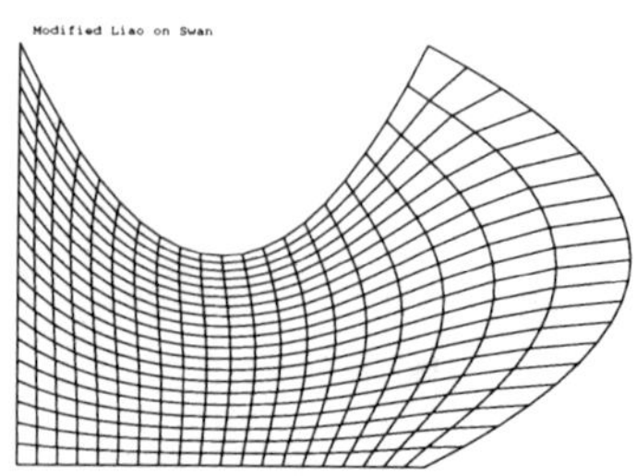

Modified Liao

Figure C.6 *Chevron grids*

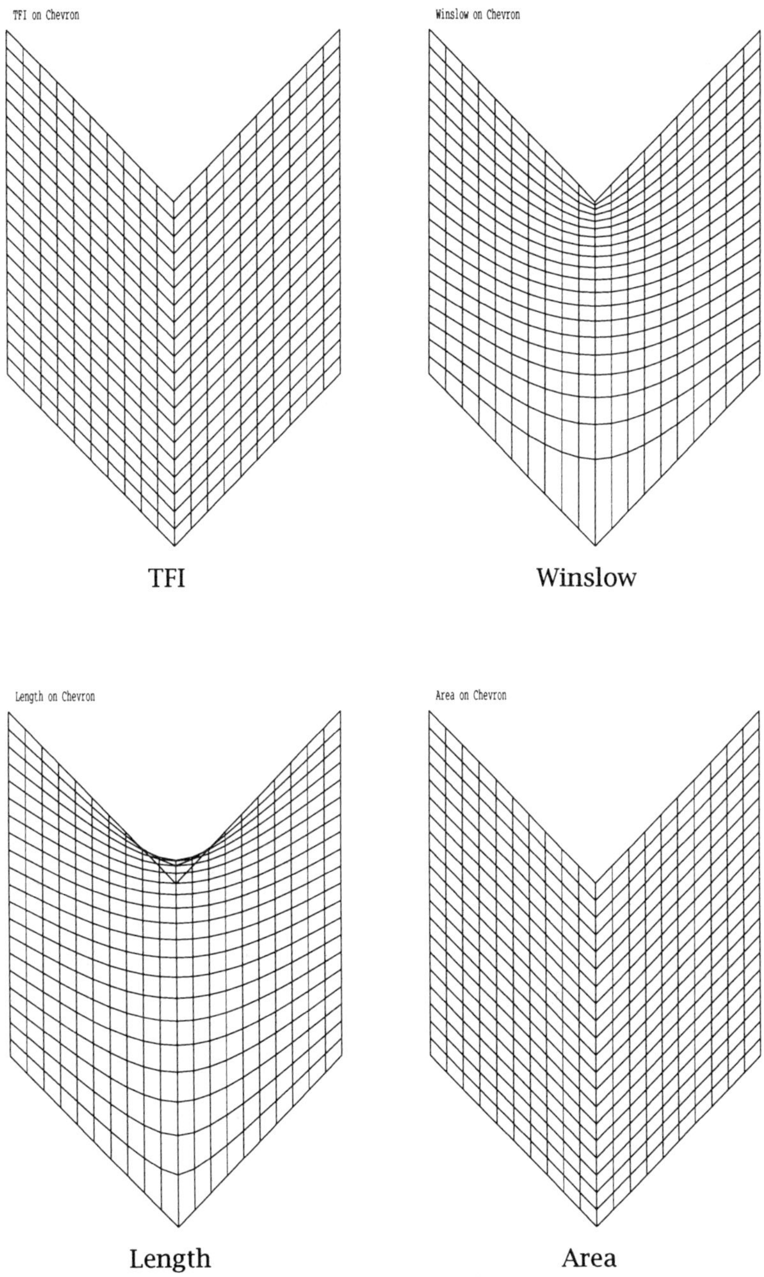

Figure C.6 *Continued*

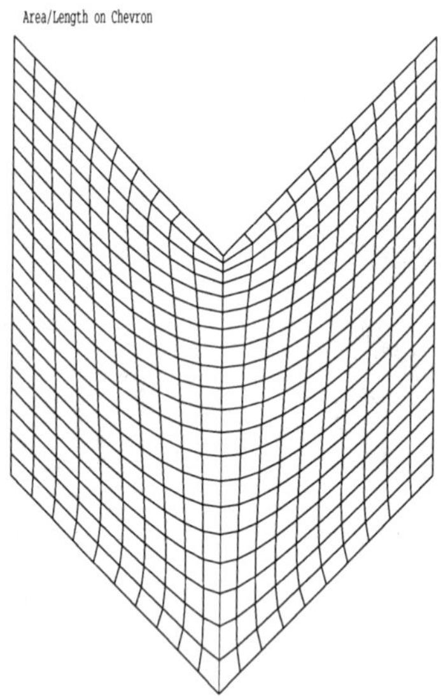

Area & Length

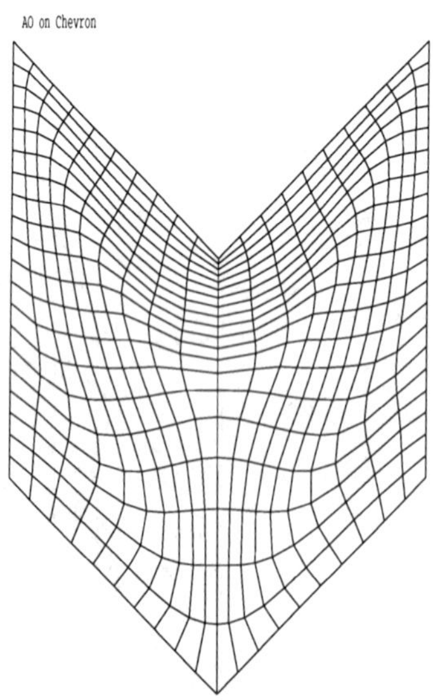

AO

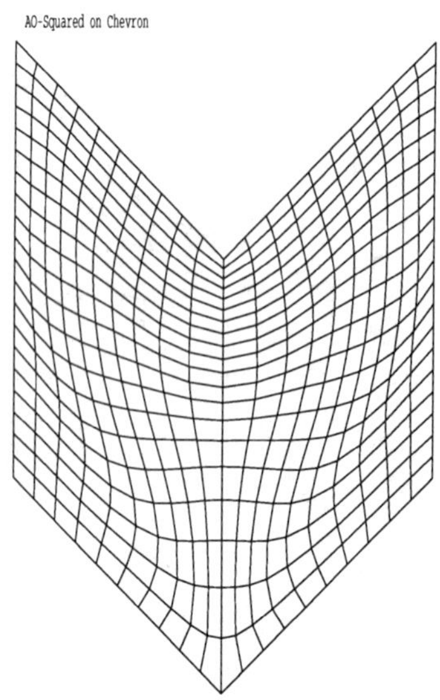

AO-Squared

Figure C.6 *Continued*

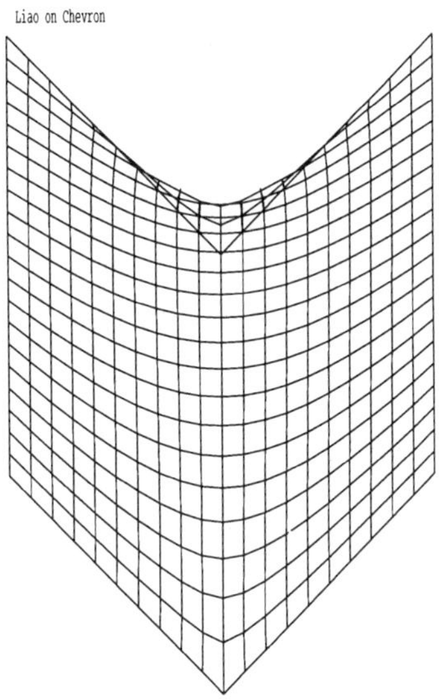

Liao

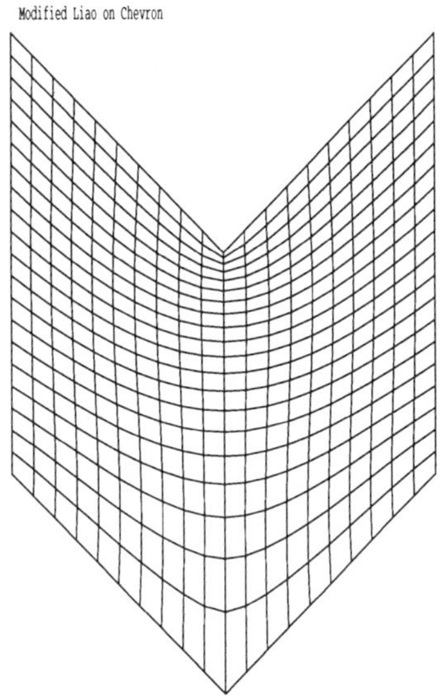

Modified Liao

Figure C.7 *Airfoil grids*

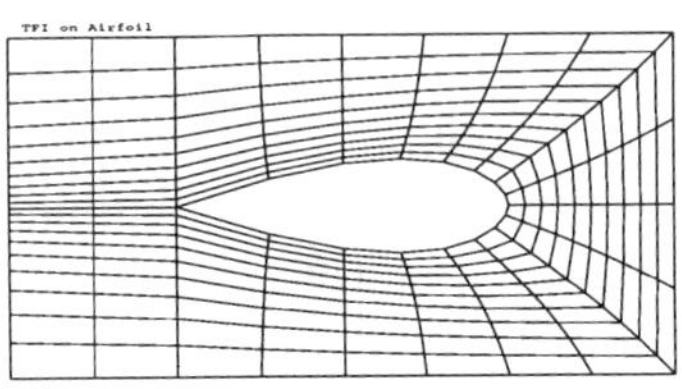

TFI

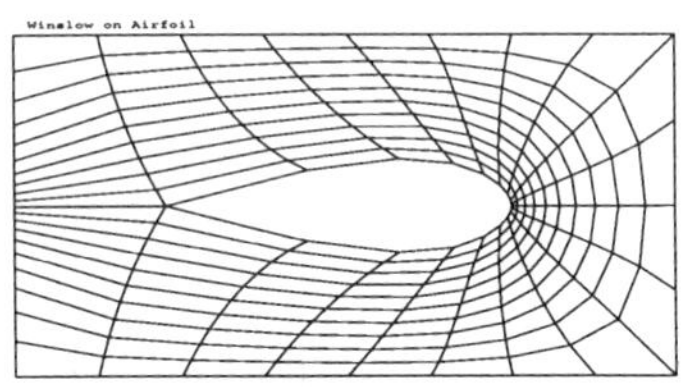

Winslow

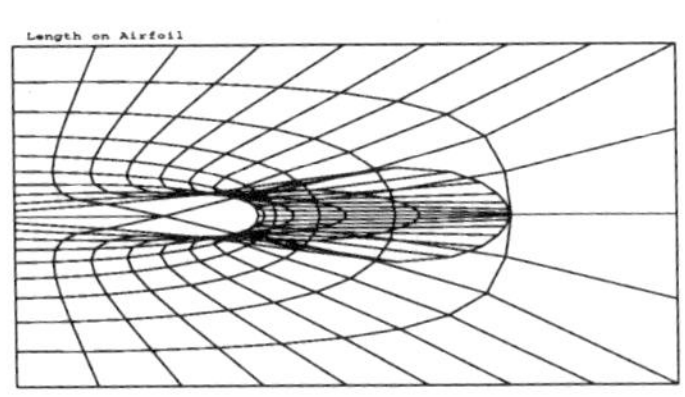

Length

Figure C.7 *Continued*

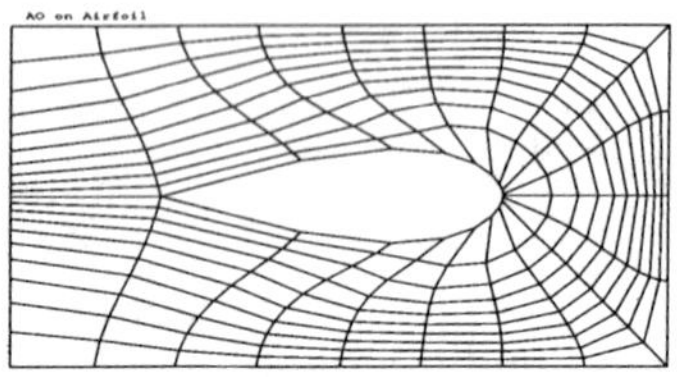

AO

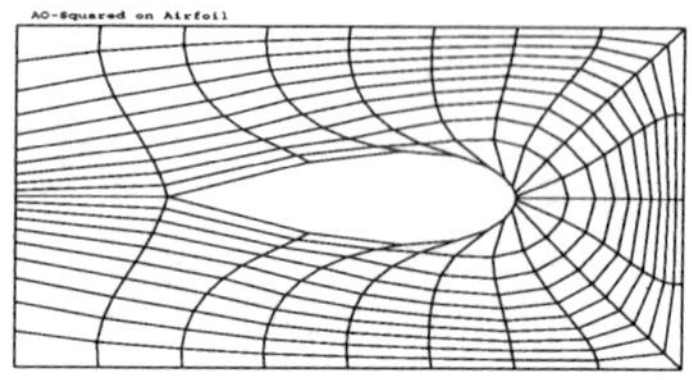

AO-Squared

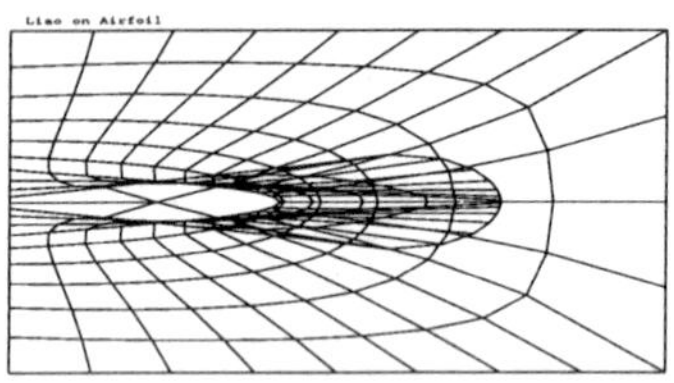

Liao

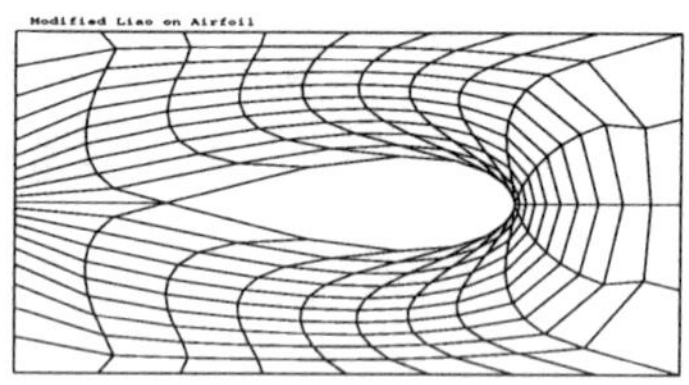

Modified Liao

Figure C.8 *Dome grids*

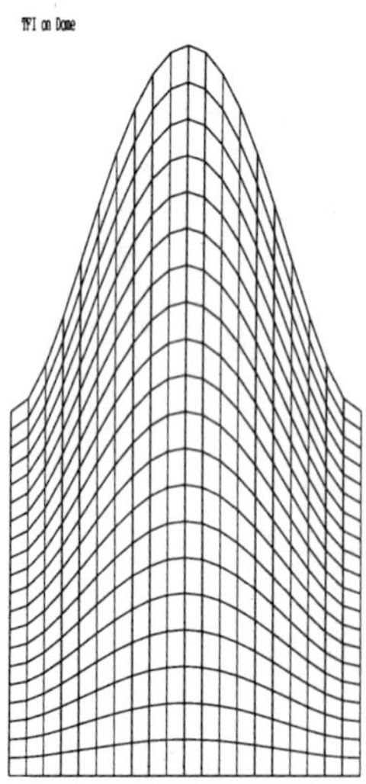

TFI

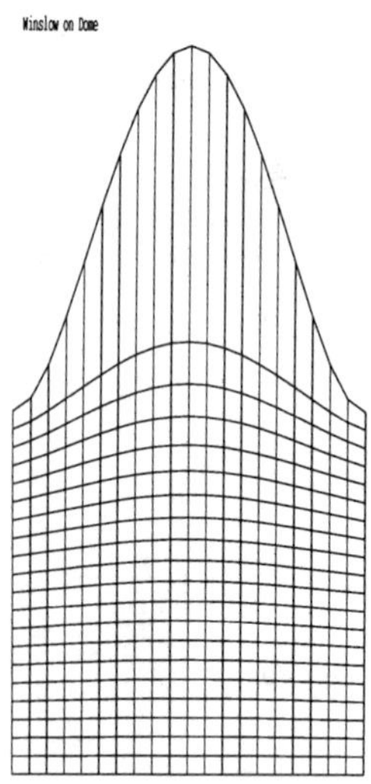

Winslow

Figure C.8 *Continued*

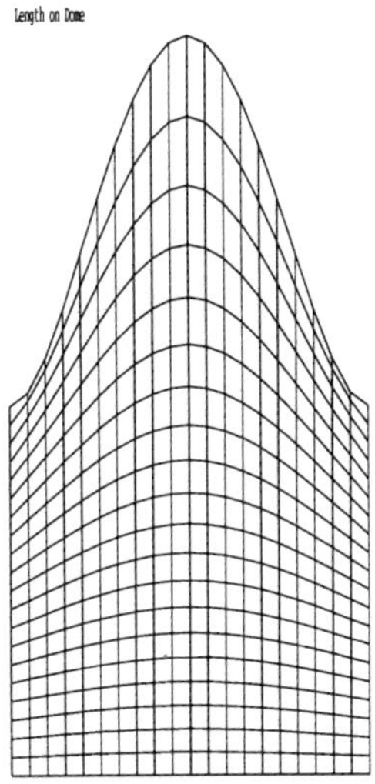

Length

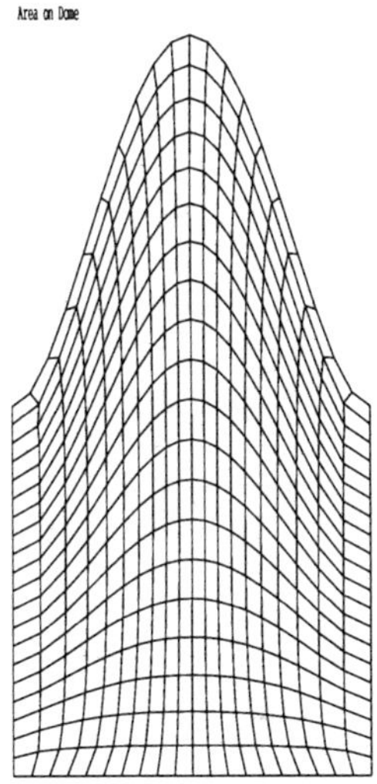

Area

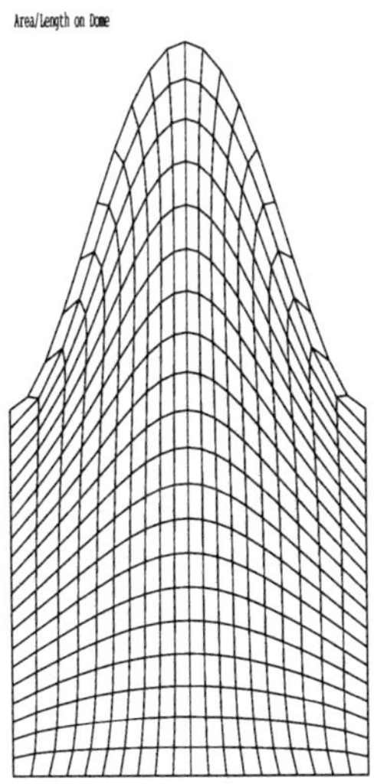

Area & Length

Figure C.8 *Continued*

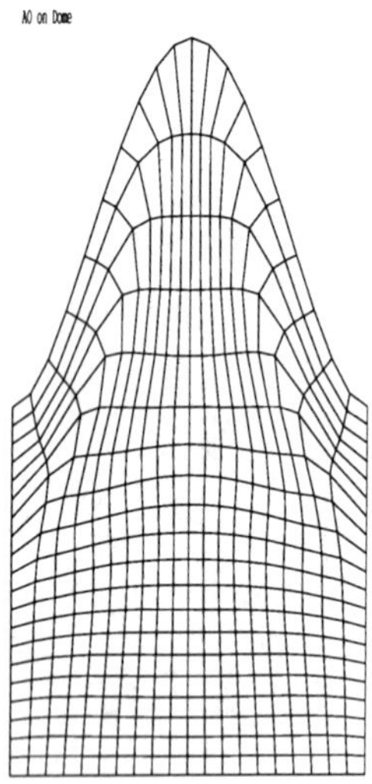

AO

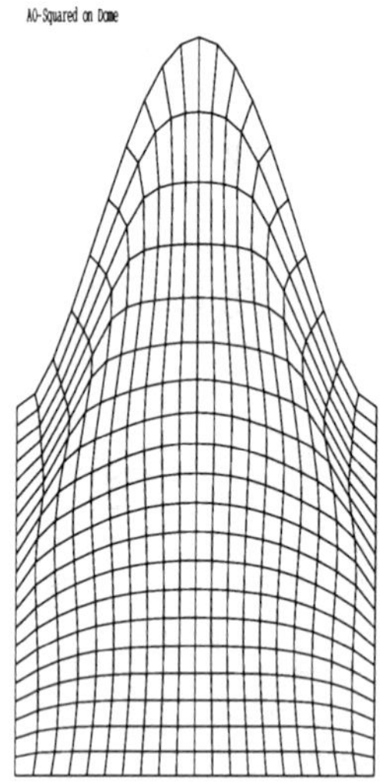

AO-Squared

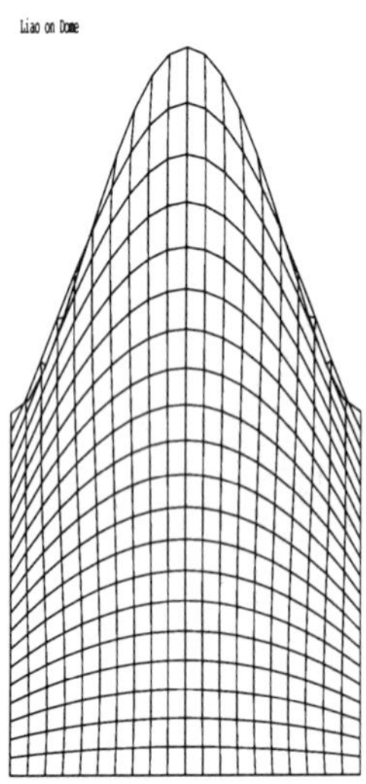

Liao

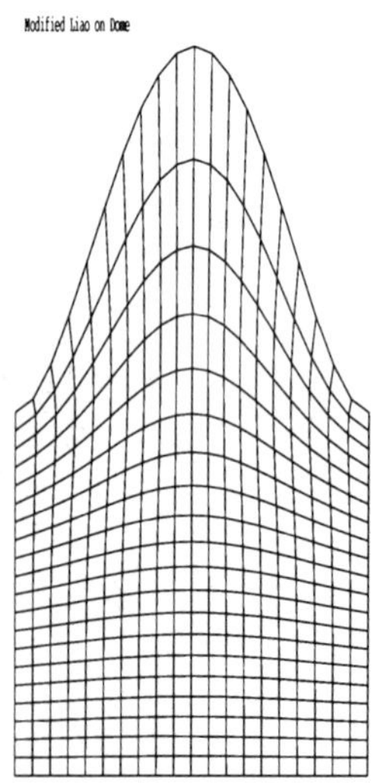

Modified Liao

Figure C.9 *Valley grids*

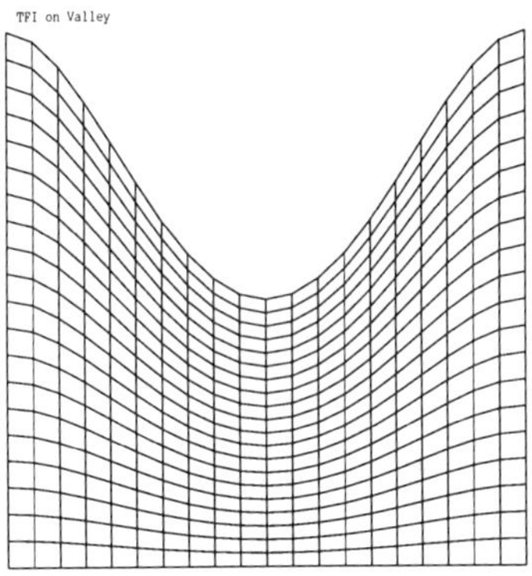

TFI

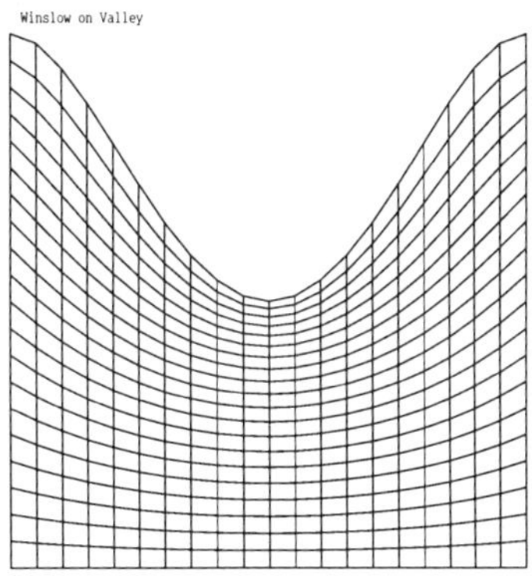

Winslow

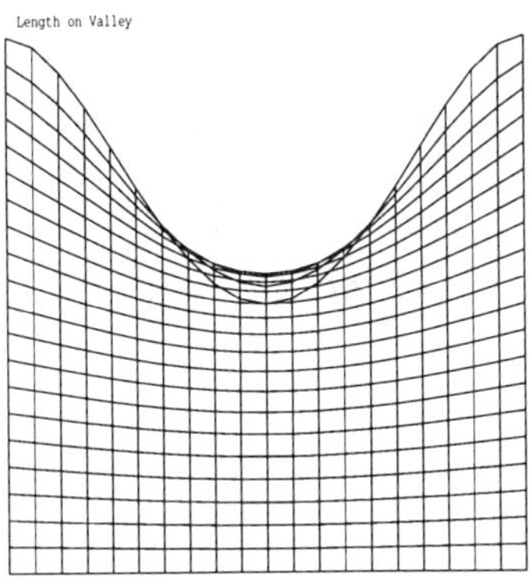

Length

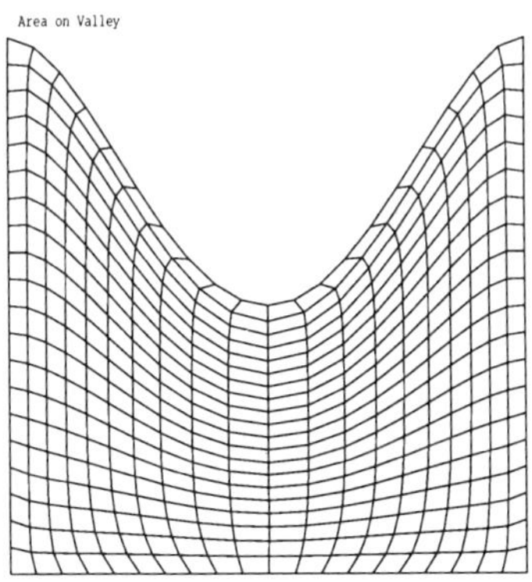

Area

Figure C.9 *Continued*

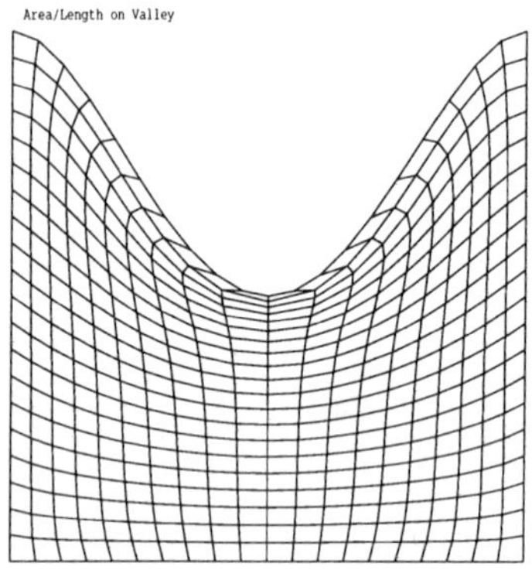

Area & Length

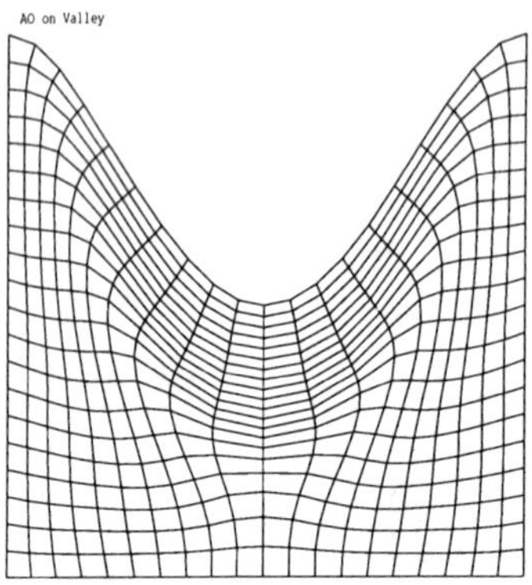

AO

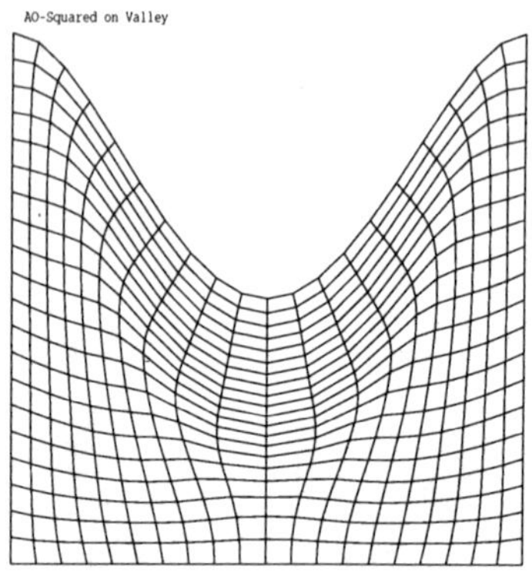

AO-Squared

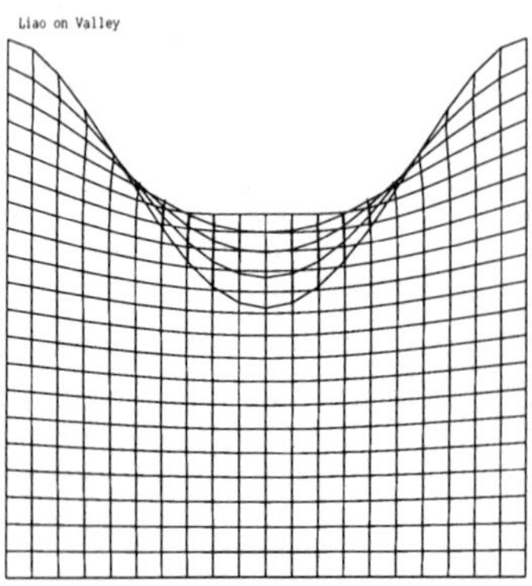

Liao

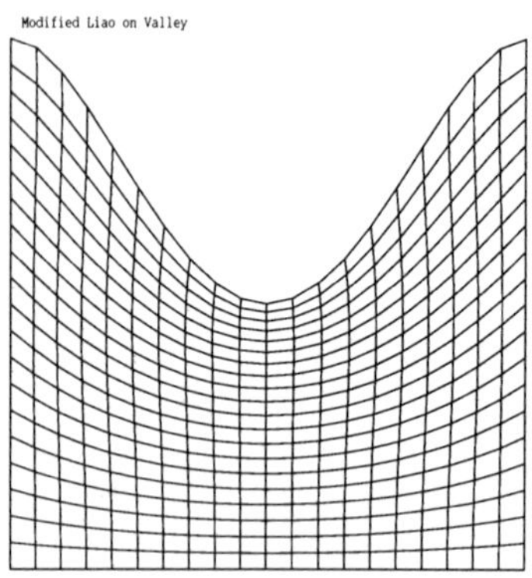

Modified Liao

Figure C.10 *Backstep grids*

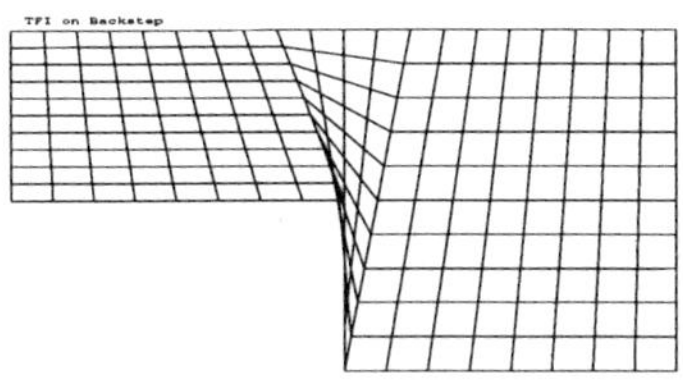

TFI

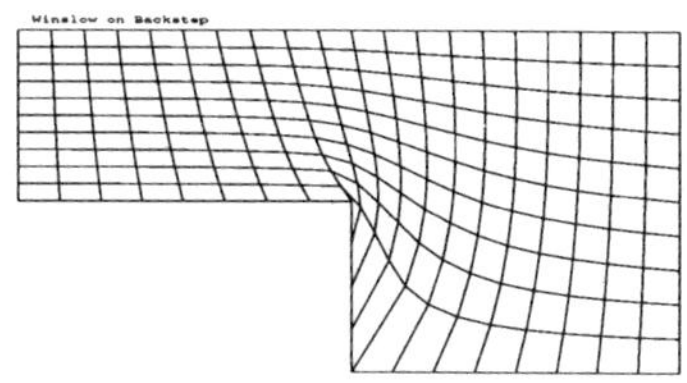

Winslow

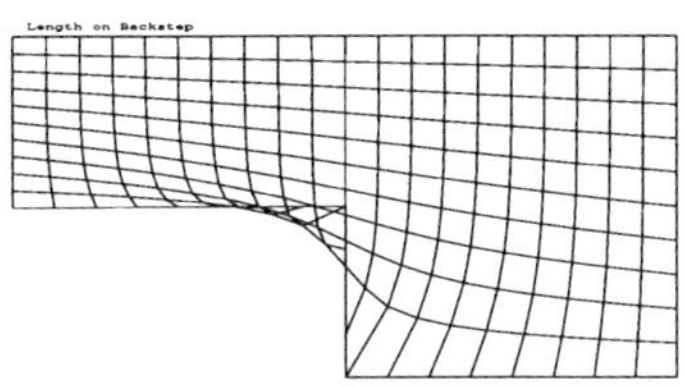

Length

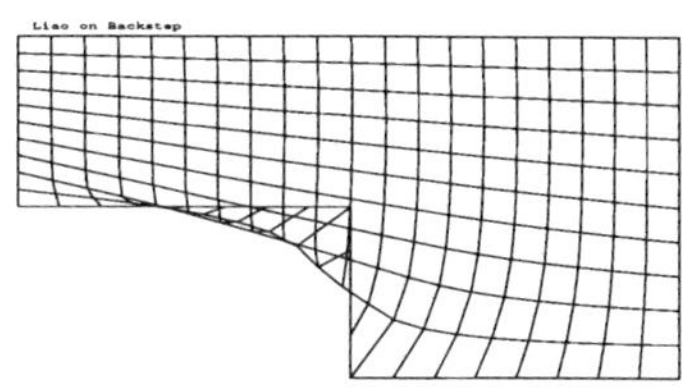

Liao

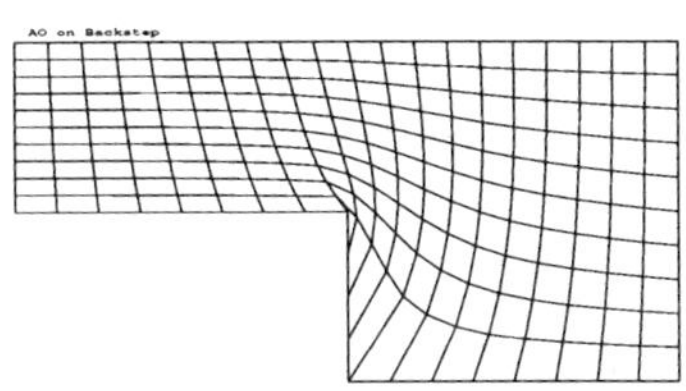

AO

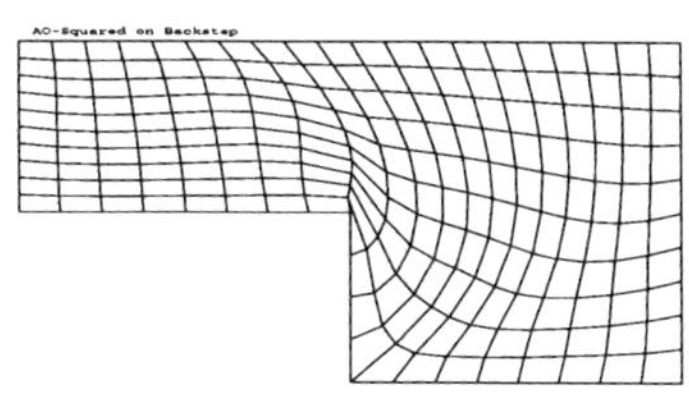

AO-Squared

Figure C.11 *Plow grids*

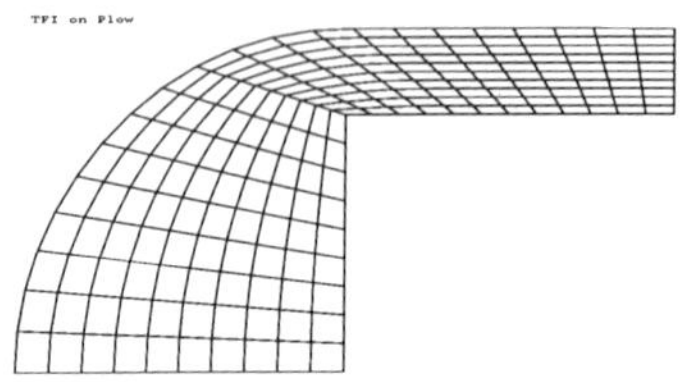

TFI

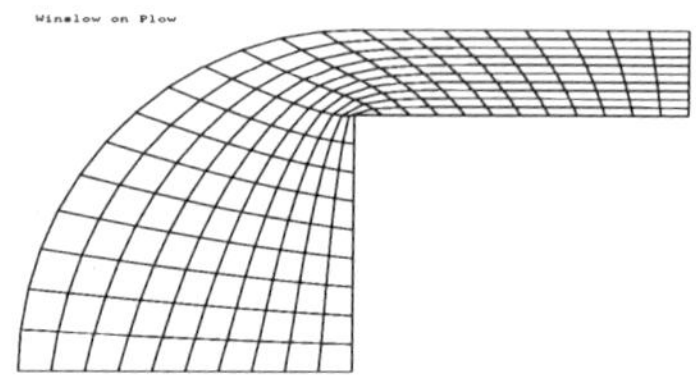

Winslow

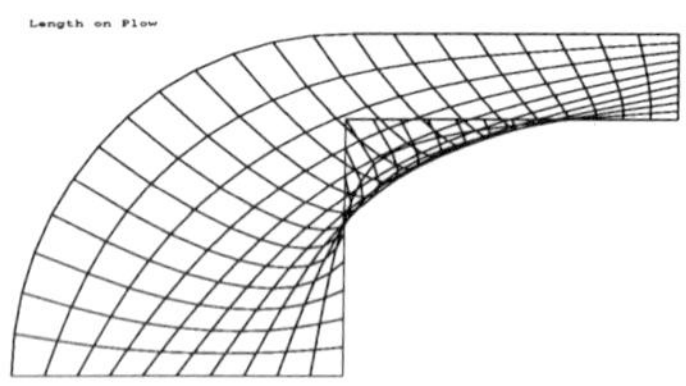

Length

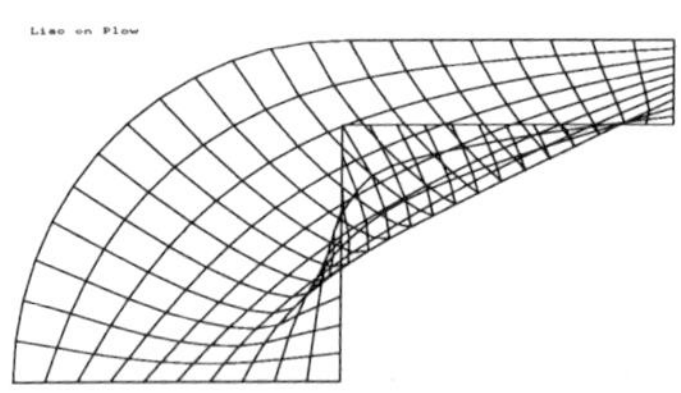

Liao

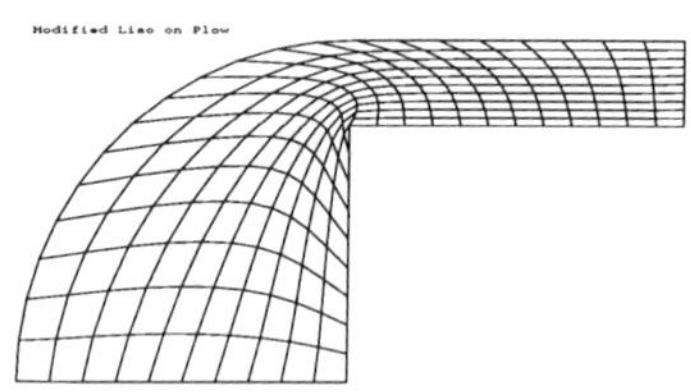

Modified Liao

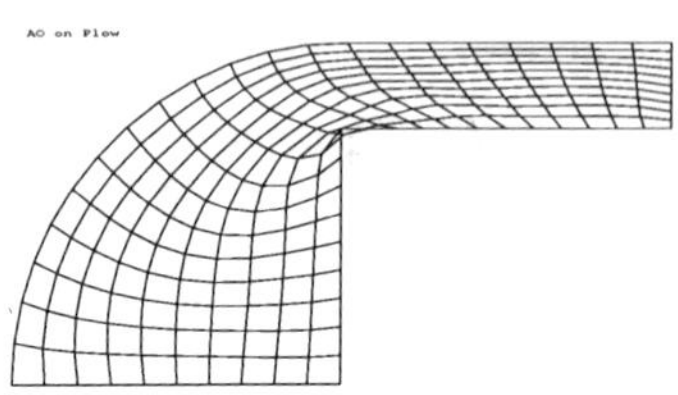

AO

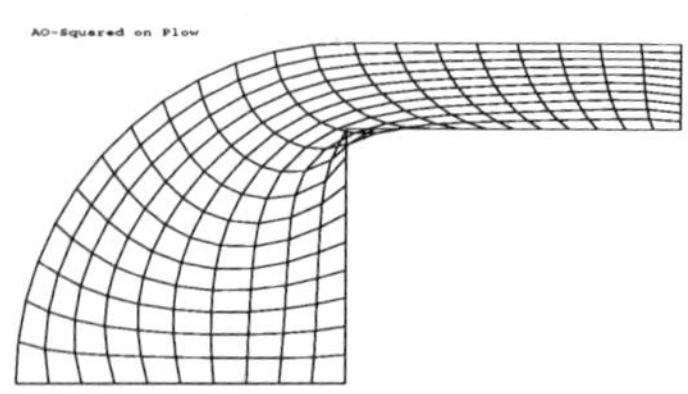

AO-Squared

Figure C.12 *C grids*

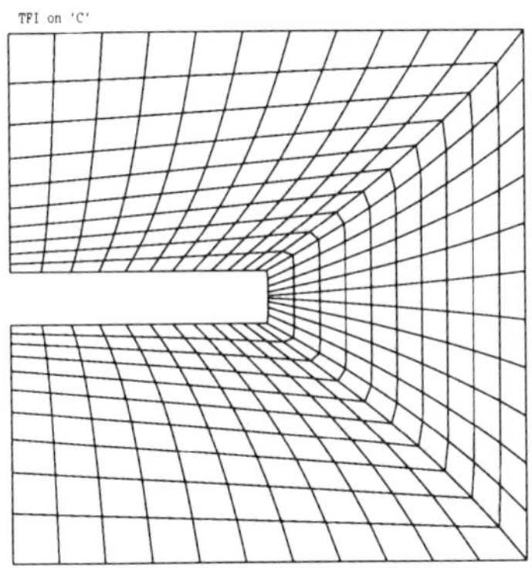

TFI

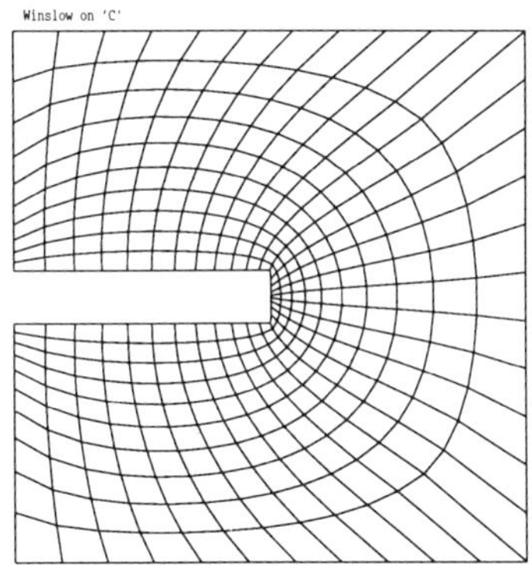

Winslow

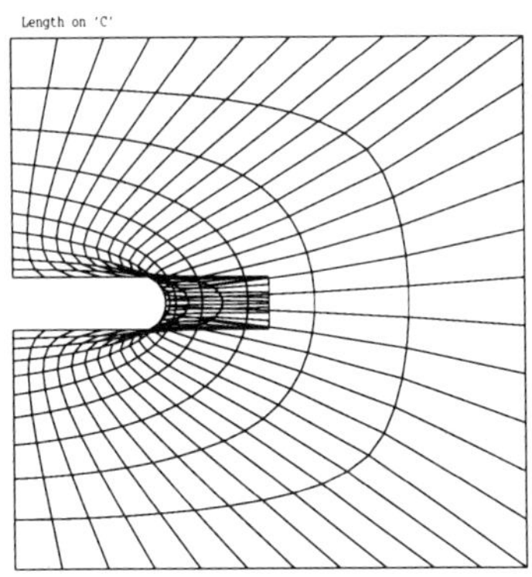

Length

Figure C.12 *Continued*

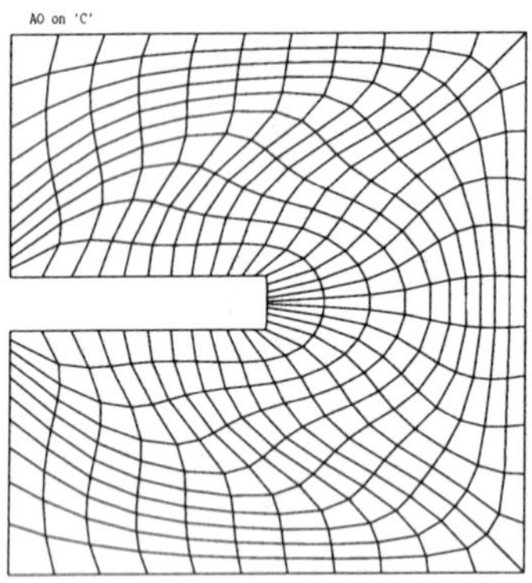

AO

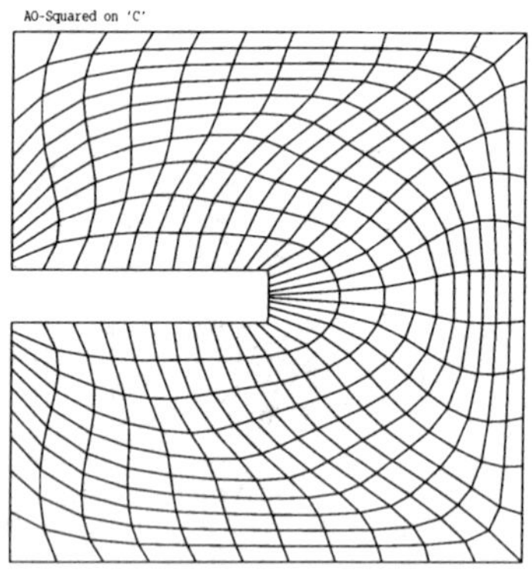

AO-Squared

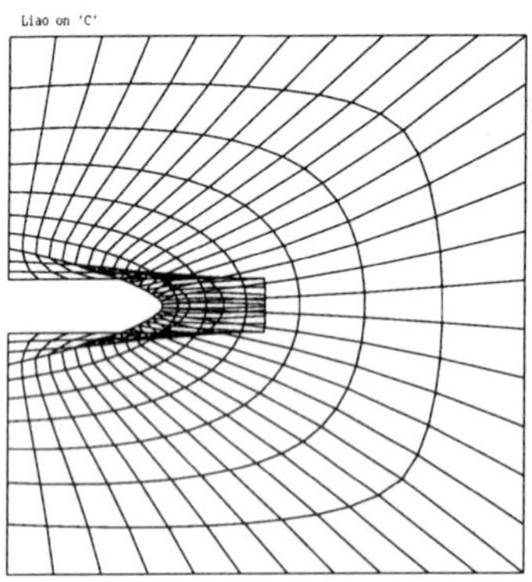

Liao

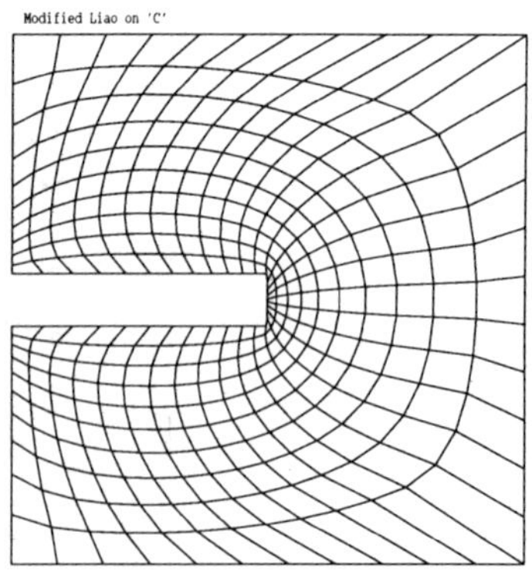

Modified Liao

Index